INTRODUCTION TO

MATHEMATICAL PROOFS

A TRANSITION

TEXTBOOKS in MATHEMATICS

Series Editor: Denny Gulick

PUBLISHED TITLES

COMPLEX VARIABLES: A PHYSICAL APPROACH WITH APPLICATIONS AND MATLAB®
Steven G. Krantz

INTRODUCTION TO ABSTRACT ALGEBRA
Jonathan D. H. Smith

INTRODUCTION TO MATHEMATICAL PROOFS: A TRANSITION
Charles E. Roberts, Jr.

LINEAR ALBEBRA: A FIRST COURSE WITH APPLICATIONS
Larry E. Knop

MATHEMATICAL AND EXPERIMENTAL MODELING OF PHYSICAL AND BIOLOGICAL PROCESSES
H. T. Banks and H. T. Tran

FORTHCOMING TITLES

ENCOUNTERS WITH CHAOS AND FRACTALS
Denny Gulick

TEXTBOOKS in MATHEMATICS

INTRODUCTION TO MATHEMATICAL PROOFS
A TRANSITION

Charles E. Roberts, Jr.

CRC Press is an imprint of the
Taylor & Francis Group an informa business
A CHAPMAN & HALL BOOK

Chapman & Hall/CRC
Taylor & Francis Group
6000 Broken Sound Parkway NW, Suite 300
Boca Raton, FL 33487-2742

Printed in the United States of America on acid-free paper
10 9 8 7 6 5 4 3 2 1

International Standard Book Number: 978-1-4200-6955-6 (Hardback)

Library of Congress Cataloging-in-Publication Data

Roberts, Charles E., 1942-
Introduction to mathematical proofs : a transition / Charles Roberts.
p. cm. -- (Textbooks in mathematics)
Includes bibliographical references and index.
ISBN 978-1-4200-6955-6 (hardcover : alk. paper)
1. Proof theory--Textbooks. 2. Logic, Symbolic and mathematical--Textbooks. I. Title. II. Series.

QA9.54.R63 2009
511.3'6--dc22 2009019726

Visit the Taylor & Francis Web site at
http://www.taylorandfrancis.com

and the CRC Press Web site at
http://www.crcpress.com

Contents

Preface

This text is written for undergraduate mathematics majors and minors who have taken only computationally oriented, problem solving mathematics courses previously. Usually these students are freshmen and sophomores. The primary objectives of the text are to teach the reader: (1) to reason logically, (2) to read the proofs of others critically, and (3) to write valid mathematical proofs. We intend to help students develop the skills necessary to write correct, clear, and concise proofs. Ultimately, we endeavor to prepare students to succeed in more advanced mathematics courses such as abstract algebra, analysis, and geometry where they are expected to write proofs and construct counterexamples instead of perform computations and solve problems. The aim of the text is to facilitate a smooth transition from courses designed to develop computational skills and problem solving abilities to courses which emphasize theorem proving.

Logic is presented in Chapter 1 because logic is the underlying language of mathematics, logic is the basis of all reasoned argument, and logic developed earliest historically. This text may well be the only place in the undergraduate mathematics curriculum where a student is introduced to the study of logic. Knowing logic should benefit students not only in future mathematics courses but in other facets of their lives as well. Formal proofs are included, because each step in a formal proof requires a justification. And students need to understand that when they write an informal proof, each statement should be justified unless the justification is apparent to the reader.

In Chapter 2, deductive mathematical systems are defined and discussed. Various proof techniques are presented and each proof technique is illustrated with several examples. Some theorems are proven using more than one proof technique, so that the reader may compare and contrast the techniques. The role of conjectures in mathematics is introduced, and proofs and disproofs of conjectures are explored. Interesting conjectures, which recently have been proved true or disproved, and conjectures, which still remain open, are stated and discussed. The integers and their properties are developed from the axioms and properties of the natural numbers; the rational numbers and their properties are derived from the integers; and, finally, the method for developing the system of real numbers from the rational numbers is described.

Elementary topics in set theory are presented in Chapter 3. A thorough understanding of basic set theory is necessary for success in advanced mathematics courses. In addition, using set notation promotes precision and clarity when communicating mathematical ideas.

Relations and functions play a major role in many branches of mathematics and the sciences. Therefore, in Chapters 4 and 5, relations and functions are defined and their various properties are examined in detail.

In Chapter 6, proof by mathematical induction, in its various forms, is introduced, and several theorems are proven using induction.

The last three chapters, which are optional, introduce the reader to the concept of cardinalities of sets (Chapter 7) and to the concepts and proofs in real analysis (Chapter 8) and in group theory (Chapter 9).

Appendix A discusses reading and writing proofs and includes some basic guidelines to follow when writing proofs. We encourage students to read Appendix A more than once during the semester and to use it as a reference when writing proofs.

Several different syllabi can be designed for this text depending upon the previous preparation and mathematical maturity of the students and the goals, objectives, and preferences of the instructor. Chapters 1 through 6 constitute the core of the course we teach during one semester. When time permits, we present some additional topics from Chapters 7, 8, and 9.

Features of the Text. This text is written in a friendly, conversational style, yet, it maintains the proper level of mathematical rigor. Most sections are of appropriate length for presentation in one lecture session. Several biographical sketches and historical comments have been included to enrich and enliven the text. Generally, mathematics is presented as a continually evolving discipline, and the material presented should fulfill the needs of students with a wide range of backgrounds. Numerous technical terms, which the student will encounter in more advanced courses, are defined and illustrated. Many theorems from different disciplines in mathematics and of varying degrees of complexity are stated and proved. Numerous examples illustrate in detail how to write proofs and show how to solve problems. These examples serve as models for students to emulate when solving exercises. Exercises of varying difficulty appear at the end of each section.

Acknowledgments. This text evolved from lecture notes for a course which I have taught at Indiana State University for a number of years. I would like to thank my students and my colleagues for their support, encouragement, and constructive criticisms. Also, I would like to thank Mr. Robert Stern of Taylor and Francis/CRC Press for his assistance in bringing this text to fruition.

Charles Roberts

Chapter 1

Logic

There are many definitions of logic; however, we will consider logic to be the study of the methods and principles used to distinguish valid reasoning from invalid reasoning. Logic is a part of mathematics; moreover, in a broad sense, it is the language of mathematics.

In this chapter, we will study elementary symbolic logic. Logic is the basis of all reasoned argument; and, therefore, logic is the basis for valid mathematical proofs. The study of logic as a body of knowledge in Western Civilization originated with Aristotle (384-322 B.C.), one of the greatest philosophers of ancient Greece. He was a student of Plato for twenty years (from 367 to 347 B.C. when Plato died). Later, Aristotle tutored Alexander the Great, and in 334 B.C. he founded his own school of philosophy in the Lyceum. After his death, Aristotle's writings on reasoning were collected together in a body of work called the *Organon*. The contents of the *Organon* is the basis for the subject of logic, although the word "logic" did not acquire its current meaning until the second century A.D. The word "logic" is a derivative of the Greek word *logos,* which translates into English as "word," "speech," or "reason."

Aristotle was the first to develop rules for correct reasoning. However, he expressed logic in ordinary language; and, consequently, it was subject to the ambiguities of natural language. At an early age, the German philosopher, mathematician, and logician Gottfried Wilhelm Leibniz (1646-1716) was not satisfied with Aristotelian logic and began to develop his own ideas. He had a lifelong goal of developing a universal language and a calculus of reasoning. His idea was that the principles of reasoning could be reduced to a formal symbolic system in which controversy (not just mathematical ones) could be settled by calculations. Thus, Leibniz envisioned an algebra or calculus of thought. He made some strides toward his goal, but his work was largely forgotten.

The English mathematician and logician August De Morgan (1806-1871) presented ideas for improving classical logic in the 1840's. The key ideas he contributed in his text *Formal Logic* (1847) include the introduction of the concept of a universe of discourse; names for contraries; disjunction, conjunction, and negation of propositions; abbreviated notation for propositions; compound names; and notation for syllogisms. De Morgan intended to improve the syllogism and use it as the main device in reasoning. In order to insure there were names for the contraries of compound names, he stated the famous De Morgan Laws. By creating some of the most basic concepts of modern logic, De Morgan contributed substantially to the change that was taking place in logic in the mid-1800's. However, his notational system was viewed as too complex, so he received little credit for the development of modern logic.

The English mathematician George Boole (1815-1864) is generally credited with founding the modern algebra of logic and, hence, symbolic logic. At the age of sixteen, Boole was an assistant teacher. In 1835, he opened his own school and began to study mathematics on his own. He never attended an institution of higher learning. He taught himself all of the higher mathematics he knew. In 1840, he began to publish papers on analysis in the *Cambridge Mathematical Journal*. In 1847, Boole published the text *The Mathematical Analysis of Logic*. Initially, Boole wanted to express all the statements of classical logic as equations, and then apply algebraic transformations to derive the known valid arguments of logic. Near the end of writing the text, Boole realized that his algebra of logic applied to any finite collection of premises with any number of symbols. Boole's logic was limited to what is presently called the **propositional calculus**. It is the propositional calculus we will study in this chapter.

1.1 Statements, Negation, and Compound Statements

In the English language, sentences are classified according to their usage. A **declarative sentence** makes a statement. An **imperative sentence** gives a command or makes a request. An **interrogative sentence** asks a question. And an **exclamatory sentence** expresses strong feeling. Consider the following sentences:

(1) Indianapolis is the capital of Indiana.

(2) Tell Tom I will be home later.

(3) What time is it?

(4) I wish you were here!

Sentence (1) is declarative, sentence (2) is imperative, sentence (3) is interrogative, and sentence (4) is exclamatory. However, the same sentence can be written to be declarative, interrogative, or exclamatory. For instance,

We won the game. [declarative]

We won the game? [interrogative]

We won the game! [exclamatory]

The declarative sentence (1) is "true" while the following declarative sentence is "false":

Minneapolis is the capital of Indiana.

Symbolic logic only applies to special declarative sentences which are statements or propositions as defined below:

DEFINITION Statement or Proposition

A **statement** or **proposition** is a declarative sentence that is either true or false, but not both true and false.

The terms true and false are left undefined, but it is assumed that their meaning is intuitively understood. Some declarative sentences might be true or false depending on the context or circumstance. Such sentences are not considered to be statements. For example, the sentences "She is hungry.", "He is handsome.", and "Chicago is far away." depend upon one's definition of "hungry," "handsome," and "far away." Consequently, such sentences are not statements, because they do not have a "truth value"—that is, because it is not possible to determine whether they are true or false. There are statements for which we do not know the truth value. For example, we do not know the truth value of Goldbach's conjecture which states:

> "Every even integer greater than two can be written as the sum of two prime numbers."

(Recall that a **prime number** is a natural number greater than one which is divisible only by itself and one. On the other hand, a **composite number** is a natural number which can be written as the product of two or more prime numbers.) To date, mathematicians have not been able to prove or disprove Goldbach's conjecture; however, it is a declarative sentence that is either true or false and not both true and false. Thus, Goldbach's conjecture is a statement or proposition in symbolic logic. Christian Goldbach made his famous conjecture in a letter written to Leonhard Euler on June 7, 1742.

EXAMPLE 1 Identifying a Statement

Determine which of the following sentences is a statement.

(a) How old are you?
(b) $x + 3 = 5$
(c) 2^{300} is a large number.
(d) Help!
(e) The author of this text was born in Washington, D. C.
(f) This sentence is false.

SOLUTION

(a) The sentence is a question (an interrogative sentence); and, therefore, it is not a statement.

(b) The declarative sentence "$x + 3 = 5$" is true for $x = 2$ and false for all other values of x, so it is not a statement.

(c) The declarative sentence "2^{300} is a large number." is not a statement, because the definition of "a large number" is not well-defined.

(d) The sentence "Help!" is exclamatory; and, hence, it is not a statement.

(e) The declarative sentence "The author of this text was born in Washington, D. C." is true or false, but not both true and false; so it is a statement even though few people would know whether the statement is true or false.

(f) The declarative sentence "This sentence is false." is an interesting sentence. If we assign the truth value "true" or the truth value "false" to this sentence, we have a contradiction. Hence, the sentence is not a statement. Because the sentence "This sentence is false." is neither "true" nor "false," it is called a **paradox**.

All statements can be divided into two types, simple and compound.

DEFINITION Simple Statement and Compound Statement

A **simple statement** (**simple proposition**) is a statement which does not contain any other statement as a component part.

Every **compound statement** (**compound proposition**) is a statement that does contain another statement as a component part.

Every statement we have examined thus far is a simple statement. Compound statements are formed from simple statements using the logical connectives "and," "or," and "not."

DEFINITION Negation of a Statement

Let P denote a statement. The **negation** of P, denoted by $\neg$ P, is the statement "not P." The negation of P is false when P is true, and the negation of P is true when P is false.

For example, the negation of the statement "Five is a prime." is the statement "Five is not a prime." And the negation of the statement "Six is an odd number." is the statement "Six is not an odd number." In English, it is also possible to indicate the negation of a statement by prefixing the statement with the phrase "it is not the case that," "it is false that," or "it is not true that." For instance, the statement "It is not true that gold is heavier than lead." is true.

DEFINITION The Conjunction of Two Statements

The **conjunction** of two statements P, Q, denoted by $P \wedge Q$, is the statement "P and Q." The conjunction of P and Q is true if and only if both P and Q are true.

Let M be the statement "It is Monday." and let R be the statement "It is raining." The statement $M \wedge R$ is "It is Monday, and it is raining." In English, several other words such as "but," "yet," "also," "still," "although," "however," "moreover," "nonetheless," and others, as well as the comma and semicolon, can mean "and" in their conjunctive sense. For instance, the statement "It is Monday; moreover, it is raining." should be translated to symbolic logic as $M \wedge R$.

DEFINITION The Disjunction of Two Statements

The **disjunction** of two statements P, Q, denoted by $P \vee Q$, is the statement "P or Q." The disjunction of P and Q is true if P is true, if Q is true, or if both P and Q are true.

In English, the word "or" has two related but distinguishable meanings. The "or" appearing in the definition of disjunction is the **inclusive or**. The **inclusive or** means "one or the other or both." In legal documents, the meaning of the inclusive "or" is often made more explicit by using the phrase "and/or." For example, the statement "This contract may be signed by John and/or Mary." means the contract is legally binding when signed by John, by Mary, or by both. On the other hand, the **exclusive or** means "one or the other but not both." For example, the statement "Ann will marry Ben or Ann will marry Ted." means either Ann will marry Ben or Ann will marry Ted but not both. In Latin there are two different words for the word "or." The word *vel* denotes the inclusive or, while the word *aut* denotes the exclusive or.

In the following two examples, we show how to write English statements in symbolic form and how to write symbolic statements in English.

EXAMPLE 2 Translate English Statements to Symbolic Form

Write the following statements in symbolic form using $\neg$, $\wedge$, and $\vee$.

(a) Madrid is the capital of Spain and Paris is the capital of France.
(b) Rome is the capital of Italy or London is the capital of England.
(c) Rome is the capital of Italy, but London is not the capital of England.
(d) Madrid is not the capital of Spain or Paris is not the capital of France.
(e) Paris is the capital of France, but London is not the capital of England or Madrid is the capital of Spain.

SOLUTION

Let M stand for the statement "Madrid is the capital of Spain."
Let P stand for the statement "Paris is the capital of France."
Let R stand for the statement "Rome is the capital of Italy."
Let L stand for the statement "London is the capital of England."

The statement (a) is written in symbolic form as $M \wedge P$.

The statement (b) written in symbolic form is $R \vee L$.

The statement (c) in symbolic form is $R \wedge (\neg L)$.

The statement (d) in symbolic form is $(\neg M) \vee (\neg P)$.

The statement (e) in symbolic form is $P \wedge ((\neg L) \vee M)$.

EXAMPLE 3 Translate Symbolic Statements to English

Let C be the statement "Today the sky is clear."
Let R be the statement "It did rain."
Let S be the statement "It did snow."
Let Y be the statement "Yesterday it was cloudy."
Write the following symbolic statements in English.

(a) $\neg Y$ (b) $R \vee S$ (c) $Y \wedge S$ (d) $C \wedge (\neg R) \wedge (\neg S)$ (e) $Y \wedge (C \vee R)$

SOLUTION

The symbolic statement (a) can be written as "Yesterday it was not cloudy."

The symbolic statement (b) can be written as "It did rain or it did snow."

The statement (c) can be written as "Yesterday it was cloudy and it did snow."

The statement (d) can be written as "Today the sky is clear and it did not rain and it did not snow."

The statement (e) can be written as "Yesterday it was cloudy, and today the sky is clear or it did rain."

EXERCISES 1.1

In exercises 1-10, determine whether the given sentence is a statement or not.

1. The integer 6 is a prime.
2. Divide 256 by 4.
3. This is a difficult problem.
4. He lives in San Francisco, California.
5. Where are you going?
6. $x^2 + 9 = 0$
7. The number π is rational.
8. Three factorial is denoted by 3!
9. That was easy!
10. George Washington never went to England.

In exercises 11-15, write the negation of each sentence.

11. The number $\sqrt{2}$ is rational.
12. Roses are not red.
13. $7 \leq 5$
14. It is false that π is rational.
15. Every even integer greater than two can be written as the sum of two prime numbers.

16. Let W represent the statement "We won the game." and let P represent the statement "There was a party."

a. Write the following statements symbolically.

(1) It is false that we won the game.

(2) We won the game, and there was a party.

(3) We did not win the game, and there was no party.

(4) There was no party; however, we won the game.

(5) We won the game, or there was no party.

b. Write the following symbolic statements in English.

(1) $\neg P$ (2) $W \vee P$ (3) $(\neg P) \vee W$ (4) $(\neg W) \wedge P$ (5) $(\neg W) \wedge (\neg P)$

17. Let T denote the statement "ABC is a triangle." and let I denote the statement "ABC is isosceles."

a. Write the following statements symbolically.

(1) ABC is not isosceles.

(2) ABC is a triangle, but ABC is not isosceles.

(3) ABC is a triangle, and ABC is isosceles.

(4) ABC is an isosceles triangle.

(5) ABC is a triangle; however, ABC is not isosceles.

b. Write the following symbolic statements in English.

(1) $\neg T$ (2) $T \vee I$ (3) $(\neg T) \wedge (\neg I)$ (4) $\neg(T \wedge I)$ (5) $(\neg T) \vee (\neg I)$

18. Let T be the statement "I drink tea for breakfast.", let S be the statement "I eat soup for lunch.", and let D be the statement "I eat dessert after dinner."

a. Write the following statements symbolically.

(1) I drink tea for breakfast and I eat soup for lunch and I eat dessert after dinner.

(2) I drink tea for breakfast and I eat soup for lunch, or I eat dessert after dinner.

(3) I drink tea for breakfast, and I eat soup for lunch or I eat dessert after dinner.

(4) I drink tea for breakfast and I eat soup for lunch, or I drink tea for breakfast and I eat dessert after dinner.

(5) I do not drink tea for breakfast or I do not eat soup for lunch or I do not eat dessert after dinner.

b. Write the following symbolic statements in English.

(1) $((\neg T) \vee S) \wedge D$ (2) $(\neg T) \vee (S \wedge D)$ (3) $(T \vee S) \wedge (S \vee D)$

(4) $\neg(T \wedge S \wedge D)$ (5) $\neg((\neg T \wedge \neg S) \vee (\neg D))$

1.2 Truth Tables and Logical Equivalences

Recall from section 1.1 that a **statement** or **proposition** is a declarative sentence that is either true or false, but not both true and false.

DEFINITION Truth Value and Truth Table

The **truth value** of a statement is true (denoted by T) if the statement is true and is false (denoted by F) if the statement is false.

A **truth table** is a table which shows the truth values of a statement for all possible combinations of truth values of its simple statement components.

The German mathematician and philosopher Friedrich Ludwig Gottlob Frege (1848-1925) is considered to be one of the founders of modern symbolic logic. Frege believed that mathematics was reducible to logic. In 1879, he published his first major work *Begriffsschrift, eine der arithmetischen nachgebildete Formelsprache des reinen Denkens* (Conceptual notation, a formal language modeled on that of arithmetic, for pure thought). In this work, Frege introduced a logical system with negation, implication, universal quantifiers, and the idea of a truth table, although it was not presented in our current notational form. Later in 1893, Frege published *Die Grundgesetze der Arithmetik, I.* (The Basic Laws of Arithmetic, I.). Here, he introduced the terms "True" and "False" and described the truth value of the statement "P implies Q" for each of the four possible combinations of truth values of P, Q. That is, he verbally described the truth table for P implies Q (which we will study in the next section); but he did not display an actual truth table.

Let P represent a statement. P may have the truth value T or the truth value F. By definition, the negation of P, $\neg$ P, is false when P is true and true when P is false. Consequently, the truth table for $\neg$ P is as follows:

Negation

P	$\neg$ P
T	F
F	T

Since this truth table explains completely the result of negating the statement P, it may be taken as the definition of $\neg$ P, the negation of P. The negation symbol "$\neg$" is a **unary logical operator**. The term "unary" means the operator acts on a single statement.

Let S be a compound statement which contains only the one simple statement P. In order to list all of the possible combinations of truth values for P, the truth table for S must have exactly two rows. In this case, **the standard truth table form** for the statement S is

P	⋯	S
T	⋯	.
F	⋯	.

The first entry in the column labeled P must be T and the second entry in the column must be F. The dots, ⋯, indicate the possible presence of columns of truth values with appropriate headings needed to "build-up" the statement S.

Next let P and Q represent two simple statements. In order to list all of the possible combinations of truth values of P, Q, the truth table for any compound statement S containing exactly the two simple statements P and Q will have exactly four rows. In this case, **the standard truth table form** for the statement S is

P	Q	⋯	S
T	T	⋯	.
T	F	⋯	.
F	T	⋯	.
F	F	⋯	.

To be in standard form, the truth value entries must be exactly as shown in the first two columns.

Since the conjunction of P, Q, the statement $P \wedge Q$, is true if and only if both P and Q are true, the truth table for $P \wedge Q$ is as follows:

Conjunction

P	Q	$P \wedge Q$
T	T	T
T	F	F
F	T	F
F	F	F

Since this truth table explains completely the result of conjuncting the statement P with the statement Q, it may be taken as the definition of $P \wedge Q$. The conjunction symbol "$\wedge$" is a **binary logical operator**. The term "binary" means the operator acts on two statement.

The disjunction symbol "$\vee$" is also a binary logical operator and its defining truth table is

Disjunction

P	Q	$P \vee Q$
T	T	T
T	F	T
F	T	T
F	F	F

EXAMPLE 1 Determining Truth Values

Given that A and B are true statements and C and D are false statements, use the definitions of $\neg$, $\wedge$, and $\vee$ to determine the truth value of each of the statements:

(a) $(\neg A) \wedge B$ (b) $\neg (A \wedge B)$ (c) $(\neg A) \vee B$
(d) $\neg (A \vee C)$ (e) $(A \vee C) \wedge (B \wedge (\neg D))$

SOLUTION

(a) Since A is true, $\neg A$ is false. Since the conjunction of a false statement, $\neg A$, and a true statement, B, is a false statement, the statement $(\neg A) \wedge B$ is false.

(b) Since A is true and B is true, their conjunction $A \wedge B$ is true. Because the negation of a true statement is a false statement,$\neg (A \wedge B)$ is a false.

(c) Since A is true, $\neg A$ is false. Because the disjunction of a false statement, $\neg A$, and a true statement, B, is a true statement, the statement $(\neg A) \vee B$ is true.

(d) Since A is true and C is false, their disjunction $A \vee C$ is true. Since the negation of a true statement is a false statement, the statement $\neg (A \vee C)$ is false

(e) Since A is true and C is false, $(A \vee C)$ is true. Since D is false, $\neg D$ is true. Since B is true and $\neg D$ is true, their conjunction $(B \wedge (\neg D))$ is true. Because $(A \vee C)$ is true and $(B \wedge (\neg D))$ is true, their conjunction $(A \vee C) \wedge (B \wedge (\neg D))$ is true.

Now let us consider the following question: "How many different truth tables are possible for statements which contain exactly one simple sentence P?" Notice that the truth table for negation contains the two different columns

(1)

T	F
F	T

The only other possible columns which can occur in a truth table for a statement which contains exactly one simple statement are

(2)

T	F
T	F

We know that the second column in (1) is the negation of the first column, since these columns came from the truth table for negation. Taking the negation of the first column in (2), we get the second column is (2). So if we could write a statement containing P which is true when P is true and also true when P is false, then we would have a statement that was always true regardless of the truth value of P. By negating that statement we would obtain a statement which is always false regardless of the truth value of P. Let us construct a truth table for the statement $P \vee (\neg P)$. As shown in (3), we start with the truth table for $\neg P$ and add a new column on the right of the table labeled $P \vee (\neg P)$.

(3)

P	$\neg$ P	P $\vee$ ($\neg$ P)
T	F	
F	T	

In the first row of the truth table (3), the statement P has truth value T and the statement $\neg P$ has the truth value F, so the disjunction $P \vee (\neg P)$ has the truth value T. In the second row of the truth table (3), the statement P has truth value F and the statement $\neg P$ has the truth value T, so the disjunction $P \vee (\neg P)$ has the truth value T. Thus, the truth table for the statement $P \vee (\neg P)$ is as shown in (4).

(4)

P	$\neg$ P	P $\vee$ ($\neg$ P)
T	F	T
F	T	T

Notice that regardless of the truth value of the statement P, the truth value of the statement $P \vee (\neg P)$ is true and the truth value of the negation of the statement $P \vee (\neg P)$ is false.

DEFINITIONS Tautology and Contradiction

A **tautology** is a statement that is true for every assignment of truth values of its component statements.

A **contradiction** is a statement that is false for every assignment of truth values of its component statements. Thus, a contradiction is the negation of a tautology.

The statement $P \vee (\neg P)$ is the simplest example of a tautology and its negation, $\neg (P \vee (\neg P))$ is an example of a contradiction, but it is not written in the simplest possible form. We will show how to simplify this expression later in this section.

DEFINITION Truth Value Equivalent or Logically Equivalent

Two statements are **truth value equivalent** or **logically equivalent** if and only if they have the same truth values for all assignments of truth values to their component statements.

It follows from the definition above that two statements which appear in the same truth table are logically equivalent if and only if their truth value columns are identical. Two statements which appear in different truth tables are logically equivalent if and only if both tables are in standard form, both tables are for statements with the same components, and their truth value columns are identical.

We need to develop an algebra for symbolic logic, so we can make calculations similar to the way in which we make algebraic calculations for real numbers. One algebraic property of real numbers is that $-(-x) = x$. So let us produce a truth value table for the statement $\neg(\neg P)$. We start our truth table with a row of column headings labeled P, $\neg P$, and $\neg(\neg P)$. In the column labeled P, we enter as the first entry T and as the second entry F. (See 5a.) Then using the negation truth table, we compute the entries for the second column labeled $\neg P$. Since negation changes the truth value T to F and the truth value F to T, the first entry in the second column is F and the second entry is T. (See 5b.) Applying negation to the second column, the column labeled $\neg P$, we obtain T for the first entry in the third column and F for the second entry. (See 5c.)

(5a)

P	¬ P	¬ (¬ P)
T F		

(5b)

P	¬ P	¬ (¬ P)
T F	F T	

(5c)

P	¬ P	¬ (¬ P)
T F	F T	T F

Observe that the truth value entries in the first column are identical to the truth value entries in the third column. Hence, we have shown that the statements P and ¬ (¬ P) are truth value equivalent. In order to denote that P and $\neg(\neg P)$ are truth value equivalent or logically equivalent, we write $P \equiv \neg(\neg P)$, which is read "P is truth value equivalent to $\neg(\neg P)$" or "P is logically equivalent to $\neg(\neg P)$" Thus, we have used a truth table to prove the Double Negation Law.

Double Negation Law: $\neg(\neg P) \equiv P$

Let t represent a statement which is a tautology, let f represent a statement which is a contradiction, and let P represent any statement. We state the following useful laws involving tautologies and contradictions. You can prove these laws by considering appropriate truth tables.

Tautology Laws:	$\neg\, \mathrm{t} \equiv \mathrm{f}$	$\mathrm{P} \wedge \mathrm{t} \equiv \mathrm{P}$	$\mathrm{P} \vee \mathrm{t} \equiv \mathrm{t}$
Contradiction Laws:	$\neg\, \mathrm{f} \equiv \mathrm{t}$	$\mathrm{P} \wedge \mathrm{f} \equiv \mathrm{f}$	$\mathrm{P} \vee \mathrm{f} \equiv \mathrm{P}$

The following six laws can easily be proven using truth tables.

Idempotent Law for Conjunction:	$P \wedge P \equiv P$
Idempotent Law for Disjunction:	$P \vee P \equiv P$
Commutative Law for Conjunction:	$P \wedge Q \equiv Q \wedge P$
Commutative Law for Disjunction:	$P \vee Q \equiv Q \vee P$
Absorption Laws:	$P \wedge (P \vee Q) \equiv P$
	$P \vee (P \wedge Q) \equiv P$

Earlier, we proved that $\neg(\neg P)$ was logically equivalent to P—that is, $\neg(\neg P) \equiv P$. We noted that this property was analogous to the algebraic property $-(-x) = x$ for real numbers. We would now like to determine what statement is logically equivalent to the negation of the conjunction of P and Q. Thus, we would like to be able to complete the statement $\neg(P \wedge Q) \equiv$ ______. A possible analogy from algebra for the real numbers might be $-(x + y) = (-x) + (-y)$. That is, we might anticipate that the statement $\neg(P \wedge Q)$ is logically equivalent to the statement $(\neg P) \wedge (\neg Q)$. To determine if this is true or not, we construct two standard form truth tables—one for $\neg(P \wedge Q)$ and one for $(\neg P) \wedge (\neg Q)$—and then see if the truth value column for $\neg(P \wedge Q)$ is identical to the truth value column for $(\neg P) \wedge (\neg Q)$ or not. The standard form truth table for $\neg(P \wedge Q)$, which appears in (6a), was constructed by adjoining a new fourth column labeled $\neg(P \wedge Q)$ on the right hand side of the conjunction truth table and then negating the truth values appearing in the third column—the column labeled $P \wedge Q$. The standard form truth table for $(\neg P) \wedge (\neg Q)$, which appears in (6b), was constructed by making column headings P, Q, $\neg$ P, $\neg$ Q, and $(\neg P) \wedge (\neg Q)$. The truth values for the columns labeled P and Q were filled in as usual. Then the truth values for the third column were calculated by negating the truth values appearing in the first column. The truth values for the fourth column were calculated by negating the truth values appearing in the second column. Finally, the truth values for the fifth column were calculated from the conjunction of the truth values appearing in the third and fourth columns.

(6a)

P	Q	$P \wedge Q$	$\neg(P \wedge Q)$
T	T	T	F
T	F	F	T
F	T	F	T
F	F	F	T

(6b)

P	Q	$\neg P$	$\neg Q$	$(\neg P) \wedge (\neg Q)$
T	T	F	F	F
T	F	F	T	F
F	T	T	F	F
F	F	T	T	T

Because the right most column of the two tables in (6a) and (6b) are not identical, $\neg(P \wedge Q)$ is not logically equivalent to $(\neg P) \wedge (\neg Q)$ as we had anticipated it might be. Assuming a statement which is logically equivalent to $\neg(P \wedge Q)$ ought to include the statement $(\neg P)$ and the statement $(\neg Q)$, we decided to compute the truth table for $(\neg P) \vee (\neg Q)$. To produce this truth table, we simply change the column heading appearing in the fifth column of (6b) from $(\neg P) \wedge (\neg Q)$ to $(\neg P) \vee (\neg Q)$ and compute the truth values for the new fifth column from the disjunction of the truth values appearing in the third and fourth columns. The required truth table appears in (7).

(7)

P	Q	¬ P	$\neg Q$	$(\neg P) \vee (\neg Q)$
T	T	F	F	F
T	F	F	T	T
F	T	T	F	T
F	F	T	T	T

Observe that the column of truth values labeled $\neg(P \wedge Q)$ in (6a) is identical to the column of truth values labeled $(\neg P) \vee (\neg Q)$ in table (7). Hence, $\neg(P \wedge Q)$ is logically equivalent to $(\neg P) \vee (\neg Q)$—that is, $\neg(P \wedge Q) \equiv (\neg P) \vee (\neg Q)$. We have just proved the first of the two De Morgan laws stated below.

De Morgan Laws:

$$\neg(P \wedge Q) \equiv (\neg P) \vee (\neg Q)$$

$$\neg(P \vee Q) \equiv (\neg P) \wedge (\neg Q)$$

EXAMPLE 2 Negating a Compound Statement

Negate the statement "I will get up and I will go to school."

SOLUTION

It is correct to say: "It is not the case that I will get up and I will go to school." However, using the De Morgan $\neg(P \wedge Q) \equiv (\neg P) \vee (\neg Q)$, we can express the negation better as "I will not get up or I will not go to school."

In order to prove logical equivalences algebraically, we need the following Rule of Substitution.

Rule of Substitution

Let P and Q be statements. Let $C(P)$ be a compound statement containing the statement P. And let $C(Q)$ be the same compound statement in which each occurrence of P is replaced by Q. If P and Q are logically equivalent, then $C(P)$ and $C(Q)$ are logically equivalent. That is,

$$\text{If } P \equiv Q, \text{then } C(P) \equiv C(Q).$$

EXAMPLE 3 Proving a Logical Equivalence Algebraically

Using stated laws and the rule of substitution, prove algebraically that $P \wedge Q \equiv \neg((\neg P) \vee (\neg Q))$

SOLUTION

1. $P \wedge Q \equiv \neg(\neg(P \wedge Q))$ — By the Double Negation Law
2. $\neg(P \wedge Q) \equiv (\neg P) \vee (\neg Q)$ — By a De Morgan Law
3. $P \wedge Q \equiv \neg((\neg P) \vee (\neg Q))$ — By the Rule of Substitution (Substituting 2. into 1.)

This example proves that it is logically possible to eliminate the conjunction operator, $\wedge$, because the operator can be expressed in terms of the negation operator, $\neg$, and the disjunction operator, $\vee$. Hence, all statements could be written using $\neg$ and $\vee$ only. If we were to write the statement "I will get up and I will go to school." using only "not" and "or" as indicated in 3, we would have to write "It is not the case that I will not get up or I will not go to school." Clearly, you can see why we prefer to use all three logical operators "not," "and," and "or." A proof similar to the one presented in example 3 can be constructed to show that $P \vee Q \equiv \neg((\neg P) \wedge (\neg Q))$. Consequently, every statement could be written in terms of $\neg$ and $\wedge$ only.

EXAMPLE 4 Simplifying a Statement Algebraically

Earlier in this section, we showed that the statement $\neg(P \vee (\neg P))$ was a contradiction. Simplify this contradiction using stated laws and the rule of substitution.

SOLUTION

1. $\neg(P \vee (\neg P)) \equiv (\neg P) \wedge (\neg(\neg P))$ — By a De Morgan Law
2. $\neg(\neg P) \equiv P$ — By the Double Negation Law
3. $\neg(P \vee (\neg P)) \equiv (\neg P) \wedge P$ — By the Rule of Substitution (Substituting 2. into 1.)

Since $\neg(P \vee (\neg P))$ is a contradiction, its logical equivalent $(\neg P) \wedge P$ is a contradiction also. The contradiction $(\neg P) \wedge P$ is called **the law of the excluded middle**. The fact that the statement $(\neg P) \wedge P$ is always false simply means that "not P" and "P" can not both be true simultaneously.

One algebraic property of the real numbers is the distributive law. The distributive law for real numbers says: "For all real numbers x, y, and z, $x \cdot (y + z) = (x \cdot y) + (x \cdot z)$." Thus, "multiplication distributes over addition." Analogously in logic, if disjunction is to distribute over conjunction, we must be able to prove $P \vee (Q \wedge R) \equiv (P \vee Q) \wedge (P \vee R)$. We have seen that a truth table for a compound statement which contains exactly one simple statement P has two rows and that a truth table for a compound statement which contains exactly two simple statements P and Q has four rows. Thus, a truth table for a compound statement which contains exactly three simple statements P, Q, and R has eight rows, since P can assume two values (T or F), Q can assume two values, R can assume two values, and $2 \times 2 \times 2 = 8$. We construct two truth tables in standard form—one for $P \vee (Q \wedge R)$ and one for $(P \vee Q) \wedge (P \vee R)$. Hence, we construct one truth table with column headings P, Q, R, $Q \wedge R$, and $P \vee (Q \wedge R)$. And, we construct a second truth table with column headings P, Q, R, $(P \vee Q)$, $(P \vee R)$, and $(P \vee Q) \wedge (P \vee R)$. In each table, we fill in the columns for P, Q, and R as shown and, then, successively calculate the columns of the table.

P	Q	R	$Q \wedge R$	$P \vee (Q \wedge R)$
T	T	T	T	T
T	T	F	F	T
T	F	T	F	T
T	F	F	F	T
F	T	T	T	T
F	T	F	F	F
F	F	T	F	F
F	F	F	F	F

P	Q	R	$P \vee Q$	$P \vee R$	$(P \vee Q) \wedge (P \vee R)$
T	T	T	T	T	T
T	T	F	T	T	T
T	F	T	T	T	T
T	F	F	T	T	T
F	T	T	T	T	T
F	T	F	T	F	F
F	F	T	F	T	F
F	F	F	F	F	F

Since the truth values appearing in the column $P \vee (Q \wedge R)$ are identical to the truth values appearing in the column $(P \vee Q) \wedge (P \vee R)$, we have proven the distributive law for disjunction: $P \vee (Q \wedge R) \equiv (P \vee Q) \wedge (P \vee R)$. The distributive law for conjunction and the associative laws for disjunction and conjunction stated below can easily be proved using truth tables.

Distributive Law for Disjunction:	$P \vee (Q \wedge R) \equiv (P \vee Q) \wedge (P \vee R)$
Distributive Law for Conjunction:	$P \wedge (Q \vee R) \equiv (P \wedge Q) \vee (P \wedge R)$
Associative Law for Disjunction:	$(P \vee Q) \vee R \equiv P \vee (Q \vee R)$
Associative Law for Conjunction:	$(P \wedge Q) \wedge R \equiv P \wedge (Q \wedge R)$

EXERCISES 1.2

In exercises 1-14 determine the truth value of the statement given that A and B are true statements and C and D are false statements.

1. $\neg A$
2. $\neg C$
3. $A \vee C$
4. $C \vee D$
5. $A \wedge B$
6. $B \wedge C$
7. $(\neg A) \wedge B$
8. $(\neg A) \vee (\neg B)$
9. $A \vee (C \wedge D)$
10. $(A \vee C) \wedge D)$
11. $A \vee (B \wedge (\neg C))$
12. $(A \vee B) \wedge (\neg D)$
13. $(\neg(A \wedge (\neg B))) \wedge ((\neg C) \vee D)$
14. $((\neg A) \vee B) \wedge (C \vee (\neg D))$

15. Given that P is a true statement, what can you say about the truth value of the following statements?

    a. $P \vee (Q \wedge R)$ b. $P \wedge R$ c. $(\neg P) \wedge (Q \wedge R)$

16. Given that P is a false statement, what can you say about the truth value of the following statements?

    a. $P \wedge R$ b. $P \vee (\neg R)$ c. $(\neg P) \vee (Q \wedge R)$

In exercises 17-21, use a truth table to prove the given logical equivalences.

17. $P \wedge P \equiv P$ idempotent law for conjunction
18. $P \vee Q \equiv Q \vee P$ commutative law for disjunction
19. $P \wedge (P \vee Q) \equiv P$ an absorption law
20. $P \wedge (Q \vee R) \equiv (P \wedge Q) \vee (P \wedge R)$ distributive law for conjunction
21. $(P \wedge Q) \wedge R \equiv P \wedge (Q \wedge R)$ associative law for conjunction

In exercises 22-27, construct a truth table for the given compound statement and identify tautologies and contradictions.

22. $(\neg P) \vee Q$
23. $\neg(P \wedge (\neg Q))$
24. $P \vee (Q \vee (\neg P))$
25. $P \wedge (\neg(Q \vee (\neg Q)))$
26. $P \vee (\neg(Q \vee (\neg Q)))$
27. $(P \wedge (Q \vee (\neg R))) \vee ((\neg P) \vee R)$

28. Find compound statements involving simple statements P, Q which have the following truth tables. For example, a statement for (f) is $\neg P$.

P	Q	(a)	(b)	(c)	(d)	(e)	(f)	(g)
T	T	T	F	T	T	T	F	F
T	F	T	T	F	T	T	F	T
F	T	T	T	T	F	T	T	F
F	F	T	T	T	T	F	T	T

P	Q	(h)	(i)	(j)	(k)	(l)	(m)	(n)
T	T	F	T	F	F	F	T	F
T	F	T	F	F	F	T	F	F
F	T	T	F	F	T	F	F	F
F	F	F	T	T	F	F	F	F

In exercises 29-34, use De Morgan's laws to negate the given statement.

29. I sweeten my tea with sugar, or I sweeten my tea with honey.

30. I sweeten my tea with sugar, but I do not sweeten my tea with honey.

31. I do not drink my coffee with sugar, and I do not drink my coffee with cream.

32. I drink my coffee with sugar; however, I do not drink my coffee with cream.

33. Although I do not go to the opera, I do go to the theater.

34. Alice did not go to France, or Alice did not go to Italy.

In exercises 35-40, use the stated laws and the rule of substitution to simplify the given expressions.

35. $\neg(P \vee (\neg Q))$
36. $\neg(P \wedge (\neg Q)) \vee Q$
37. $P \wedge ((\neg P) \vee Q)$
38. $Q \vee (P \wedge (\neg Q))$
39. $(P \wedge Q) \vee (P \wedge (\neg Q))$
40. $(P \wedge Q) \vee ((\neg Q) \wedge (P \vee R))$

As noted earlier, in English, the word "or" has two related but distinguishable meanings. The "inclusive or," $\vee$, is the "or" used most generally in mathematics and it means "one or the other or both". The "exclusive or" means "one or the other but not both". We will let $\triangledown$ denote the "exclusive or" which is defined by the truth table

Exclusive Disjunction

P	Q	P $\triangledown$ Q
T	T	F
T	F	T
F	T	T
F	F	F

In exercises 41-44 identify the disjunction as inclusive or exclusive.

41. Ted will walk to Mary's house or Ted will drive to Mary's house.

42. I will eat dinner or I will go to a movie.

43. Dinner starts with soup or dinner starts with a salad.

44. Coffee is served after dinner with sugar or coffee is served after dinner with cream.

45. Construct a standard form truth table for $(P \vee Q) \wedge (\neg(P \wedge Q))$. Compare this truth table with the exclusive disjunction table. What do you conclude?

46. (i) Construct standard form truth tables for the following:

    (a) $(\neg P) \triangledown Q$ (b) $P \triangledown (\neg Q)$ (c) $\neg(P \triangledown Q)$ (d) $(\neg P) \triangledown (\neg Q)$

    (ii) Which of the expressions in (i) are logically equivalent and which are equivalent to $P \triangledown Q$?

1.3 Conditional and Biconditional Statements

Statements of the form "If P, then Q." occur often and are very important in mathematics. The statement "If P, then Q." is called a **conditional statement**, the statement P is called the **hypothesis** or **antecedent** of the conditional statement, and the statement Q is called the **conclusion** or **consequent** of the conditional statement. For example, in the conditional statement "If n is a prime number greater than two, then n is odd.", the hypothesis is "n is a prime number greater than two" and the conclusion is

"n is odd." Of course, conditional statements occur in every day life as well. For instance, you might recall conditional statements such as "If you clean your room, then you may go to a movie." or "If you mow the grass, I will pay you twenty-five dollars."

A conditional statement asserts that its hypothesis implies its conclusion. The conditional statement, itself, does not assert that its hypothesis is true, but only that if its hypothesis is true, then its conclusion is true. Furthermore, the conditional statement does not assert that its conclusion is true, but only that its conclusion is true if its hypothesis is true. Thus, when the hypothesis and conclusion of a conditional statement are both true, we want the conditional statement to be true. And when the hypothesis is true and the conclusion is false, we want the conditional statement to be false. We denote the statement "If P, then Q." symbolically by $P \Rightarrow Q$ which is read "P implies Q" or "If P, then Q." Just as with negation, conjunction, and disjunction, the conditional statement is defined by its truth table. From the discussion above, the first two rows of the standard truth table for $P \Rightarrow Q$ should be as follows.

Conditional

P	Q	P ⇒ Q
T	T	T
T	F	F
F	T	
F	F	

The essential meaning of the conditional statement appears in the partial truth table above. Earlier, we indicated that in English the word "or" has two meanings—the meaning of the "inclusive or" and the meaning of the "exclusive or." Since we have two missing truth values in the truth table for the conditional statement, there are four possible ways to complete the truth table. The question is: "For use in mathematical discussions, what truth value assignments should we make in the last two rows of the truth table?" Observe that when the hypothesis, P, is true, the truth value of the conditional statement $P \Rightarrow Q$ is identical to the truth value of the conclusion, Q. Also observe that when P is true, the truth value of the statement $(\neg P) \vee Q$ is identical to the truth value of Q. Hence, we define $P \Rightarrow Q$ to be logically equivalent to $(\neg P) \vee Q$. That is, we take $P \Rightarrow Q$ to be an abbreviation for $(\neg P) \vee Q$. Below, we have constructed the standard truth table for $(\neg P) \vee Q$ and have attached at the right a column labeled $P \Rightarrow Q$. Since we have defined the conditional statement $P \Rightarrow Q$ to be logically equivalent to the statement $(\neg P) \vee Q$, the column of truth values for $P \Rightarrow Q$ is identical to the column of truth values for $(\neg P) \vee Q$.

Conditional

P	Q	$\neg P$	$(\neg P) \vee Q$	$P \Rightarrow Q$
T	T	F	T	T
T	F	F	F	F
F	T	T	T	T
F	F	T	T	T

Let P stand for the statement "3 = 4" which is false, and let Q stand for the statement "$\sqrt{2}$ is rational." which is false. It is easier to see that the conditional statement $P \Rightarrow Q$, "If 3 = 4, then $\sqrt{2}$ is rational.", is true when it is written in the logically equivalent form $(\neg P) \vee Q$, "Not 3 = 4 or $\sqrt{2}$ is rational." In the form "Not 3 = 4 or $\sqrt{2}$ is rational." it is clear that "Not 3 = 4" is true and, therefore, the conjunction "Not 3 = 4 or $\sqrt{2}$ is rational." is true.

We have defined $P \Rightarrow Q$ to be logically equivalent to $(\neg P) \vee Q$, and we can see from the truth table for the conditional statement that it may be defined as follows also.

DEFINITION The Conditional Statement

Given two statements P and Q, the **conditional statement** $P \Rightarrow Q$ (read "P implies Q") is the statement "If P, then Q." The conditional statement $P \Rightarrow Q$ is true unless P is true and Q is false, in which case it is false.

Later, in constructing some proofs and counterexamples, we will need the negation of the conditional statement. Since the columns of truth values for $P \Rightarrow Q$ and $(\neg P) \vee Q$ are identical, the columns of truth values for their negations will be identical also. Hence,

1. $\neg(P \Rightarrow Q) \equiv \neg((\neg P) \vee Q)$ — By definition of logically equivalence
2. $\neg((\neg P) \vee Q) \equiv (\neg(\neg P)) \wedge (\neg Q)$ — By a De Morgan Law
3. $\neg(P \Rightarrow Q) \equiv (\neg(\neg P)) \wedge (\neg Q)$ — By the Rule of Substitution (Substituting 2. into 1.)
4. $\neg(\neg P) \equiv P$ — By the Double Negation Law
5. $\neg(P \Rightarrow Q) \equiv P \wedge (\neg Q)$ — By the Rule of Substitution (Substituting 4. into 3.)

Thus, we have proven the following logical equivalence for the negation of the conditional statement.

Negation of the Conditional Statement
$\neg(P \Rightarrow Q) \equiv P \wedge (\neg Q)$

There are many different ways to express the conditional statement, $P \Rightarrow Q$, in words. The following is a nonexhaustive list.

Alternative Expressions for the Conditional Statement $P \Rightarrow Q$	
If P, then Q	Q, if P
P implies Q	Q is implied by P
P only if Q	Q provided P
P is sufficient for Q	Q is necessary for P

EXAMPLE 1 Analyzing Conditional Statements and Writing Their Negations

For each of the following conditional statements:

(1) identify the hypothesis and conclusion;

(2) determine the truth value of the hypothesis, the conclusion, and the conditional statement; and

(3) write the negation of the conditional statement.

(a) If $1 + 1 = 2$, then $1 + 2 = 3$.

(b) If $1 + 1 = 2$, then $1 + 2 = 4$.

(c) The number $\sqrt{2}$ is rational, if $2 + 2 = 5$.

(d) The number $\sqrt{2}$ is irrational, if $2 + 2 = 5$.

(e) The moon is made of green cheese is necessary for the golden gate bridge to be in California.

(f) The moon is made of green cheese is sufficient for the golden gate bridge to be in California.

SOLUTION

(a) The hypothesis is "$1 + 1 = 2$" which is true and the conclusion is "$1+2 = 3$" which is true. Since the hypothesis is true and the conclusion is true, the conditional statement "If $1 + 1 = 2$, then $1 + 2 = 3$." is true. The negation of a conditional statement is the conjunction of the hypothesis and the negation of the conclusion. Thus, the negation of "If $1 + 1 = 2$, then $1 + 2 = 3$." is "$1 + 1 = 2$ and not $1 + 2 = 3$."

(b) The hypothesis is "$1 + 1 = 2$" which is true and the conclusion is "$1+2 = 4$" which is false. Since the hypothesis is true and the conclusion is false, the conditional statement "If $1+1 = 2$, then $1+2 = 4$." is false. The negation of the conditional statement "If $1+1 = 2$, then $1+2 = 4$." is "$1 + 1 = 2$ and not $1 + 2 = 4$."

(c) The hypothesis is "$2 + 2 = 5$" which is false and the conclusion is "the number $\sqrt{2}$ is rational" which is false. Since the hypothesis is false, the conditional statement "The number $\sqrt{2}$ is rational, if $2 + 2 = 5$." is true. The negation of this conditional statement is "$2 + 2 = 5$ and the number $\sqrt{2}$ is not rational."

(d) The hypothesis is "$2 + 2 = 5$" which is false and the conclusion is "the number $\sqrt{2}$ is irrational" which is true. Since the hypothesis is false, the conditional statement "The number $\sqrt{2}$ is irrational, if $2 + 2 = 5$." is true. The negation of this conditional statement is "$2 + 2 = 5$ and the number $\sqrt{2}$ is not irrational."

(e) The hypothesis is "The golden gate bridge is in California." which is true and the conclusion is "The moon is made of green cheese." which is false. Since the hypothesis is true and the conclusion is false, the conditional statement "The moon is made of green cheese is necessary for the golden gate bridge to be in California." is false. The negation of this conditional statement is "The golden gate bridge is in California and the moon is not made of green cheese."

(f) The hypothesis is "The moon is made of green cheese." which is false and the conclusion is "The golden gate bridge is in California." which is true. Since the hypothesis is false, the conditional statement "The moon is made of green cheese is sufficient for the golden gate bridge to be in California." is true. The negation of this conditional statement is "The moon is made of green cheese and the golden gate bridge is not in California."

Every conditional statement has associated with it three other statements: the **converse**, the **inverse**, and **contrapositive**.

DEFINITION Converse, Inverse, and Contrapositive

The **converse** of $P \Rightarrow Q$ is $Q \Rightarrow P$.

The **inverse** of $P \Rightarrow Q$ is $(\neg P) \Rightarrow (\neg Q)$.

The **contrapositive** of $P \Rightarrow Q$ is $(\neg Q) \Rightarrow (\neg P)$.

EXAMPLE 2 The Converse, Inverse, and Contrapositive of a Conditional Statement

Write the converse, inverse, and contrapositive of the conditional statement "If we win the game, we will celebrate."

SOLUTION

The converse is "If we celebrate, we will win the game."

The inverse is "If we do not win the game, we will not celebrate."

The contrapositive is "If we do not celebrate, we will not win the game."

We constructed the following truth table for the conditional statement, $P \Rightarrow Q$; its converse, $Q \Rightarrow P$; its inverse, $(\neg P) \Rightarrow (\neg Q)$; and its contrapositive, $(\neg Q) \Rightarrow (\neg P)$.

P	Q	$P \Rightarrow Q$	$Q \Rightarrow P$	$\neg P$	$\neg Q$	$(\neg P) \Rightarrow (\neg Q)$	$(\neg Q) \Rightarrow (\neg P)$
T	T	T	T	F	F	T	T
T	F	F	T	F	T	T	F
F	T	T	F	T	F	F	T
F	F	T	T	T	T	T	T

Examining the columns of truth values, we find that $P \Rightarrow Q$ is logically equivalent to $(\neg Q) \Rightarrow (\neg P)$, since their columns of truth values are identical and that $Q \Rightarrow P$ is logically equivalent to $(\neg P) \Rightarrow (\neg Q)$ since their columns of truth values are identical.

DEFINITION The Biconditional Statement

Given two statements P and Q, the **biconditional statement** $P \Leftrightarrow Q$ is the statement "P if and only if Q." The biconditional statement $P \Leftrightarrow Q$ is true when P and Q have the same truth values and false when P and Q have different truth values.

In mathematics, the phrase "if and only if" is often abbreviated by "iff" and sometimes "if and only if" is expressed in the alternate form "is necessary and sufficient for." The truth table for the biconditional statement appears below.

Biconditional

P	Q	P ⇔ Q
T	T	T
T	F	F
F	T	F
F	F	T

EXAMPLE 3 Constructing a Truth Table for $(P \Rightarrow Q) \wedge (Q \Rightarrow P)$

Construct the standard truth table for $(P \Rightarrow Q) \wedge (Q \Rightarrow P)$ and adjoin the column of truth values for the biconditional statement $P \Leftrightarrow Q$ at the right. What can you conclude about $(P \Rightarrow Q) \wedge (Q \Rightarrow P)$ and $(P \Leftrightarrow Q)$?

SOLUTION

We make column headings P, Q, $P \Rightarrow Q$, $Q \Rightarrow P$, $(P \Rightarrow Q) \wedge (Q \Rightarrow P)$, and $(P \Leftrightarrow Q)$. We fill in the appropriate truth value columns for P and Q, and then we determine and enter the remaining truth values by column from left to right. Thus, we obtain the following standard truth table.

P	Q	$P \Rightarrow Q$	$Q \Rightarrow P$	$(P \Rightarrow Q) \wedge (Q \Rightarrow P)$	$(P \Leftrightarrow Q)$
T	T	T	T	T	T
T	F	F	T	F	F
F	T	T	F	F	F
F	F	T	T	T	T

Since the truth value columns for $(P \Rightarrow Q) \wedge (Q \Rightarrow P)$ and $(P \Leftrightarrow Q)$ are identical, $(P \Rightarrow Q) \wedge (Q \Rightarrow P)$ is logically equivalent to $(P \Leftrightarrow Q)$—that is, $(P \Rightarrow Q) \wedge (Q \Rightarrow P) \equiv (P \Leftrightarrow Q)$.

In algebra, we have an established hierarchy for operations. For algebraic expression with no parentheses, the operation of exponentiation is performed first and from left to right; next, the operations of multiplication and division are performed from left to right; and finally, the operations of addition and subtraction are performed from left to right. Thus, it is understood that $x + y * z$ written without any additional parentheses means $x + (y * z)$ and not $(x+y)*z$. The hierarchy for connectives in symbolic logic for expressions with no parentheses is: negation $\neg$ is performed first, next conjunction $\wedge$ and disjunction $\vee$ are performed, and lastly the conditional $\Rightarrow$ and biconditional $\Leftrightarrow$ connectives are performed. Hence, by this convention $\neg P \Rightarrow Q \wedge R$ means $(\neg P) \Rightarrow (Q \wedge R)$. However, without parentheses the meaning of the expression $P \wedge Q \vee R$ is ambiguous. Does it mean $(P \wedge Q) \vee R$ or $P \wedge (Q \vee R)$? So in this case, a set of parentheses is necessary to indicate which expression is intended.

EXERCISES 1.3

In exercises 1-8, (1) identify the hypothesis and the conclusion; (2) determine the truth value of the hypothesis, the conclusion, and the conditional statement; and (3) write the negation of the conditional statement.

1. If New York City is on the East Coast, then Los Angeles is on the West Coast.
2. New York City is on the East Coast, provided Los Angeles is on the West Coast.
3. The number π is rational, if the number $\sqrt{2}$ is irrational.
4. The number π is rational, only if the number $\sqrt{2}$ is irrational.
5. $2^3 > 3^2$ is implied by $2 < 3$.
6. $2^3 > 3^2$ is necessary for $2 < 3$.
7. $2^3 > 3^2$ is sufficient for $2 < 3$.
8. $2 < 3$ implies $2^3 > 3^2$.

In exercises 9-16 write the converse, inverse, and contrapositive of statements 1-8 respectively.

In exercises 17-20 determine the truth value of the given biconditional statement.

17. The Leaning Tower of Pisa is in Germany if and only if the Eiffel Tower is in Spain.

18. The Leaning Tower of Pisa is in Italy if and only if the Eiffel Tower is in France.

19. The number 3 is even if and only if the number 4 is even.

20. The number $\sqrt{2}$ is irrational iff the number π is rational.

21. Given that the truth value of the implication $(P \vee Q) \Rightarrow (\neg R)$ is false and that the truth value of P is false, what are the truth values of Q and R?

In exercises 22-33 construct truth tables for the given statement. Identify tautologies and contradictions.

22. $P \Rightarrow (P \vee Q)$
23. $P \Rightarrow (P \wedge Q)$
24. $(P \vee Q) \Rightarrow P$
25. $(P \wedge Q) \Rightarrow P$
26. $(P \wedge Q) \Rightarrow (P \vee Q)$
27. $(P \vee Q) \Rightarrow (P \wedge Q)$
28. $(P \wedge (P \Rightarrow Q)) \Rightarrow Q$
29. $(Q \wedge (P \Rightarrow Q)) \Rightarrow P$
30. $((\neg Q) \wedge (P \Rightarrow Q)) \Rightarrow (\neg P)$
31. $(P \vee Q) \Leftrightarrow P$
32. $P \Leftrightarrow (P \wedge (P \vee Q))$
33. $(P \Rightarrow Q) \Leftrightarrow (Q \Rightarrow P)$

In exercises 34-39 simplify each statement by replacing conditional statement such as $H \Rightarrow C$ by the logically equivalent statement $(\neg H) \vee C$ and using stated laws and the rule of substitution.

34. $(\neg P) \Rightarrow Q$
35. $P \Rightarrow (Q \vee R)$
36. $\neg((\neg P) \Rightarrow (\neg Q))$
37. $(P \Rightarrow Q) \Rightarrow P$
38. $(P \Rightarrow Q) \Rightarrow Q$
39. $(P \Rightarrow Q) \wedge (Q \Rightarrow P)$

1.4 Logical Arguments

Proofs play a major role in mathematics, and deductive reasoning is the foundation on which proofs rest. In mathematics, as in law, a logical argument is a claim that from certain **premises** (statements that are assumed to be true) one can infer a certain **conclusion** (statement) is true. Logic is concerned with the connections between statements and with what deductions can be made, assuming that the premises are true. Let the symbol $\therefore$ stand for the word "therefore." The symbolic form of a logical argument written in horizontal form is

$$P_1, \ldots, P_n \quad \therefore C$$

and written in vertical form is

$$\begin{array}{c} P_1 \\ \vdots \\ P_n \\ \hline \therefore C \end{array}$$

where $P_1, \ldots, P_n$ are premises and C is the conclusion. In logic, we study the methods and principles used in distinguishing valid ("correct") arguments from invalid ("incorrect") arguments. Notice that statements are said to be true or false, while arguments are said to be valid or invalid. We make the following definitions regarding arguments.

DEFINITION Valid and Invalid Arguments

An argument $P_1, \ldots, P_n \therefore C$ with premises $P_1, \ldots, P_n$ and conclusion C is **valid** if and only if $P_1 \wedge P_2 \wedge \ldots \wedge P_n \Rightarrow C$ is a tautology.

If an argument is not valid, it is called **invalid**.

Thus, an argument is valid if and only if when the premises are all true, the conclusion must be true. Stated another way, an argument is valid if and only if it is not possible for the conclusion to be false unless at least one of the premises is false. Validity concerns the relationship between the premises and the conclusion, and not the truth values of the premises and conclusions.

Let us consider the argument

The sun is shining.

Therefore, the sun is shining or it is raining.

Let P denote the premise "The sun is shining." And let Q denote the statement "It is raining." Then this argument can be written symbolically as

$$\frac{P}{\therefore P \vee Q}$$

We construct the following standard truth table for $P \Rightarrow (P \vee Q)$.

P	Q	$P \vee Q$	$P \Rightarrow (P \vee Q)$
T	T	T	T
T	F	T	T
F	T	T	T
F	F	F	T

Observe that the statement $P \Rightarrow (P \vee Q)$ is a tautology and, consequently,

$$\frac{P}{\therefore P \vee Q}$$

is a valid argument. Thus, we have just proven the rule of disjunction.

Rule of Disjunction
The argument $P \quad \therefore P \vee Q$ is a valid argument known as the **rule of disjunction**.

Now, let us consider the symbolic argument

$$\begin{array}{l} P \\ \hline \therefore P \wedge Q \end{array}$$

We construct the following standard truth table for $P \Rightarrow (P \wedge Q)$.

P	Q	$P \wedge Q$	$P \Rightarrow (P \wedge Q)$
T	T	T	T
T	F	F	F
F	T	F	T
F	F	F	T

Observe that the statement $P \Rightarrow (P \wedge Q)$ is not a tautology and; therefore,

$$\begin{array}{l} P \\ \hline \therefore P \wedge Q \end{array}$$

is an invalid argument. In order to show that an argument is invalid, it is not necessary to construct an entire truth table. It is sufficient to show how to choose the truth values of the component statements in such a way that the premises are true and the conclusion is false. Thus, to show that the argument $P \therefore P \wedge Q$ is invalid, we need to show only the first three columns of row two of the proceeding truth table. That is,

P	Q	$P \wedge Q$
T	F	F

shows that the argument $P \therefore P \wedge Q$ is invalid.

The following three arguments are valid.

Rule of Conjunction

The argument $P, \ Q \ \therefore P \wedge Q$ is a valid argument known as the **rule of conjunction**.

Rule of Conjunctive Simplification

The argument $P \wedge Q \ \therefore P$ is a valid argument called the **rule of conjunctive simplification**.

Rule of Disjunctive Syllogism

The argument $P \vee Q, \ \neg P \ \therefore Q$ is a valid argument called the **rule of disjunctive syllogism**.

Truth tables which verify that these arguments are valid appear below.

Rule of Conjunction

P	Q	$P \wedge Q$	$(P \wedge Q) \Rightarrow (P \wedge Q)$
T	T	T	T
T	F	F	T
F	T	F	T
F	F	F	T

Rule of Conjunctive Simplification

P	Q	$P \wedge Q$	$(P \wedge Q) \Rightarrow P$
T	T	T	T
T	F	F	T
F	T	F	T
F	F	F	T

Rule of Disjunctive Syllogism

P	Q	$\neg P$	$P \vee Q$	$(P \vee Q) \wedge (\neg P)$	$(P \vee Q) \wedge (\neg P) \Rightarrow Q$
T	T	F	T	F	T
T	F	F	T	F	T
F	T	T	T	T	T
F	F	T	F	F	T

When an argument contains three or more different simple statements, it becomes tedious to determine the validity of the argument by making a truth table. A more convenient way to prove an argument is valid is to deduce its conclusion from its premises by a sequence of elementary arguments which are known to be valid. Hence, a **formal proof** of a given argument is a sequence of statements such that each statement is either: (1) a premise of the argument, or (2) a statement which follows from preceding statements by an elementary valid argument, or (3) a statement which is logically equivalent to a preceding statement in the sequence, or (4) the last statement in the sequence—the conclusion of the argument. The proper way to write a formal proof is to list the premises and the sequence of statements deduced from them along with a justification for each statement. The justification of each deduced statement specifies the preceding statement or statements from which it was deduced and the rule of inference which was used. Thus far, we have proved by the use of truth tables that four elementary arguments are valid—namely, the rule of disjunction, the rule of conjunction, the rule of conjunctive simplification, and the rule of disjunctive syllogism.

EXAMPLE 1 A Formal Proof of a Valid Argument

Give a formal proof that the argument $P \Rightarrow Q,\ P\ \therefore Q$ is a valid argument.

SOLUTION

1. $P \Rightarrow Q$ premise
2. P premise
3. $(\neg P) \vee Q$ 1, logical equivalence (definition of $\Rightarrow$)
4. $\neg(\neg P)$ 2, logical equivalence (double negation law)
5. Q 3, 4, disjunctive syllogism

Example 1 is a formal proof of the following rule known as the rule of detachment or modus ponens

Rule of Detachment or Modus Ponens

The argument $P \Rightarrow Q, \;\; P \;\; \therefore Q$ is a valid argument known as the **rule of detachment** or **modus ponens**.

EXAMPLE 2 A Formal Proof of a Valid Argument

Give a formal proof that the argument $P \Rightarrow Q, \;\; \neg Q \;\; \therefore \neg P$ is a valid argument.

SOLUTION

1. $P \Rightarrow Q$ premise
2. $\neg Q$ premise
3. $(\neg Q) \Rightarrow (\neg P)$ 1, logical equivalence (the contrapositive of $P \Rightarrow Q$)
4. $\neg P$ 2, 3, rule of detachment

Example 2 is a formal proof of the rule of contrapositive inference or modus tollens.

Rule of Contrapositive Inference or Modus Tollens

The argument $P \Rightarrow Q, \;\; \neg Q \;\; \therefore \neg P$ is a valid argument known as the **rule of contrapositive inference** or **modus tollens.**

The rule of detachment and the rule of contrapositive inference are both valid arguments which contain exactly two simple statements, one of which is the conditional statement $P \Rightarrow Q$. Two commonly used invalid arguments which contain exactly two simple statements, one of which is the conditional statement $P \Rightarrow Q$, are the **Fallacy of the Converse** and the **Fallacy of the Inverse.**

Two Invalid Arguments

The argument $P \Rightarrow Q, \quad Q \quad \therefore P$ is a invalid argument called the **fallacy of the converse.**

The argument $P \Rightarrow Q, \quad \neg P \quad \therefore \neg Q$ is a invalid argument called the **fallacy of the inverse.**

Recall that an argument is valid if and only if when the premises are all true, the conclusion must be true. Thus, in order to prove that the fallacy of the converse is an invalid argument, all we need to do is determine truth values for P and Q such that both premises $P \Rightarrow Q$ and Q have the truth value T and the conclusion P has the truth value F. Assigning P the truth value F and Q the truth value T achieves the required result. Likewise, to prove that the fallacy of the inverse is an invalid argument, we need to determine truth values for P and Q such that both premises $P \Rightarrow Q$ and $\neg P$ have the truth value T and the conclusion $\neg Q$ has the truth value F. Again, assigning P the truth value F and Q the truth value T achieves the required result.

In the following two examples, we determine the validity of two arguments.

EXAMPLE 3 Determining the Validity of an Argument

Write the following argument in symbolic form. Determine if the argument is valid or invalid and state the name of the argument form.

I took a nap, or I watched television.
I did not watch television.

Therefore, I took a nap.

SOLUTION

Let N denote the statement "I took a nap." And let T denote the statement "I watched television." Then the symbolic form of the argument is

$$\begin{array}{l} N \vee T \\ \neg T \\ \hline \therefore N \end{array}$$

This argument has the form of the disjunctive syllogism and it is a valid argument.

EXAMPLE 4 Determining the Validity of an Argument

Write the following argument in symbolic form. Determine if the argument is valid or invalid and state the name of the argument form.

If I went out to eat dinner, then I ate dessert.

I ate dessert.

Therefore, I went out to eat dinner.

SOLUTION

Let D denote the statement "I went out to eat dinner." And let P denote the statement "I ate dessert." Then the symbolic form of the given argument is

$$\begin{array}{l} D \Rightarrow P \\ P \\ \hline \therefore D \end{array}$$

This argument has the form of the fallacy of the converse and it is an invalid argument.

The conditional statement has the following transitive property.

Rule of Transitive Inference or Hypothetical Syllogism

The argument $P \Rightarrow Q,\ \ Q \Rightarrow R\ \therefore P \Rightarrow R$ is a valid argument called the **rule of transitive inference** or **hypothetical syllogism**.

The following truth table establishes the validity of the rule of transitive inference.

Let A denote the statement $(\mathrm{P} \Rightarrow \mathrm{Q}) \wedge (\mathrm{Q} \Rightarrow \mathrm{P})$ and let B denote the statement $[(\mathrm{P} \Rightarrow \mathrm{Q}) \wedge (\mathrm{Q} \Rightarrow \mathrm{P})] \Rightarrow (\mathrm{P} \Rightarrow \mathrm{R})$.

Rule of Transitive Inference

P	Q	R	P ⇒ Q	Q ⇒ R	A	P ⇒ R	B
T	T	T	T	T	T	T	T
T	T	F	T	F	F	F	T
T	F	T	F	T	F	T	T
T	F	F	F	T	F	F	T
F	T	T	T	T	T	T	T
F	T	F	T	F	F	T	T
F	F	T	T	T	T	T	T
F	F	F	T	T	T	T	T

In the following example we show in more detail how to analyze an argument and how to construct a formal proof.

EXAMPLE 5 Analyzing an Argument and Constructing a Formal Proof

Write the following argument in symbolic form and construct a formal proof of its validity.

If I study, I make good grades.
If I do not study, I have fun.

Therefore, I make good grades or I have fun.

SOLUTION

Let S denote the statement "I study." Let G denote the statement "I make good grades." And let F denote the statement "I have fun." Then the symbolic form of the given argument is

$$(1) \qquad \begin{array}{l} S \Rightarrow G \\ (\neg S) \Rightarrow F \\ \hline \therefore G \vee F \end{array}$$

Both premises are conditional statements. One contains S as a hypothesis and the other contains $\neg S$ as a hypothesis. Because a conditional statement and its contrapositive are logically equivalent, we could replace the first or second premise by its contrapositive. Since the conclusion contains the statement F and since the second premise contains the statement F as a conclusion, we decided to replace the first premise $S \Rightarrow G$ by its contrapositive $(\neg G) \Rightarrow (\neg S)$. From the true statements $(\neg G) \Rightarrow (\neg S)$ and $(\neg S) \Rightarrow F$ we deduce by the rule of transitive inference $(\neg G) \Rightarrow F$. From the definition of the conditional statement, $(\neg G) \Rightarrow F$ is logically equivalent to the statement $(\neg(\neg G)) \vee F$.

Since $(\neg(\neg G)) \equiv G$, we obtain the desired conclusion $G \vee F$. Hence, we have developed the following formal proof that (1) is a valid argument.

1. $S \Rightarrow G$	premise
2. $(\neg S) \Rightarrow F$	premise
3. $(\neg G) \Rightarrow (\neg S)$	1, logical equivalence (contrapositive of 1)
4. $(\neg G) \Rightarrow F$	2, 3, transitivity of inference
5. $(\neg(\neg G)) \vee F$	4, logical equivalence (definition of $\Rightarrow$)
6. $(\neg(\neg G)) \equiv G$	logical equivalence (double negation law)
7. $G \vee F$	5, 6, rule of substitution

In addition to using symbolic logic to prove that an argument is valid, symbolic logic may be used to deduce a valid consequence or consequences from a collection of premises. When possible, one tries to use all of the given premises to deduce a valid conclusion. When it is not possible to use all the premises to deduce a valid conclusion, then one generally uses as large a subcollection of premises as possible to deduce a valid conclusion.

EXAMPLE 6 Deducing a Valid Conclusion

Deduce a valid conclusion for the following argument.

Ursula went to the conference by train or by airplane.

If Ursula went to the conference by train or drove her own car, then she arrived late and she missed the opening ceremony.

Ursula did not arrive late.

Therefore, ______________________

SOLUTION

Let T denote the statement "Ursula went to the conference by train." Let A denote the statement "Ursula went to the conference by airplane." Let C denote the statement "Ursula drove her own car." Let L denote the statement "Ursula arrived late." And let M denote the statement "Ursula missed the opening ceremony." Then the symbolic form of the given argument is

$T \vee A$
$(T \vee C) \Rightarrow (L \wedge M)$
$\neg L$

$\therefore$ ______________

The third premise $\neg L$ is true by assumption. Hence, L is false and $L \wedge M$ is false. Consequently, $\neg(L \wedge M)$ is true. From $\neg(L \wedge M)$ and the second premise, $(T \vee C) \Rightarrow (L \wedge M)$, it follows using the rule of contrapositive inference that $\neg(T \vee C)$. Then by a De Morgan law, we have $(\neg T) \wedge (\neg C)$ and by the rule of conjunctive simplification, we get $(\neg T)$. From $(\neg T)$ and the first premise, $T \vee A$, we conclude A using the rule of disjunctive syllogism. Hence, "Ursala went to the conference by airplane." is a valid conclusion, which we deduced using all of the given premises.

EXAMPLE 7 Deducing a Valid Conclusion

Deduce a valid conclusion for the following argument.

If Rob studies medicine, Rob prepares to earn a good living.

If Rob studies humanities, Rob prepares to live the good life.

If Rob prepares to earn a good living or Rob prepares to live a good life, then Rob's years at the university are well spent.

Rob's years at the university are not well spent.

Therefore, ______________

SOLUTION

Let M denote the statement "Rob studies medicine." Let E denote the statement "Rob prepares to earn a good living." Let H denote the statement "Rob studies humanities." Let L denote the statement "Rob prepares to live a good life." And let U denote the statement "Rob's years at the university are well spent." Then the symbolic form of the given argument is

$M \Rightarrow E$
$H \Rightarrow L$
$(E \vee L) \Rightarrow U$
$\neg U$

$\therefore$ ______________

Applying the rule of contrapositive inference (modus tollens) to the third and fourth premises, we infer $\neg(E \vee L)$ which is logically equivalent to $(\neg E) \wedge (\neg L)$. Using the rule of conjunctive simplification twice, yields $\neg E$ and $\neg L$. From the first premise $M \Rightarrow E$ and $\neg E$, we conclude by modus tollens $\neg M$. Likewise, from $\neg L$ and the second premise $H \Rightarrow L$ we conclude $\neg H$ by modus tollens. Hence, two valid conclusions we can deduce are $\neg M$ ("Rob does not study medicine.") and $\neg H$ ("Rob does not study humanities.").

It should be noted we can deduce three valid conclusions from the first two premises alone—namely, $(M \vee H) \Rightarrow (E \vee L)$, $(M \wedge H) \Rightarrow (E \wedge L)$, and $(M \wedge H) \Rightarrow (E \vee L)$. However, in doing so, we have not made use of the remaining two premises.

EXERCISES 1.4

In exercises 1-5, construct an appropriate truth table in standard form to determine if the given argument is valid or invalid.

1. $P \wedge Q,\ P \Rightarrow \neg Q \therefore P \wedge (\neg Q)$
2. $P,\ (\neg P) \Rightarrow Q \therefore \neg Q$
3. $P \wedge Q,\ R \Rightarrow P \therefore Q \vee R$
4. $\neg P,\ Q \Rightarrow (P \wedge R) \therefore \neg Q$
5. $P \Rightarrow Q,\ Q \Rightarrow R,\ \neg Q \therefore \neg R$

In exercises 6-10, prove each argument is invalid by choosing truth values for the component statements in such a way that all the premises are true and the conclusion is false.

6. $P \vee (\neg Q),\ \neg Q \therefore P$
7. $(\neg P) \Rightarrow Q,\ P \therefore \neg Q$
8. $P \Rightarrow Q,\ (\neg P) \Rightarrow Q,\ Q \therefore P$
9. $P \Rightarrow Q,\ Q \Rightarrow R,\ R \therefore Q$
10. $P \Rightarrow Q,\ R \Rightarrow S,\ Q \vee R \therefore P \vee S$

In exercises 11-18, write each argument in symbolic form by letting M denote the statement "I go to a movie." and P denote the statement "I eat popcorn." Determine if the argument is valid or invalid and state the name of the argument form.

11. I go to a movie.

    Therefore, I go to a movie, or I eat popcorn.

12. I go to a movie, and I eat popcorn.

    Therefore, I eat popcorn.

13. I go to a movie.
    I eat popcorn.

    Therefore, I go to a movie, and I eat popcorn.

14. If I go to a movie, then I eat popcorn.
    I eat popcorn.

    Therefore, I go to a movie.

15. If I go to a movie, I eat popcorn.
    I go to a movie.

    Therefore, I eat popcorn.

16. I go to a movie, or I eat popcorn.
    I do not eat popcorn.

    Therefore, I go to a movie.

17. If I go to a movie, I eat popcorn.
    I do not eat popcorn.

    Therefore, I do not go to a movie.

18. If I go to a movie, I eat popcorn.
    I do not go to a movie.

    Therefore, I do not eat popcorn.

In exercises 19-24, a formal proof of the validity of an argument is given. State a justification for each step in the proof.

19. $A \wedge B \therefore A \vee B$

1. $A \wedge B$	premise
2. A	1,
3. B	1,
4. $A \vee B$	2, 3,

20. $C \vee (D \wedge E),\ \neg E \therefore C$

1.	$C \vee (D \wedge E)$	premise
2.	$\neg E$	premise
3.	$(C \vee D) \wedge (C \vee E)$	1,
4.	$C \vee E$	3,
5.	C	2, 4,

21. $F \Rightarrow G,\ G \Rightarrow H,\ F \wedge I \therefore H \wedge I$

1.	$F \Rightarrow G$	premise
2.	$G \Rightarrow H$	premise
3.	$F \wedge I$	premise
4.	$F \Rightarrow H$	1, 2,
5.	F	3,
6.	H	4, 5,
7.	I	3,
8.	$H \wedge I$	6, 7,

22. $J \Rightarrow K,\ J \Rightarrow L \therefore J \Rightarrow (K \wedge L)$

1.	$J \Rightarrow K$	premise
2.	$J \Rightarrow L$	premise
3.	$(\neg J) \vee K$	1,
4.	$(\neg J) \vee L$	2,
5.	$[(\neg J) \vee K] \wedge [(\neg J) \vee L]$	3, 4,
6.	$(\neg J) \vee (K \wedge L)$	5,
7.	$J \Rightarrow (K \wedge L)$	6,

23. $M \Rightarrow (N \Rightarrow O) \therefore N \Rightarrow (M \Rightarrow O)$

1.	$M \Rightarrow (N \Rightarrow O)$	premise
2.	$(\neg M) \vee (N \Rightarrow O)$	1,
3.	$(N \Rightarrow O) \equiv (\neg N) \vee O$	
4.	$(\neg M) \vee ((\neg N) \vee O)$	2, 3,
5.	$((\neg M) \vee (\neg N)) \vee O$	4,
6.	$((\neg M) \vee (\neg N)) \equiv ((\neg N) \vee (\neg M))$	
7.	$((\neg N) \vee (\neg M)) \vee O$	5, 6,
8.	$(\neg N) \vee ((\neg M) \vee O)$	7,
9.	$((\neg M) \vee O) \equiv (M \Rightarrow O)$	
10.	$(\neg N) \vee (M \Rightarrow O)$	8, 9,
11.	$N \Rightarrow (M \Rightarrow O)$	10,

24. a. The following is a formal proof of a valid argument known as the **constructive dilemma:**

$$P \Rightarrow Q, \quad R \Rightarrow S, \quad P \vee R \quad \therefore\ Q \vee S$$

1.	$P \Rightarrow Q$	premise
2.	$R \Rightarrow S$	premise
3.	$P \vee R$	premise
4.	$\neg P \vee Q$	1,
5.	$(\neg P \vee Q) \vee S$	4,
6.	$\neg P \vee (Q \vee S)$	5,
7.	$\neg R \vee S$	2,
8.	$(\neg R \vee S) \vee Q$	7,
9.	$\neg R \vee (S \vee Q)$	8,
10.	$(S \vee Q) \equiv (Q \vee S)$	
11.	$\neg R \vee (Q \vee S)$	9,10
12.	$[\neg P \vee (Q \vee S)] \wedge [\neg R \vee (Q \vee S)]$	6,11,
13.	$[(\neg P) \wedge (\neg R)] \vee (Q \vee S)$	12,
14.	$[(\neg P) \wedge (\neg R)] \equiv \neg(P \vee R)$	
15.	$\neg(P \vee R) \vee (Q \vee S)$	13,14
16.	$(P \vee R) \Rightarrow (Q \vee S)$	15,
17.	$Q \vee S$	3,16,

b. The following is a formal proof of a valid argument known as the **destructive dilemma:**

$$P \Rightarrow Q, \quad R \Rightarrow S, \quad \neg Q \vee \neg S \quad \therefore\ \neg P \vee \neg R$$

(**HINT:** The constructive dilemma (see part a.) occurs as one justification.)

1.	$P \Rightarrow Q$	premise
2.	$R \Rightarrow S$	premise
3.	$\neg Q \vee \neg S$	premise
4.	$\neg Q \Rightarrow \neg P$	1,
5.	$\neg S \Rightarrow \neg R$	2,
6.	$\neg P \vee \neg R$	3, 4, 5

In exercises 25-33, construct a formal proof of the validity of each argument.

25. $A \therefore B \Rightarrow A$
26. $\neg C \therefore C \Rightarrow D$
27. $E \wedge F \therefore E \vee F$
28. $G \Rightarrow H \therefore G \Rightarrow (H \vee I)$
29. $J \Rightarrow (K \wedge L) \therefore J \Rightarrow K$
30. $M \Rightarrow N \therefore M \wedge O \Rightarrow N$
31. $(P \vee Q) \Rightarrow R \therefore P \Rightarrow R$
32. $S \Rightarrow (T \vee U),\ \neg T \therefore S \Rightarrow U$
33. $V \Rightarrow W,\ V \vee W \therefore W$

In exercises 34-38, write each argument in symbolic form using the letters indicated and construct a formal proof of the validity of each argument.

34. If Jerry uses an artificial lure (L), then if the fish are biting (B), then Jerry catches the legal limit of fish (F). Jerry uses an artificial lure, but Jerry does not catch the legal limit of fish. Therefore, the fish are not biting.

35. If Alex attends class (A) or Bob attends class (B), then Charles does not attend class ($\neg C$). Bob attends class or Charles attends class. If Bob attends class or Alex does not attend class, then Don attends class (D). Alex attends class. Therefore, Bob does not attend class or Don attends class.

36. If Emery studies (S), Emery will graduate (G). If Emery graduates, Emery will travel (T) or Emery will work for his uncle (U). Emery studies, but Emery does not work for his uncle. Therefore, Emery will travel.

37. I will take a vacation (V), provided I have time (T) and I have money (M). I have time or I have aspirations (A). Therefore, if I do not have aspirations, I have no money or I will take a vacation.

38. If Robin goes to the state park (P), Robin hikes (H) and Robin fishes (F). Robin did not hike or Robin did not fish or Robin did camp (C). Robin did not camp. Therefore, Robin did not go to the state park.

In exercises 39-43, write each argument in symbolic form using the letters indicated and deduce a valid conclusion for the argument.

39. Elsa will attend (A), if she receives the e-mail (E), provided she is interested (I). Although Elsa did not attend, she is interested. Therefore, ______________________________.

40. If the supply of gold remains fixed (F) and the use of gold increases (I), then the price of gold will rise (R). The price of gold does not rise. Therefore, ______________________________.

41. If Alice attends the meeting (A), Betty attends the meeting (B). If Betty attends the meeting, Carol will not attend the meeting ($\neg C$). If Carol attends the meeting, Donna does not attend the meeting ($\neg D$). If Betty attends the meeting, Eve does not attend the meeting ($\neg E$). If Donna does not attend the meeting, Fay attends the meeting (F). Eve does not attend the meeting or Fay does not attend the meeting. Therefore, ______________________________.

42. If Imogene goes to the picnic (P), then Imogene wears blue jeans (J). If Imogene wears blue jeans, then Imogene does not attend both the banquet (B) and the dance (D). Imogene attended the dance. If Imogene did not attend the banquet, then she has her banquet ticket (T), but she does not have her ticket. Therefore, __________________________.

43. He will be interested (I) if and only if he is an acquaintance (A) or he is curious (C). If he is in management (M) or he is a shareholder (S), he is curious. He is a shareholder, but he is not an acquaintance. Therefore, __________________________.

1.5 Open Statements and Quantifiers

Chapter 3 is devoted to the study of sets; however, at this point we need to introduce some very basic concepts related to sets and some notation used in conjunction with sets. The German mathematician Georg Cantor (1845-1918) initiated the theory of sets in the 1870s and 1880s. According to Cantor,

> "A **set** is any collection of definite, distinguishable objects of our intuition or our intellect to be conceived as a whole."

The objects of a set are called the **elements** or **members** of the set. Every day we use words which convey the meaning of a set. For example, when we say "The committee recommended a course in logic be required for graduation.", we are considering the committee to be a set and the members of the committee are the members (elements) of the set. When we mention a "flock" of geese, we are regarding a particular set of geese as a single entity and any individual goose in that flock is a member of that set. Thus, the essential point of Cantor's concept of a set is that the collection of objects is to be regarded as a single entity. The word "definite" in Cantor's concept of a set means given a set and an object, it is possible to determine whether the object belongs to the set or not. And the word "distinguishable" means given any two pair of objects qualified to appear as elements of a set, one must be able to determine whether the objects are the same or different. As a result, a set is completely determined by its members.

Generally, we will denote sets by capital letters and will use the conventional notation "$x \in A$" to denote that "x is an element of the set A" or "x is an member of the set A." Also, we will use the notation "$x,\ y \in A$" to indicate "x and y are elements of the set A." The Italian mathematician Giuseppe Peano (1858-1932) introduced the symbol $\in$ for "is an element of" in 1889. The symbol comes from the first letter of the Greek word for "is." In order to indicate that "x is not an element of the set A", we will write "$x \notin A$". For example, let A denote the set of letters in the English alphabet. Then $b \in A$; $e, f \in A$; $a, e, i, o, u \in A$; but $5 \notin A$.

Sets are usually described by one of two notations—**roster notation** (also called **enumeration notation**) or **set-builder notation**. In roster notation, the elements of the set are enclosed in braces, { }, and separated by commas. The order in which the elements are listed within the braces is immaterial. For example, the set of one digit prime numbers may be written using roster notation as $\{2, 3, 5, 7\}$, or $\{5, 7, 2, 3\}$, or $\{7, 3, 5, 2\}$, etc. These are just different representations of the same set. Of course, roster notation is appropriate for finite sets. However, this notation can also be used to represent infinite sets. For instance, we may represent the set of positive integers, P, by $P = \{1, 2, 3, \ldots\}$. When representing sets in this manner the dots ... (called ellipsis) indicate that the pattern used to obtain the elements listed previously is to be followed to obtain the remaining elements of the set. This convention may be used to represent finite sets which have a relatively large number of elements as well. For example, $F = \{2, 4, 6, \ldots, 100\}$ is a representation of the set of all positive even integers less than or equal to 100.

Throughout the text, we will let **N** denote the set of **natural numbers**, which are also sometimes called the counting numbers or the positive integers. Thus,

$$\mathbf{N} = \{1, 2, 3, \ldots\}$$

The set of **integers** will be represented by **Z**, so

$$\mathbf{Z} = \{\ldots, -2, -1, 0, 1, 2, \ldots\}$$

The symbol **Z** comes from the German word for number, *Zählen*. We will use **Q** to denote the set of **rational numbers**. Recall that a rational number is any number of the form p/q where p and q are integers and $q \neq 0$. The set of **real numbers** will be denoted by **R**.

Using set-builder notation the set, F, of all even integers less than or equal to 100 would be written as

$$F = \{x \mid x \text{ is an even integer less than or equal to } 100\}$$

The vertical bar, $|$, in the definition above is read "such that." Hence, the set-builder definition of the set F given above is read "F equals the set of all x such that x is an even integer less than or equal to 100." Another example of a set specified using set-builder notation is

$$O = \{x \mid x \text{ is an odd integer}\}$$

Observe that set-builder notation is very appropriate for representing both infinite sets and finite sets with a relatively large number of elements.

The set with no elements is called the **empty set** or **null set**. The empty set is symbolized by $\emptyset$. The symbol $\emptyset$ is last letter of the Danish-Norwegian alphabet. **The empty set is not denoted by { }. The empty set is not the set** $\{0\}$. The set $\{0\}$ is a set with one element, namely 0. Thus,

$0 \in \{0\}$. **And the empty set is not the set** $\{\emptyset\}$. The set $\{\emptyset\}$ is a set with one element, namely $\emptyset$. That is, $\emptyset \in \{\emptyset\}$.

Many sentences in mathematics involve one or more variables, and, therefore, are not statements.

DEFINITIONS Variable, Universal Set, and Open Statement

A **variable** is a symbol, say x, which represents an unspecified object from a given set U.

The set of values, U, that can be assigned to the variable x is called the **universe, universal set,** or **universe of discourse**.

An **open statement in one variable** is a sentence that involves one variable and that becomes a statement (a declarative sentence that is true or false) when values from the universal set are substituted for the variable. An open statement in the variable x is denoted by $P(x)$.

It should be noted that an open statement in one variable is not a statement, because the sentence is not true or false until a specific value from the universal set is substituted into the sentence making it a statement—that is, making it true or false. Thus, the truth value of an open statement remains "open" until a specific value for the variable is substituted into the statement. Let $P(x)$ denote the open statement "The natural number x is a prime number." Observe that $P(x)$ is neither true nor false; the statement $P(5)$ is a true; and the statement $P(4)$ is a false.

DEFINITION Truth Set of an Open Statement

The **truth set of an open statement** is the set of all values from the universal set that make the open statement a true statement.

Let $P(x)$ be an open statement with the specified nonempty universal set U. Then in set-builder notation, the truth set of $P(x)$ is the set

$$T_P = \{x \in U \mid P(x)\}$$

The following example illustrates that the truth set of an open statement depends on the choice of the universal set. Let $P(x)$ be the open statement "$x^2 < 9$." Then $\{x \in \mathbf{N} \mid x^2 < 9\} = \{1, 2, 3\}$, $\{x \in \mathbf{Z} \mid x^2 < 9\} = \{-3, -2, -1, 0, 1, 2, 3\}$, and $\{x \in \mathbf{R} \mid x^2 < 9\} = \{x \in \mathbf{R} \mid -3 < x < 3\}$.

EXAMPLE 1 Finding Truth Sets

Find the following truth sets.

a. $\{x \in \mathbf{N} \mid 2x^2 - x = 0\}$ b. $\{x \in \mathbf{Z} \mid 2x^2 - x = 0\}$
c. $\{x \in \mathbf{Q} \mid 2x^2 - x = 0\}$

SOLUTION

Factoring the equation $2x^2 - x = 0$ which appears in a., b., and c., we find $x(2x - 1) = 0$. Recall from algebra that in the set of real numbers, if $ab = 0$, then either $a = 0$ or $b = 0$. Since $x(2x - 1) = 0$, it follows that $x = 0$ or $2x - 1 = 0$. Solving the last equation, we find $x = 1/2$. Thus, the two real roots of the equation $2x^2 - x = 0$ are $x = 0$ and $x = 1/2$. Therefore, the required truth sets are as follows:

a. $\{x \in \mathbf{N} \mid 2x^2 - x = 0\} = \emptyset$

b. $\{x \in \mathbf{Z} \mid 2x^2 - x = 0\} = \{0\}$

c. $\{x \in \mathbf{Q} \mid 2x^2 - x = 0\} = \{0, \frac{1}{2}\}$

Often we want to indicate how many values of the variable x make the open statement $P(x)$ true. Specifically, we would like to know if $P(x)$ is true for every x in the universe U or if $P(x)$ is true for at least one $x \in U$. Thus, we introduce two quantifiers.

DEFINITIONS Universal Quantifier and Existential Quantifier

The symbol $\forall$ is called the **universal quantifier** and represents the phrase "for all," "for each," or "for every." The statement $(\forall x \in U)(P(x))$ is read "for all $x \in U$, $P(x)$" and is true precisely when the truth set $T_P = \{x \in U \mid P(x)\} = U$.

The symbol $\exists$ is called the **existential quantifier** and represents the phrase "there exists," "there is," or "for some." The statement $(\exists x \in U)(P(x))$ is read "there exists an $x \in U$ such that $P(x)$" and is true precisely when the truth set $T_P = \{x \in U \mid P(x)\} \neq \emptyset$.

Several comments are in order. First of all, to prove that the statement $(\exists x \in U)(P(x))$ is true, it is necessary to find only a single value of $x \in U$ for which $P(x)$ is true. On the other hand, to prove that the statement

$(\forall x \in U)(P(x))$ is true, it is necessary to prove that $P(x)$ is true for all $x \in U$. Second, the conditional statement

$$(\forall x \in U)(P(x)) \Rightarrow (\exists x \in U)(P(x))$$

is true, since an open statement which is true for all values of x in a universe U is true for some (any) value of $x \in U$. On the contrary, the conditional statement

$$(\exists x \in U)(P(x)) \Rightarrow (\forall x \in U)(P(x))$$

is false, since an open statement can be true for some x in a universe and false for other x in the universe. And finally, the truth value of a quantified statement depends on the universe as the following examples show. Let $P(x)$ denote the statement "$x^2 > 0$.". The quantified statement $(\forall x \in \mathbf{N})(P(x))$ is true; while the quantified statement $(\forall x \in \mathbf{Z})(P(x))$ is false, because $P(0)$ is false. Now let $Q(x)$ represent the statement "$x \leq 0$." The quantified statement $(\exists x \in \mathbf{N})(Q(x))$ is false; while the quantified statement $(\exists x \in \mathbf{Z})(Q(x))$ is true, because $Q(0)$ is true.

EXAMPLE 2 Writing English Sentences Symbolically

Translate the following English sentences into symbolic statements containing one quantifier. Indicate the truth value of each statement.

a. For every natural number x, $2x + 1 > 0$.
b. For every integer x, $2x + 1 > 0$.
c. There exists an integer x such that $2x + 1 < 0$.
d. There exists a natural number x such that $x^2 + x + 41$ is a prime number.
e. For every natural number x, $x^2 + x + 41$ is a prime number.

SOLUTION

The corresponding symbolic statements are as follows.

a. $(\forall x \in \mathbf{N})(2x + 1 > 0)$. This statement is true, since for $x \in \mathbf{N}$, $x > 0$; therefore, $2x > 0$ and $2x + 1 > 1 > 0$.

b. $(\forall x \in \mathbf{Z})(2x + 1 > 0)$. This statement is false, since $-1 \in \mathbf{Z}$ and $2(-1) + 1 = -2 + 1 = -1 < 0$.

c. $(\exists x \in \mathbf{Z})(2x + 1 < 0)$. This statement is true, since it is true for $x = -1$. See the computation in part b. In fact, this statement is true for all integers $x \leq -1$.

d. $(\exists x \in \mathbf{N})(x^2 + x + 41$ is a prime number.) This statement is true, since for $x = 1$, $1^2 + 1 + 41 = 43$ which is a prime number.

e. $(\forall x \in \mathbf{N})(x^2 + x + 41$ is a prime number.) This statement is false. Can you find one specific natural number x such that $x^2 + x + 41$ is not a prime number?

The symbolic statement $(\forall x \in A)(P(x))$ which is read "For all $x \in A$, $P(x)$." can also be stated as "For all x, if $x \in A$, then $P(x)$." and symbolized by $(\forall x)((x \in A) \Rightarrow P(x))$. Likewise, the symbolic statement $(\exists x \in A)(P(x))$ can be written as $(\exists x)((x \in A) \Rightarrow P(x))$ which is read "There exists an x such that if $x \in A$, then $P(x)$." The following example illustrates these usages.

EXAMPLE 3 Translating Symbolic Statements with One Quantifier into English

In this example, let the universe of discourse be the set of integers, $\mathbf{Z}$. Translate the following symbolic statements which involve one quantifier into English and indicate the truth value of each statement.

a. $(\forall x)((x \in \mathbf{N}) \Rightarrow (x \in \mathbf{Z}))$

b. $(\forall x)((x \in \mathbf{Z}) \Rightarrow (x \in \mathbf{N}))$

c. $(\exists x)((x \in \mathbf{Z}) \wedge (x \in \mathbf{N}))$

d. $(\exists x)((x \in \mathbf{Z}) \wedge (x \notin \mathbf{N}))$

e. $(\exists x)((x \in \mathbf{N}) \wedge (x \notin \mathbf{Z}))$

f. $(\forall x)((x$ is a prime$) \Rightarrow (x$ is not a composite$))$

SOLUTION

a. $(\forall x)((x \in \mathbf{N}) \Rightarrow (x \in \mathbf{Z}))$ translates into English as "For all x, if x is a natural number, then x is an integer." This translation can be shortened to "Every natural number is an integer." The symbolic statement and the equivalent English translations are true.

b. $(\forall x)((x \in \mathbf{Z}) \Rightarrow (x \in \mathbf{N}))$ translates as "For all x, if x is an integer, then x is a natural number." This translation can be shortened to "Every integer is a natural number." This statement is false, because $x = -1$ makes the statement $x \in \mathbf{Z}$ true and the statement $x \in \mathbf{N}$ false. Hence, for $x = -1$, the conditional statement $(x \in \mathbf{Z}) \Rightarrow (x \in \mathbf{N})$ is false.

c. $(\exists x)((x \in \mathbf{Z}) \wedge (x \in \mathbf{N}))$ means "There is a number x which is an integer and a natural number." This symbolic statement and its translation are true, since 1 is both an integer and a natural number.

d. $(\exists x)((x \in \mathbf{Z}) \wedge (x \notin \mathbf{N}))$ translates into English as "For some x, x is an integer and x is not a natural number." A condensed translation is "Some integer is not a natural number." This symbolic statement and its translations are all true, since $x = -1$ is an integer which is not a natural number.

e. $(\exists x)((x \in \mathbf{N})) \wedge (x \notin \mathbf{Z}))$ translates as "For some x, x is a natural number and x is not an integer." A shortened translation is "Some natural number is not an integer." This symbolic statement and its translations are all false.

f. $(\forall x)((x \text{ is a prime}) \Rightarrow (x \text{ is not a composite}))$ translates as "For all x, if x is a prime, then x is not a composite." A shortened translation is "No prime is a composite." This symbolic statement and its translations are all true.

Traditional logic emphasized four basic types of statements involving a single quantifier: the universal affirmative, the universal denial, the particular affirmative, and the particular denial. Examples of these four types of statements appear in example 3. We now provide a general summary for these types of statements. Let a universe for the variable x be specified and let $P(x)$ and $Q(x)$ be appropriate statements. Then the symbolic statement and English translation of the four statements of traditional logic are as follows.

Statement Type	Symbolic Statement	English Sentence
1. Universal Affirmative	$(\forall x)(P(x) \Rightarrow Q(x))$	All $P(x)$ are $Q(x)$.
2. Universal Denial	$(\forall x)(P(x) \Rightarrow (\neg Q(x)))$	No $P(x)$ are $Q(x)$.
3. Particular Affirmative	$(\exists x)(P(x) \wedge Q(x))$	Some $P(x)$ are $Q(x)$.
4. Particular Denial	$(\exists x)(P(x) \wedge (\neg Q(x)))$	Some $P(x)$ are not $Q(x)$.

If the statement $(\exists x \in U)(P(x))$ is true, then we know that **there is at least one** x in the universe U such that $P(x)$ is true. However, in mathematics, it is often the case that **there exists exactly one** x in the universe for which $P(x)$ is true. For example, for the set of integers there exists exactly one additive identity—namely, the number 0. (That is, the number 0 is the unique number in Z such that $x + 0 = 0 + x = x$ for all $x \in Z$.) For the set of natural numbers there exists exactly one multiplicative identity—namely, the number 1. (The number 1 is the unique number in N such that $x \cdot 1 = 1 \cdot x = x$

for all $x \in N$.) To indicate that there exists a unique element in a universe with a specific property, we define the unique existential quantifier, $\exists!$, as follows.

DEFINITION Unique Existential Quantifier

The symbol $\exists!$ is called the **unique existential quantifier** and represents the phrase "there exists a unique," or "there exists exactly one." The statement $(\exists! x \in U)(P(x))$ is read "there exists a unique $x \in U$ such that $P(x)$" or "there exists exactly one $x \in U$ such that $P(x)$." The statement $(\exists! x \in U)(P(x))$ is true precisely when the truth set $T_P = \{x \in U \mid P(x)\}$ has exactly one element.

It follows directly from the definitions of the quantifiers $\exists$ and $\exists!$ that the conditional statement $(\exists! x \in U)(P(x)) \Rightarrow (\exists x \in U)(P(x))$ is true, while the conditional statement $(\exists x \in U)(P(x)) \Rightarrow (\exists! x \in U)(P(x))$ is false.

EXAMPLE 4 Determining the Truth Value of Statements Containing the Unique Existential Quantifier

Find the truth value of the following statements which contain the unique existential quantifier.

a. $(\exists! x \in \mathbf{N})(|x + 4| = 1)$ b. $(\exists! x \in \mathbf{N})(|x - 4| = 5)$
c. $(\exists! x \in \mathbf{N})(|x - 4| = 3)$

SOLUTION

Notice that an equation of the form $|y| = a$ where a is positive appears in a., b., and c. Recall from algebra that in the set of real numbers, if $|y| = a$ where a is positive, then either $y = a$ or $y = -a$.

a. It follows from the discussion above, that if $|x + 4| = 1$, then either $x + 4 = 1$ or $x + 4 = -1$. Adding -4 to both of the last two equations, we find if $|x + 4| = 1$, then either $x = -3$ or $x = -5$. Since $-3 \notin \mathbf{N}$ and $-5 \notin \mathbf{N}$, the statement $(\exists! x \in \mathbf{N})(|x + 4| = 1)$ is false, because the truth set of $\{x \in \mathbf{N} \mid |x + 4| = 1\}$ is the empty set, which contains no element.

b. If $|x - 4| = 5$, then either $x - 4 = 5$ or $x - 4 = -5$. Adding 4 to both of the last two equations, we find if $|x - 4| = 5$, then either $x = 9$ or $x = -1$. Since $9 \in \mathbf{N}$ and $-1 \notin \mathbf{N}$, the statement $(\exists! x \in \mathbf{N})(|x + 4| = 1)$ is true, because the truth set of $\{x \in \mathbf{N} \mid |x - 4| = 5\} = \{9\}$, which contains exactly one element.

c. If $|x-4|=3$, then either $x-4=3$ or $x-4=-3$. Adding 4 to both of the last two equations, we find if $|x-4|=3$, then either $x=7$ or $x=1$. Since $7 \in \mathbf{N}$ and $1 \in \mathbf{N}$, the statement $(\exists! x \in \mathbf{N})(|x+4|=1)$ is false, because the truth set of $\{x \in \mathbf{N} \mid |x-4|=3\} = \{7,1\}$, which contains two elements.

Observe that if the universal set in a., b., and c. were changed from the set of natural numbers, **N**, to the set of integers, **Z**, the set of rational numbers, **Q**, or the set of real numbers, **R**, then all three statements a., b., and c. containing the unique existential quantifier would be false, since each corresponding truth set would contain exactly two elements.

Example 4 illustrates that the statement $(\exists! x \in U)(P(x))$ can be proven to be false by showing that the truth set of $\{x \in U \mid P(x)\}$ is either the empty set or a set with two or more elements. In order to prove that the statement $(\exists! x \in U)(P(x))$ is true, it is necessary to show that the truth set of $\{x \in U \mid P(x)\}$ contains exactly one element.

In mathematics, it is very important to be able to negate quantified statements such as definitions and theorems. However, before we can negate quantified statements it is necessary to define equivalence for two quantified statements.

DEFINITIONS Equivalent Quantified Statements

Let $P(x)$ and $Q(x)$ be two quantified statements with nonempty universe U.

Two statements $P(x)$ and $Q(x)$ are **equivalent in the universe** U if and only if $P(x)$ and $Q(x)$ have the same truth value for all $x \in U$. That is, $P(x)$ and $Q(x)$ are **equivalent in the universe** U if and only if $(\forall x \in U)(P(x) \Leftrightarrow Q(x))$.

The two quantified statements $P(x)$ and $Q(x)$ are **equivalent** if and only if they are equivalent in every universe U.

Let $A(x)$ denote the statement "$x^2 = x$" and let $B(x)$ denote the statement "$x = 1$." The quantified statement $(\forall x)(A(x))$ is equivalent to the quantified statement $(\forall x)(B(x))$ in the universe of natural numbers, **N**, since in **N**

$$\{x \mid x^2 = x\} = \{1\} = \{x \mid x = 1\}.$$

However, the quantified statement $(\forall x)(A(x))$ is not equivalent to the quantified statement $(\forall x)(B(x))$ in the universe of integers, **Z**, since in **Z**

$$\{x \mid x^2 = x\} = \{0, 1\} \neq \{1\} = \{x \mid x = 1\}.$$

Consequently, the quantified statements $(\forall x)(A(x))$ and $(\forall x)(B(x))$ are not equivalent.

Let $P(x)$ and $Q(x)$ be open sentences in x with nonempty universe U. Since the conditional statement $P \Rightarrow Q$ is logically equivalent to $\neg P \vee Q$,

$$(\forall x \in U)((P(x) \Rightarrow Q(x)) \Leftrightarrow (\neg P(x) \vee Q(x))) \quad \text{for any nonempty universe } U.$$

Hence, the quantified statements $(\forall x)((P(x) \Rightarrow Q(x))$ and $(\forall x)(\neg P(x) \vee Q(x))$ are equivalent. Other important equivalent statements are

1. $(\forall x)((P(x) \wedge Q(x))$ and $(\forall x)(Q(x) \wedge P(x))$
2. $(\forall x)((P(x) \vee Q(x))$ and $(\forall x)(Q(x) \vee P(x))$
3. $(\forall x)((P(x) \Rightarrow Q(x))$ and $(\forall x)(\neg Q(x) \Rightarrow \neg P(x))$
4. $(\forall x)(\neg(P(x) \vee Q(x)))$ and $(\forall x)((\neg P(x)) \wedge (\neg Q(x)))$
5. $(\forall x)(\neg(P(x) \wedge Q(x)))$ and $(\forall x)((\neg P(x)) \vee (\neg Q(x)))$

Additional equivalent pairs of quantified statements may be obtained by replacing each occurrence of $\forall$ in the above statements by $\exists$.

Now we consider the negation of the quantified statement $(\forall x)(P(x))$. Let a nonempty universe U be given. The negation of $(\forall x)(P(x))$ is

$$\begin{aligned}
\neg(\forall x)(P(x)) \text{ is true} &\Leftrightarrow (\forall x)(P(x)) \text{ is false} \\
&\Leftrightarrow \{x \in U \mid P(x)\} \neq U \\
&\Leftrightarrow \{x \in U \mid \neg P(x)\} \neq \emptyset \\
&\Leftrightarrow (\exists x)(\neg P(x))
\end{aligned}$$

Thus, $\neg(\forall x)(P(x))$ is equivalent to $(\exists x)(\neg P(x))$. In a like manner, it can be shown that $\neg(\exists x)(P(x))$ is equivalent to $(\forall x)(\neg P(x))$. Consequently, we have the following theorem regarding the negation of quantified statements.

THEOREM The Negation of Quantified Statements

Let $P(x)$ be an open statement in the variable x. Then $\neg(\forall x)(P(x))$ is equivalent to $(\exists x)(\neg P(x))$ and $\neg(\exists x)(P(x))$ is equivalent to $(\forall x)(\neg P(x))$.

Henceforth, we will use the symbol $\equiv$ to denote the phrase "is equivalent to." Thus, the conclusion of the last theorem may be written as $\neg(\forall x)(P(x)) \equiv (\exists x)(\neg P(x))$ and $\neg(\exists x)(P(x)) \equiv (\forall x)(\neg P(x))$.

EXAMPLE 5 Negating a Quantified Statement

Negate the universal affirmative statement, $(\forall x)(P(x) \Rightarrow Q(x))$, and translate the negation into English.

SOLUTION

The negation of the universal affirmative statement is

$$\begin{aligned}
\neg(\forall x)(P(x) \Rightarrow Q(x)) &\equiv \neg(\forall x)((\neg P(x)) \vee Q(x)) && \text{Definition of Conditional Statement} \\
&\equiv (\exists x)\neg((\neg P(x)) \vee Q(x)) && \text{Negation of Quantified Statement Theorem} \\
&\equiv (\exists x)(\neg(\neg P(x)) \wedge (\neg Q(x))) && \text{De Morgan's Law} \\
&\equiv (\exists x)(P(x) \wedge (\neg Q(x))) && \text{Double Negation}
\end{aligned}$$

The statement $(\exists x)(P(x) \wedge (\neg Q(x)))$ is the particular denial statement and its English translation is "Some $P(x)$ are not $Q(x)$. Thus, the negation of the universal affirmative statement is the particular denial statement and by double negation, the negation of the particular denial statement is the universal affirmative statement. As one might anticipate, the negation of the universal denial statement is the particular affirmative statement and the negation of the particular affirmative statement is the universal denial statement.

Thus far, we have discussed only quantified statements in one variable. Sentences in mathematics often contain quantified statements in two or more variables. Moreover, the variables may not appear explicitly in the sentence; and, initially, it may not be apparent exactly how many variables the sentence actually contains. To make matters even more challenging, phrases such as "for all," "for each," "for every," "there exists," "there is," or "for some" may not appear in the sentence either. Thus, it is often difficult to translate English sentences regarding mathematics into symbolic statements. On the other hand, it is usually fairly easy to translate quantified symbolic statements from mathematics into English.

First, let us consider the sentence, "For every real number there is a natural number greater than the real number." This sentence contains the quantifiers "for every" and "there is" and the key words "real number" and "natural number." Thus, we are reasonably certain there are two variables in this sentence. Let us use r for the variable denoting a real number and n for the variable denoting a natural number. Then the translation of the given sentence into a symbolic statement in mathematics is

$$(\forall r \in \mathbf{R})(\exists n \in \mathbf{N})(n > r)$$

The negation of this statement is

$$\begin{aligned} \neg((\forall r \in \mathbf{R})(\exists n \in \mathbf{N})(n > r)) &\equiv (\exists r \in \mathbf{R})(\neg((\exists n \in \mathbf{N})(n > r))) \\ &\equiv (\exists r \in \mathbf{R})(\forall n \in \mathbf{N})(\neg(n > r)) \\ &\equiv (\exists r \in \mathbf{R})(\forall n \in \mathbf{N})(n \leq r) \end{aligned}$$

Translating the last quantified statement into English, we find that the negation of sentence "For every real number there is a natural number greater than the real number." is the sentence "There exists a real number such that all natural numbers are less than or equal to the real number." Clearly, the negation of the original sentence is false, since it says that the natural numbers are bounded above by some real number. Consequently, the original statement is true.

The following examples illustrate that the order of the quantifiers in a statement is very important. Let us consider the following five symbolic statements with the universe of discourse being the integers, **Z**.

1. $(\forall x)(\forall y)(x + y = 0)$
2. $(\forall y)(\forall x)(x + y = 0)$
3. $(\forall x)(\exists y)(x + y = 0)$
4. $(\exists y)(\forall x)(x + y = 0)$
5. $(\exists x)(\exists y)(x + y = 0)$

The first statement $(\forall x)(\forall y)(x + y = 0)$ says "For all integers x and for all integers y, $x + y = 0$. This statement is false, since $x = 2 \in \mathbf{Z}$ and $y = 3 \in \mathbf{Z}$, but $x + y = 2 + 3 = 5 \neq 0$. The second statement has the same meaning as the first. It says "For all integers y and for all integers x, $x + y = 0$." and it is false. Thus, we note that $(\forall x)(\forall y)(x + y = 0) \equiv (\forall y)(\forall x)(x + y = 0)$. The third statement $(\forall x)(\exists y)(x + y = 0)$ says "For every integer x there exists an integer y such that $x + y = 0$." or "Every integer has an additive inverse." Given an integer x, the integer $y = -x$ satisfies the equation $x + y = 0$. Thus, the third statement is true. Notice that in the fourth statement the order of the quantifiers are reversed from the order in which they appear in the third statement. The fourth statement says "There exists an integer y such that for all integers x, $x + y = 0$." This statement is false, since for $x = 2$, y must be -2 in order to satisfy $x + y = 0$, while for $x = 3$, y must be -3 in order to satisfy $x + y = 0$ and, of course, $-2 \neq -3$. Since the third statement is true and the fourth statement is false, we see that $(\forall x)(\exists y)(x + y = 0)$

is not equivalent to $(\exists y)(\forall x)(x+y=0)$. The fifth statement says "There exists an integer x and there exists an integer y such that $x+y=0$." We can easily verify that this statement is true by choosing $x=2$ and $y=-2$. If we reverse the quantifiers appearing in the fifth statement, we obtain the statement $(\exists y)(\exists x)(x+y=0)$ which is equivalent to the fifth statement.

The examples presented above illustrate the following facts which we state without proof. Let x and y be distinct variables and let $P(x,y)$ be an open statement in x and y. Then

$$(\forall x)(\forall y)P(x,y) \equiv (\forall y)(\forall x)P(x,y)$$

$$(\exists x)(\exists y)P(x,y) \equiv (\exists y)(\exists x)P(x,y)$$

and

$$(\exists x)(\forall y)P(x,y) \Rightarrow (\forall y)(\exists x)P(x,y)$$

EXAMPLE 6 Translating a Sentence Into a Quantified Statement

Translate the sentence "Any nonzero real number has a multiplicative inverse." into a quantified symbolic statement.

SOLUTION

An initial translation is $(\forall x \in \mathbf{R})((x \neq 0) \Rightarrow (x \text{ has a multiplicative inverse}))$. At first glance, it appears that the quantified statement contains only one variable, namely x. However, by definition, a real number y is a multiplicative inverse of a nonzero real number x if and only if $xy=1$. Using this definition, our translation into a quantified symbolic statement in two variables, becomes

$$(\forall x \in \mathbf{R})((x \neq 0) \Rightarrow ((\exists y \in \mathbf{R})(xy=1))).$$

EXERCISES 1.5

In exercises 1-10, write the truth set of the given set. When possible, use roster notation.

1. $\{x \in \mathbf{N} \mid x \text{ is an even prime number}\}$
2. $\{x \in \mathbf{N} \mid x \text{ is a multiple of } 4\}$
3. $\{x \in \mathbf{Z} \mid x \text{ is a multiple of } 4\}$
4. $\{x \in \mathbf{N} \mid x < 5\}$
5. $\{x \in \mathbf{Z} \mid x < 5\}$
6. $\{x \in \mathbf{N} \mid 3x > x\}$
7. $\{x \in \mathbf{Z} \mid 3x > x\}$
8. $\{x \in \mathbf{N} \mid 3x < x\}$
9. $\{x \in \mathbf{Z} \mid 3x < x\}$
10. $\{x \in \mathbf{N} \mid \sqrt{x} \in \mathbf{N}\}$

In exercises 11-20, let the universal set, U, be the set of all triangles. Let $E(x)$ denote the open statement "x is an equilateral triangle.", let $I(x)$ denote "x is an isosceles triangle.", and let $R(x)$ denote "x is a right triangle." Translate each English sentence into a symbolic statement with one quantifier and indicate the truth value of each statement.

11. All isosceles triangles are equilateral triangles.
12. All equilateral triangles are isosceles triangles.
13. All isosceles triangles are not right triangles.
14. Some isosceles triangles are equilateral triangles.
15. Some isosceles triangles are not right triangles.
16. Some right triangles are equilateral triangles.
17. Some right triangles are isosceles triangles.
18. No equilateral triangle is a right triangle.
19. No right triangle is an isosceles triangle.
20. No equilateral triangle is not an isosceles triangle.

21. Write the negation of exercises 11-20.

In exercises 22-31, translate the given symbolic statement into English and indicate the truth value of the statement.

22. $(\forall x \in \mathbf{R})(2x > 0)$
23. $(\exists x \in \mathbf{N})(2x > 0)$
24. $(\forall x \in \mathbf{N})((x \text{ is a prime}) \Rightarrow (x \text{ is odd}))$
25. $(\exists x \in \mathbf{N})(2x \leq 0)$
26. $(\forall x \in \mathbf{N})(((x \text{ is a prime}) \wedge (x \neq 2)) \Rightarrow (x \text{ is odd}))$
27. $(\exists! x \in \mathbf{Z})(x^2 = 0)$
28. $(\exists! x \in \mathbf{N})((x \text{ is a prime}) \wedge (x \text{ is not odd}))$
29. $(\exists! x \in \mathbf{Z})(x^2 = x)$
30. $(\exists! x \in \mathbf{R})(e^x = 1)$
31. $(\exists! x \in \mathbf{R})(x = \sqrt{7})$

32. An equivalent form of the unique existential quantifier $(\exists! x \in U)(P(x))$ is (*) $\quad (\exists x \in U)(P(x) \wedge ((\forall y \in U)(P(y) \Rightarrow (y = x))))$
Write the negation of (*) and observe that the unique existential quantifier $(\exists! x \in U)(P(x))$ is false when, for every x, either $P(x)$ is false or for some y, $P(y)$ is true for $y \neq x$.

In exercises 33-36, write the given English sentence symbolically and indicate the truth value of the statement.

33. All primes are odd.
34. Some primes are even.
35. There exists a unique even prime.
36. There is a unique smallest natural number.

37. Write the negation of exercises 22-31 and 33-36.

In exercises 38-52, translate each sentence into a quantified symbolic statement and indicate the truth value of the statement.

38. There exist natural numbers m and n such that m is greater than n.

39. There exists a natural number m such that for all natural numbers n, m is greater than n.

40. For all natural numbers m there exists a natural number n such that m is greater than n.

41. For all natural numbers m and n, m is greater than n.

42. For every integer x there exists an integer y such that $y = 2x$.

43. For every integer x there exists an integer y such that $x = 2y$.

44. There exists an integer y such that $y = 2x$ for every integer x.

45. There exists a unique rational number x such that $x + y = 0$ for all rational numbers y.

46. There exists a unique rational number x such that $xy = 1$ for all rational numbers y.

47. For all rational numbers x there exists a unique rational number y such that $x + y = 0$.

48. For all nonzero rational numbers x there exists a unique rational number y such that $xy = 1$.

49. For every positive real number x there exists a natural number n such that $\frac{1}{n} < x$.

50. For all real numbers x, y, and z, $(x + y) + z = x + (y + z)$.

51. For all real numbers x, y, and z, if $x < y$, then $x + z < y + z$.

52. For all real numbers x there exists a unique real number y such that $xy = yz$ for all real numbers z.

1.6 Chapter Review

Definitions

A **statement** or **proposition** is a declarative sentence that is either true or false, but not both true and false.

A **simple statement** (**simple proposition**) is a statement which does not contain any other statement as a component part.

Every **compound statement** (**compound proposition**) is a statement that does contain another statement as a component part.

Let P denote a statement. The **negation** of P, denoted by $\neg$ P, is the statement "not P." The negation of P is false when P is true, and the negation of P is true when P is false.

The **conjunction** of two statements P, Q, denoted by P $\wedge$ Q, is the statement "P and Q." The conjunction of P and Q is true if and only if both P and Q are true.

The **disjunction** of two statements P, Q, denoted by P $\vee$ Q, is the statement "P or Q." The disjunction of P and Q is true if P is true, if Q is true, or if both P and Q are true.

The **truth value** of a statement is true (denoted by T) if the statement is true and is false (denoted by F) if the statement is false.

A **truth table** is a table which shows the truth values of a statement for all possible combinations of truth values of its simple statement components.

A **tautology** is a statement that is true for every assignment of truth values of its component statements.

A **contradiction** is a statement that is false for every assignment of truth values of its component statements. Thus, a contradiction is the negation of a tautology.

Two statements are **truth value equivalent** or **logically equivalent** if and only if they have the same truth values for all assignments of truth values to their component statements.

The statement "If P, then Q." is called a **conditional statement**, the statement P is called the **hypothesis** or **antecedent** of the conditional statement, and the statement Q is called the **conclusion** or **consequent** of the conditional statement.

Given two statements P and Q, the **conditional statement** $P \Rightarrow Q$ (read "P implies Q") is the statement "If P, then Q." The conditional statement $P \Rightarrow Q$ is true unless P is true and Q is false, in which case it is false.

Alternative expressions for the conditional statement $P \Rightarrow Q$ are

If P, then Q	Q, if P
P implies Q	Q is implied by P
P only if Q	Q provided P
P is sufficient for Q	Q is necessary for P

The **converse** of $P \Rightarrow Q$ is $Q \Rightarrow P$.

The **inverse** of $P \Rightarrow Q$ is $(\neg P) \Rightarrow (\neg Q)$.

The **contrapositive** of $P \Rightarrow Q$ is $(\neg Q) \Rightarrow (\neg P)$.

Given two statements P and Q, the **biconditional statement** $P \Leftrightarrow Q$ is the statement "P if and only if Q." The biconditional statement $P \Leftrightarrow Q$ is true when P and Q have the same truth values and false when P and Q have different truth values.

An argument $P_1, \ldots, P_n \therefore C$ with premises $P_1, \ldots, P_n$ and conclusion C is **valid** if and only if $P_1 \wedge P_2 \wedge \ldots \wedge P_n \Rightarrow C$ is a tautology.

If an argument is not valid, it is called **invalid**.

Sets are usually described by one of two notations—**roster notation** (also called **enumeration notation**) or **set-builder notation**. In roster notation, the elements of the set are enclosed in braces, { }, and separated by commas.

The set with no elements is called the **empty set** or **null set**.

A **variable** is a symbol, say x, which represents an unspecified object from a given set U.

The set of values, U, that can be assigned to the variable x is called the **universe, universal set,** or **universe of discourse**.

An **open statement in one variable** is a sentence that involves one variable and that becomes a statement (a declarative sentence that is true or false) when values from the universal set are substituted for the variable. An open statement in the variable x is denoted by $P(x)$.

The **truth set of an open statement** is the set of all values from the universal set that make the open statement a true statement.

The symbol $\forall$ is called the **universal quantifier** and represents the phrase "for all," "for each," or "for every." The statement $(\forall x \in U)(P(x))$ is read "for all $x \in U$, $P(x)$" and is true when the truth set $T_P = \{x \in U \mid P(x)\} = U$.

The symbol $\exists$ is called the **existential quantifier** and represents the phrase "there exists," "there is," or "for some." The statement $(\exists x \in U)(P(x))$ is read "there exists an $x \in U$ such that $P(x)$" and is true when the truth set $T_P = \{x \in U \mid P(x)\} \neq \emptyset$.

The symbol $\exists!$ is called the **unique existential quantifier** and represents the phrase "there exists a unique," or "there exists exactly one." The statement $(\exists! x \in U)(P(x))$ is read "there exists a unique $x \in U$ such that $P(x)$" or "there exists exactly one $x \in U$ such that $P(x)$." The statement $(\exists! x \in U)(P(x))$ is true precisely when the truth set $T_P = \{x \in U \mid P(x)\}$ has exactly one element.

Two statements $P(x)$ and $Q(x)$ are **equivalent in the universe** U if and only if $P(x)$ and $Q(x)$ have the same truth value for all $x \in U$. That is, $P(x)$ and $Q(x)$ are **equivalent in the universe** U if and only if $(\forall x \in U)(P(x) \Leftrightarrow Q(x))$.

The two quantified statements $P(x)$ and $Q(x)$ are **equivalent** if and only if they are equivalent in every universe U.

Useful Laws

Let t represent a statement which is a tautology, let f represent a statement which is a contradiction, and let P represent any statement.

Double Negation Law: $\neg(\neg P) \equiv P$

Tautology Laws: $\neg\, \mathrm{t} \equiv \mathrm{f}$ $\quad \mathrm{P} \wedge \mathrm{t} \equiv \mathrm{P}$ $\quad \mathrm{P} \vee \mathrm{t} \equiv \mathrm{t}$

Contradiction Laws: $\neg\, \mathrm{f} \equiv \mathrm{t}$ $\quad \mathrm{P} \wedge \mathrm{f} \equiv \mathrm{f}$ $\quad \mathrm{P} \vee \mathrm{f} \equiv \mathrm{P}$

Idempotent Law for Conjunction: $P \wedge P \equiv P$

Idempotent Law for Disjunction: $P \vee P \equiv P$

Commutative Law for Conjunction: $P \wedge Q \equiv Q \wedge P$

Commutative Law for Disjunction: $P \vee Q \equiv Q \vee P$

Absorption Laws: $P \wedge (P \vee Q) \equiv P$

$P \vee (P \wedge Q) \equiv P$

De Morgan Laws: $\neg(P \wedge Q) \equiv (\neg P) \vee (\neg Q)$

$\neg(P \vee Q) \equiv (\neg P) \wedge (\neg Q)$

Law of the Excluded Middle: $(\neg P) \wedge P$

Distributive Law for Disjunction: $P \vee (Q \wedge R) \equiv (P \vee Q) \wedge (P \vee R)$

Distributive Law for Conjunction: $P \wedge (Q \vee R) \equiv (P \wedge Q) \vee (P \wedge R)$

Associative Law for Disjunction: $(P \vee Q) \vee R \equiv P \vee (Q \vee R)$

Associative Law for Conjunction: $(P \wedge Q) \wedge R \equiv P \wedge (Q \wedge R)$

Negation of the Conditional Statement $\neg(P \Rightarrow Q) \equiv P \wedge (\neg Q)$

Rules of Logic

Rule of Substitution

Let P and Q be statements. Let $C(P)$ be a compound statement containing the statement P. And let $C(Q)$ be the same compound statement in which each occurrence of P is replaced by Q. If P and Q are logically equivalent, then $C(P)$ and $C(Q)$ are logically equivalent. That is,

$$\text{If } P \equiv Q, \text{then } C(P) \equiv C(Q).$$

The argument $P \quad \therefore P \vee Q$ is a valid argument known as the **rule of disjunction.**

The argument $P,\ Q \quad \therefore P \wedge Q$ is a valid argument known as the **rule of conjunction.**

The argument $P \wedge Q \quad \therefore P$ is a valid argument called the **rule of conjunctive simplification.**

The argument $P \vee Q,\ \neg P \quad \therefore Q$ is a valid argument called the **rule of disjunctive syllogism.**

The argument $P \Rightarrow Q,\quad P \ \therefore Q$ is a valid argument known as the **rule of detachment** or **modus ponens.**

The argument $P \Rightarrow Q,\quad \neg Q \ \therefore \neg P$ is a valid argument known as the **rule of contrapositive inference** or **modus tollens.**

Two Invalid Arguments

The argument $P \Rightarrow Q,\quad Q \ \therefore P$ is a invalid argument called the **fallacy of the converse.**

The argument $P \Rightarrow Q,\quad \neg P \quad \therefore \neg Q$ is a invalid argument called the **fallacy of the inverse.**

The argument $P \Rightarrow Q, \quad Q \Rightarrow R \quad \therefore P \Rightarrow R$ is a valid argument called the **rule of transitive inference** or **hypothetical syllogism**.

THEOREM The Negation of Quantified Statements

Let $P(x)$ be an open statement in the variable x. Then $\neg(\forall x)(P(x))$ is equivalent to $(\exists x)(\neg P(x))$ and $\neg(\exists x)(P(x))$ is equivalent to $(\forall x)(\neg P(x))$.

Statements of Traditional Logic

Statement Type	Symbolic Statement	English Sentence
1. Universal Affirmative	$(\forall x)(P(x) \Rightarrow Q(x))$	All $P(x)$ are $Q(x)$.
2. Universal Denial	$(\forall x)(P(x) \Rightarrow (\neg Q(x)))$	No $P(x)$ are $Q(x)$.
3. Particular Affirmative	$(\exists x)(P(x) \wedge Q(x))$	Some $P(x)$ are $Q(x)$.
4. Particular Denial	$(\exists x)(P(x) \wedge (\neg Q(x)))$	Some $P(x)$ are not $Q(x)$.

Review Exercises

In exercises 1-5, determine if the given sentence is a statement or not.

1. Multiply 8 by 7.
2. What time is it?
3. The number 47 is a composite number.
4. The number $\sqrt{11}$ is an irrational number.
5. Alice is very intelligent.

6. Let T denote the statement "I pass the test." Let C be the statement "I pass the course." And, let D stand for the statement "I make the dean's list."

   a. Write the following statements symbolically.

      (1) I did not make the dean's list.

      (2) I passed the test, but I did not pass the course.

      (3) Either I pass the test, or I do not pass the course.

(4) If I pass the test, then I pass the course.

(5) If I pass the test and I pass the course, then I make the dean's list.

(6) I make the dean's list if and only if I pass the course.

b. Write the following statements in English.

(1) $C \wedge D$ (2) $(T \wedge C) \wedge (\neg D)$ (3) $C \Leftrightarrow T$ (4) $(\neg C) \Rightarrow (\neg D)$

(5) $(T \Rightarrow C) \wedge (C \Rightarrow D)$

In exercises 7-13, write the negation of the given statement.

7. The number π is rational.
8. I did not go to the game.
9. It is cloudy, or the sun is shining.
10. I went home, but I did not read the newspaper.
11. If $2 + 3 = 4$, then $5 + 6 = 7$.
12. The number π is even if and only if the number e is odd.
13. If I finish my homework, I play tennis unless it rains.

14. Construct a truth table for the following statements. Identify tautologies and contradictions.

a. $P \Leftrightarrow (\neg P)$ b. $P \Rightarrow (P \vee Q)$ c. $P \Rightarrow (P \wedge Q)$

d. $(P \Rightarrow Q) \vee (Q \Rightarrow P)$ e. $P \Rightarrow [(\neg P) \Rightarrow (Q \wedge (\neg Q))]$

f. $\{[(P \wedge Q) \Rightarrow R] \Rightarrow (P \Rightarrow R)\} \Leftrightarrow (P \Rightarrow Q)$

15. Rewrite each of the following statements in the conditional form $P \Rightarrow Q$. That is, rewrite each statement in the form "If P, then Q."

a. You may vote provided you are old enough.

b. The gasoline engine is running implies there is fuel in the tank.

c. You may run for the United States Senate only if you are at least thirty-five years of age.

d. Rain is necessary for a garden.

e. A triangle is equilateral is sufficient for the triangle to be isosceles.

f. Today is Friday is implied by yesterday being Thursday.

16. Write the negation of the statements in exercise 15.

17. Write the converse, inverse, and contrapositive of the statements in exercise 15.

18. Determine the truth value of the following biconditional statements.

    a. The number $\sqrt{3}$ is negative if and only if the number $-\sqrt{3}$ is positive.

    b. $-1 > 0$ iff $0 < 1$.

19. Which of the following arguments are valid?

    a. $P \therefore P \wedge Q$

    b. $P \vee Q, \neg P \therefore Q$

    c. $P \Rightarrow Q, \neg Q \therefore \neg P$

    d. $P \Rightarrow Q, Q \therefore P$

    e. $P \Rightarrow Q, R \Rightarrow Q \therefore P \Leftrightarrow R$

20. Write the following argument in symbolic form using the letters indicated and deduce a valid conclusion for the argument.

    If Allen attends the meeting (A), then Barbar attends the meeting (B). If Allen and Barbar attend the meeting, then Carly will be elected (C) or Dave will be elected (D). If Carly or Dave is elected, Earl will resign (E). If Allen's attendance implies Earl will not resign, then Fae will be the new chairperson (F). Therefore, ________________.

21. Complete the given logical equivalences.

    a. $\neg(P \wedge Q) \equiv$

    b. $P \vee (Q \wedge R) \equiv$

    c. $(\neg P) \wedge (\neg Q) \equiv$

    d. $(P \wedge Q) \vee (P \wedge R) \equiv$

    e. $\neg(P \Rightarrow Q) \equiv$

    f. $P \vee (P \wedge Q) \equiv$

22. Write the truth set of each of the following sets.

    a. $\{x \in \mathbf{N} \mid x^2 + 3x = 0\}$

    b. $\{x \in \mathbf{Z} \mid x^2 + 3x = 0\}$

    c. $\{x \in \mathbf{N} \mid 2x^2 + 3x + 1 = 0\}$

    d. $\{x \in \mathbf{Z} \mid 2x^2 + 3x + 1 = 0\}$

    e. $\{x \in \mathbf{Q} \mid 2x^2 + 3x + 1 = 0\}$

In exercises 23-27, let $M(x)$ denote the open statement "x is a mathematician." and let $P(x)$ denote the open statement "x is a philosopher." Write each statement symbolically.

23. No mathematician is a philosopher.

24. Some philosophers are mathematicians.

25. All mathematicians are not philosophers.

26. No philosophers are not mathematicians.

27. Some mathematicians are philosophers.

28. Write the negation of exercises 23-27 both symbolically and in English.

In exercises 29-30, translate each sentence into a quantified statement.

29. For any integer x there exists a unique integer y such that $x + y = 0$.

30. There exists a natural number x such that $xy = 0$ for all natural numbers y.

31. There exists a unique real number x such that $x + y = x$ for all real numbers y.

32. Write the negation of exercises 29-31.

Chapter 2

Deductive Mathematical Systems and Proofs

This chapter is devoted to the study of deductive mathematical systems and elementary proof techniques. In section 2.1, we define deductive mathematical systems and discuss the Euclidean and non-Euclidean geometries, the system of natural numbers, and the system of integers. In section 2.2, basic techniques for proving conditional statements such as direct proof, proof by contraposition, proof by contradiction, and proof by cases are presented. We also discuss proving biconditional statements, proving a statement by contradiction, and proving statements which contain quantifiers. Next, we present some well-known mathematical conjectures and show how to prove and disprove conjectures. And finally, we define and examine the system of rational numbers and the system of real numbers.

2.1 Deductive Mathematical Systems

A deductive mathematical system (an axiomatic theory) consists of the following six elements:

1. An underlying language—English, in our case.
2. A deductive logic system—which we introduced in chapter 1.
3. A list of undefined terms.
4. A list of formally defined technical terms, called definitions.
5. A list of statements which are assumed to be true, called postulates or axioms.
6. A list of deduced statements, called theorems.

If, on one of the first days of class, we were to ask beginning geometry students the definition of a "point," we would receive several responses. Among those responses we would most likely find: "A point is the intersection of two lines." If a little later in the class, we ask the same students to define a "line," we would probably receive one or both of these responses: "A line is the shortest distance between two points." "A line is determined by two points." That is, beginning geometry students would want to define a "point" in terms of "lines" and to define a "line" in terms of "points." The students would be guilty of providing circular definitions. However, they have no choice but to do so, if they believe all terms should be defined. Suppose we have an English dictionary which includes every word in the English language. And suppose we start by looking up a single word. It will be defined in terms of other English words, all of which are found in the dictionary. Included with the definition of the word is a list of synonyms—words having the same or almost the same meaning. Next, if we look up one of the synonyms, we will find a new list of synonyms for the second word. If we continue the process in this manner, then eventually, because there are only a finite number of words in the dictionary, we will look up a word that we have looked up before—it may not be the word with which we started the process, but it will be a word we have looked up before. Consequently, because there are only a finite number of words, it is impossible to define every word and also avoid circularity of definitions. Thus, by necessity, an abstract mathematical system requires some undefined terms. In geometry, those undefined terms may be words such as, "point," "line," "distance," "on," etc.

A **definition** is an agreement to use a symbol, a word, or a short phrase to substitute for something else, usually for some expression that is too long to write easily or conveniently. Hence, in mathematics, a definition is simply an agreed upon shorthand. For example, we agreed earlier that a **prime number** is a natural number greater than one which is divisible only by itself and one. Of course, this definition requires us to know the definitions of a "natural number," of "greater than," and of "divisible." Usually, a definition is made to introduce a new concept in terms of the undefined terms and previously given definitions. Consequently, by substitution every definition could be written in terms of undefined terms only; however, such expressions would be lengthy and tedious to read.

An **axiom** or **postulate** is a statement that is assumed to be true. The axioms of a deductive mathematical system are the statements from which all other statements of the system can be derived. The Greeks were the first to develop the concept of a logical discourse based on a set of assumed axioms. According to Aristotle, "every demonstrative science must start from indemonstrative principles; otherwise, the steps of demonstration would be endless." A mathematical system is **consistent** if and only if contradictory statements are not implied by the axioms of the system. A set of axioms is **inconsistent** if it is possible to deduce from the axioms that some statement is both true and false. An inconsistent mathematical system is worthless, because the negation of any statement which can be deduced can be deduced also. Hence, the one requirement that is made of a set of axioms is that it be consistent. For aesthetic purposes, mathematicians attempt to keep both the number of undefined terms and the number of axioms of a system to a minimum.

A **theorem** is a true statement that has been proved by a valid argument. A **proof** is a logically valid deduction of a theorem from the premises of the theorem, the axioms, or previously proven theorems. A **formal proof** of a theorem is a finite sequence of statements $S_1, S_2, \ldots, S_k$ such that each statement S is a premise of the theorem to be established, an axiom of the system, follows from one or more of the preceeding S's by a logical rule of inference, or follows from one or more of the preceeding S's by a previously proven theorem. Hence, a **theorem** or **provable statement** is the last statement of some proof. Later in this chapter, we will present several techniques for proving theorems.

Euclidean and Non-Euclidean Geometry Systems

Thales of Miletus (c. 640-546 B.C.) is considered to be the father of Greek mathematics. Early in his life, he traveled to Egypt and learned the practical geometry that the Egyptians knew. Thales is the first person to conceive of the idea of geometry as a deductive science, with a succession of propositions, each resting upon the axioms and postulates and upon the propositions which had been proven previously. A noted pupil of Thales was Pythagoras (c. 572-501 B.C.) who was born on the island of Samos, not far from Miletus. To acquaint himself with all learning of the time, he traveled to Greece, Egypt, Babylonia, and India. Pythagoras and his disciples proved many theorems in geometry, the most famous of which bears his name. He devoted much attention to the study of areas, volumes, proportions, and regular solids. During the next three centuries, hundreds of Greek mathematicians pursued Thales' vision of geometry by verifying geometric relationships, proving theorems, and discovering constructions. According to the French mathematician and philosopher Henri Poincaré (1854-1912): "Science is built up with facts, as a house is with stones. But a collection of facts is no more a science than a heap of stones is a home." Prior to the time of Thales and Pythagoras, geometry consisted of a collection of disorganized facts. The Egyptians knew geometrical truths thousands of years earlier and the Babylonians even earlier. However, their knowledge consisted mainly of a collection of empirical results for surveying and constructing. The Greeks introduced order into geometry and converted it from a collection of isolated facts into an organized science. The organization and systematization of geometry was begun by Thales and Pythagoras, continued by their successors, and culminated with the publication in about 300 B.C. of the treatise, *Elements*, by Euclid of Alexandria (c. 325 B.C. to c. 265 B.C.). Euclid's book has been the basis for the study of geometry ever since. It has even been said that next to the *Bible*, the *Elements* may be the most translated, published, and studied book in the Western world.

The *Elements* is based on twenty-three definitions, five postulates, and five axioms. Euclid divided the statements he assumed to be true into two sets—postulates and axioms. It is not clear why he did this; but, perhaps, he thought some statements were more general or more "self-evident" than others. Currently, we call statements that are assumed to be true either axioms or postulates. The postulates of Euclid were

1. A straight line may be drawn between any two points.
2. Any terminated straight line may be produced indefinitely.
3. About any point as center, a circle with any radius may be described.
4. All right angles are equal.
5. If two straight lines lying in a plane are met by another line, making the sum of the internal angles on one side less than two right angles, then those straight lines will meet, if sufficiently produced, on the side on which the sum of the angles is less than two right angles.

The first three postulates are postulates regarding construction and the fourth postulate says that space is homogeneous. The fifth postulate is called the "parallel postulate." In the *Elements*, Euclid proves twenty-eight propositions (theorems) using only the first four postulates. He needed the parallel postulate to prove the twenty-ninth proposition. The collection of statements of geometry that can be proven using only the first four postulates is known as **absolute geometry** or **neutral geometry**. The fifth century commentator Proclus (410-485) wrote of the parallel postulate: "This ought even to be struck out of the Postulates altogether; for it is a theorem involving many difficulties ... ". Although Proclus did not deny the truth of the fifth postulate, he denied that it was "self-evident." He even presented a false proof of his own that the fifth postulate could be deduced from the first four postulates. The history of mathematics is filled with attempts to prove that the parallel postulate is not a postulate but a theorem. For centuries mathematicians tried to show the fifth postulate was a theorem by either (1) deducing it from the first four postulates or (2) deducing it from the first four postulates and an additional "self-evident" postulate. Attempts of the first kind failed, because it is not possible to deduce the fifth postulate from the first four. However, this fact was not proven until the mid-nineteenth century. Attempts of the second kind failed because every proposed alternate postulate which was strong enough to permit the deduction of the parallel postulate was no more "self-evident" than the original fifth postulate.

In 1685, Giovanni Girolamo Saccheri (1667-1733) entered the Jesuit Order. He was ordained a priest in 1694 and taught at Jesuit colleges throughout Italy. From 1699 until his death, he held the chair of mathematics at Pavia. In 1697, Saccheri replaced the parallel postulate with a contrary assumption. He intended to derive a contradiction from Euclid's first four postulates and his assumed postulate; thereby, deducing the fifth postulate. Instead, he succeeded in proving many theorems. Consequently, Saccheri was the first to postulate and develop a system of non-Euclidean geometry. However, he did not realize this fact! He considered the results to be absurd, since they violated ordinary geometrical intuition. In the first half of the nineteenth century, Gauss, Bolyai, Lobachevsky, and Riemann developed the non-Euclidean geometries.

The System of Natural Numbers

Mathematicians produce new objects to study in a variety of ways. One way to produce new objects is through the specification of a set of axioms—which in essence provide "implicit definitions" of the objects. All other facts about the new objects are derived through logical deduction. A second way to produce new objects is to construct the objects from already existing objects. During the nineteenth century, mathematicians carried out the "arithmetization of analysis." In 1888, the German mathematician Richard Dedekind (1831-1916) developed the set of natural numbers axiomatically. In 1889, the Italian mathematician and logician Giuseppe Peano (1858-1932) presented an alternate axiomatic development of the set of natural numbers. Other mathematicians showed how to construct the integers from the natural numbers, the rational numbers from the integers, the real numbers from the rational numbers, and the complex numbers from the real numbers.

Peano's axiomatic system for the natural numbers has three undefined terms: "1","number," and "successor." Four of Peano's five axioms are

P1. 1 is a number.

P2. Every number n has a unique successor, n', which is a number.

P3. If m and n are numbers and if $m' = n'$, then $m = n$.

P4. 1 is not the successor of any number.

Axioms P1 and P4 together say that 1 is the "first" or "initial" element in the set of natural numbers. Axiom P2 says that 1 has exactly one successor. What Peano intends by axiom P2 is that the successor of n will be obtained from n by adding 1—that is, Peano intends for $n' = n + 1$. So the unique successor of 1 is $1 + 1 = 2$, and the unique successor of 2 is 3, and so forth. The elements of the set of natural numbers essentially form a chain. The first link in the chain in the number 1, the second link in the chain is the number 2, and so on. The second axiom, P2, does not allow the chain to "branch" or "divide" as more links are added, since only one unique link can be added at an existing link. The third axiom, P3, prohibits the chain from "looping back onto itself," because P3 says that every natural number except 1 has a unique "predecessor."

Peano knew that there were an infinite number of sets in addition to the set of natural numbers that satisfied the axioms P1 through P4. But he also knew that all of these sets were "equivalent to" the set of natural numbers.

For example, if we define the successor of n to be $n' = n/2$, then we obtain the sequence of numbers $1, \frac{1}{2}, \frac{1}{4}, \frac{1}{8} \ldots$ which satisfies axioms P1 through P4.

Later, we will state and discuss axiom P5 (which leads to the principle of mathematical induction). From axioms P1 through P5 the following well-known theorems (properties) for natural numbers can be proven.

BASIC THEOREMS FOR THE NATURAL NUMBERS

Closure Properties for Addition and Multiplication

N1 $\quad \forall m, n \in \mathbf{N},\ m + n \in \mathbf{N}$

N2 $\quad \forall m, n \in \mathbf{N},\ mn \in \mathbf{N}$

Commutative Properties for Addition and Multiplication

N3 $\quad \forall m, n \in \mathbf{N},\ m + n = n + m$

N4 $\quad \forall m, n \in \mathbf{N},\ mn = nm$

Associative Properties for Addition and Multiplication

N5 $\quad \forall m, n, p \in \mathbf{N},\ m + (n + p) = (m + n) + p$

N6 $\quad \forall m, n, p \in \mathbf{N},\ m(np) = (mn)p$

Distributive Properties of Multiplication over Addition

N7 $\quad \forall m, n, p \in \mathbf{N},\ m(n + p) = mn + mp$

N8 $\quad \forall m, n, p \in \mathbf{N},\ (m + n)p = mp + np$

Existence of a Multiplicative Identity Property

N9 $\quad (\exists 1 \in \mathbf{N})(\forall m \in \mathbf{N}),\ 1m = m$

The System of Integers

One of the first things of a mathematical nature a young child learns to do is count: "one, two, three, ..." That is, the child learns the names of the elements of the set of natural numbers. Then a child learns to add single digit natural numbers such as $2+3$ by "counting on their fingers." Some time later a child learns how to subtract 3 from 5 by asking and answering the question: "What number do I add to 3 to get 5?" Hence, at an early age we learn that subtraction is defined in terms of addition.

DEFINITION Subtraction

Let S be a set of numbers and let m and n be elements of S. The number $x = m - n$, read "m minus n" or "n subtracted from m," is the number x which makes the statement $m = n + x$ true, if there be such a number x.

Thus, $x = 3-1$ is the number which makes $3 = 1+x$ a true statement. (Recall that you learned how to add first and then how to subtract. Furthermore, you learned to "check" your subtraction result by addition.)

Theorem N1 for the natural numbers states that the natural numbers are closed under addition. That is, if we add two natural numbers, we get another natural number. Is the set of natural numbers closed under the operation of subtraction? The answer is no. Ask a child: "If you have two apples and you give both of them away, how many apples do you have left?" The child will most likely answer "none." Consequently, if we want a set of numbers which contains the elements of the set of natural numbers and which is closed under the operation of subtraction, we must enlarge the set of natural numbers by including a number which represents the concept of "none." The number we need to include is called "zero" and it is represented by the symbol "0." Let $\mathbf{W} = \{0, 1, 2, 3, \ldots\}$. $\mathbf{W}$ contains the elements of the set of natural numbers and the element 0. We "invented" the number 0 to have the property that for all $m \in \mathbf{W}$, $m - m = 0$, or stated in terms of addition, 0 has the property that for all $m \in \mathbf{W}$, $m = m + 0$. Therefore, the number 0 is called the **additive identity for the set W**, because when one adds 0 to any number $m \in \mathbf{W}$ the result is m.

If we replace the set $\mathbf{N}$ by the set $\mathbf{W}$ in theorems N1 through N9 and name the new theorems W1 through W9 respectively, we can prove that the set $\mathbf{W}$ satisfies theorems W1 through W9 and the additional following theorem:

Existence of an Additive Identity Property
W10 $(\exists 0 \in \mathbf{W})(\forall m \in \mathbf{W}),\ m + 0 = m$

Is the set $\mathbf{W}$ closed under the operation of subtraction? That is, for all $m, n \in \mathbf{W}$, is the number $x = (m - n) \in \mathbf{W}$? The answer is no, because the number $(2 - 3) \notin \mathbf{W}$ since there is no number $x \in \mathbf{W}$ which makes the statement $2 = 3 + x$ true. The question we need to answer next is "What new numbers do we need to combine with the elements of $\mathbf{W}$ in order to obtain a new set which is closed under the operation of subtraction and, in addition, satisfies theorems W1 through W10 when the set $\mathbf{W}$ is replaced by the new set? Let S be a set which contains the elements of $\mathbf{W}$ and all of the other new numbers we need to "invent" so that S is closed under subtraction. That is, we want S to have the property that if $m, n \in S$, then $(m - n) \in S$. Since $0 \in \mathbf{W}$, $0 \in S$. And for every $m \in \mathbf{N}$, $m \in S$. Thus, for every $m \in \mathbf{N}$ we need a new number $x \in S$ such that $0 - m = x \in S$. Because subtraction is defined in terms of addition, for every $m \in \mathbf{N}$ we need a new number $x \in S$ such that $0 = m + x$ is a true statement. We define that number x to be $-m$, which is read "minus m" or "negative m." Hence, for subtraction to be closed in S, the set of numbers we need to include in S in addition to the numbers in $\mathbf{W}$ is $\{\ldots, -3, -2, -1\}$. Therefore, the "smallest set" which includes the elements of the natural numbers and which is closed under subtraction is the set of integers

$$Z = \{\ldots, -3, -2, -1, 0, 1, 2, 3, \ldots\}.$$

Notice that the integers include the natural numbers (which are also called the **positive integers**), zero, and the **negative integers** (the set $\{\ldots, -3, -2,$

$-1\}$). If we replace the set **N** by the set **Z** in theorems N1 through N9 and call the new theorems Z1 through Z9, if we replace the set **W** by the set **Z** in theorem W10 and call the new theorem Z10, we can prove that the set of integers, **Z**, satisfies theorems Z1 through Z10 and satisfies the additional theorem Z11 stated below.

BASIC THEOREMS FOR THE INTEGERS

Closure Properties for Addition and Multiplication

Z1 $\forall m, n \in \mathbf{Z},\ m + n \in \mathbf{Z}$

Z2 $\forall m, n \in \mathbf{Z},\ mn \in \mathbf{Z}$

Commutative Properties for Addition and Multiplication

Z3 $\forall m, n \in \mathbf{Z},\ m + n = n + m$

Z4 $\forall m, n \in \mathbf{Z},\ mn = nm$

Associative Properties for Addition and Multiplication

Z5 $\forall m, n, p \in \mathbf{Z},\ m + (n + p) = (m + n) + p$

Z6 $\forall m, n, p \in \mathbf{Z},\ m(np) = (mn)p$

Distributive Properties of Multiplication over Addition

Z7 $\forall m, n, p \in \mathbf{Z},\ m(n + p) = mn + mp$

Z8 $\forall m, n, p \in \mathbf{Z},\ (m + n)p = mp + np$

Existence of a Multiplicative Identity Property

Z9 $(\exists 1 \in \mathbf{Z})(\forall m \in \mathbf{Z}),\ 1m = m$

Existence of an Additive Identity Property

Z10 $(\exists 0 \in Z)(\forall m \in \mathbf{Z}),\ m + 0 = m$

Existence of Additive Inverses Property

Z11 $(\forall m \in \mathbf{Z})(\exists (-m) \in \mathbf{Z}),\ m + (-m) = 0$

In summary, the natural numbers are closed under the operations of addition and multiplication (theorems N1 and N2) and satisfy theorems N3 through N9. We enlarged the set of natural numbers by including the numbers $0, -1, -2, \ldots$, and obtained the set of integers. The integers satisfy the extended theorems Z1 through Z11. Thus, the integers are closed under the operations of addition, multiplication, and subtraction. However, subtraction is not commutative nor associative as are addition and multiplication.

EXAMPLE 1 Justification of the Steps in a Formal Proof

Provide a justification for each step in the following formal proof. Each of the missing justifications should be one of the theorems Z1 through Z11 or "substitution of s_i into s_j" where s_i and s_j are prior statements in the proof.

Theorem 2.1 For every integer m, $m0 = 0$.

(Or written symbolically, $\forall m \in \mathbf{Z}$, $m0 = 0$.)

Proof: For all $m \in \mathbf{Z}$

1. $mm = mm$	1. The relation = is reflexive*.
2. $m + 0 = m$	2.
3. $m(m + 0) = mm$	3.
4. $m(m + 0) = mm + m0$	4.
5. $mm + m0 = mm$	5.
6. $m0 = 0$	6.

SOLUTION

The justifications are as follows: 2. Theorem Z10 3. Substitution of 2 into 1. 4. Theorem Z7 5. Substitution of 4 into 3. 6. Theorem Z10.

Notice that since multiplication is commutative in the integers (theorem Z4), $\forall m \in \mathbf{Z}$, $0m = m0 = 0$.

*Our first justification is the fact that the relation = is reflexive—that is, $a = a$. However, we do not study relations until chapter 4.

EXERCISES 2.1

1. Is the set $\{1, 2, \ldots, 10\}$ closed with respect to addition? multiplication? subtraction? division?

2. Is the set of even integers $\{\ldots, -4, -2, 0, 2, 4, \ldots\}$ closed with respect to addition? multiplication? subtraction? division?

3. Is the set of odd integers $\{\ldots, -3, -1, 1, 3, \ldots\}$ closed with respect to addition? multiplication? subtraction? division?

4. For the integers, give examples to show that

    a. the commutative property of subtraction $m - n = n - m$ is false.

    b. the associative property of subtraction $m - (n - p) = (m - n) - p$ is false.

    c. the commutative property of division $m \div n = n \div m$ is false.

    d. the associative property of division $m \div (n \div p) = (m \div n) \div p$ is false.

5. For the set of natural numbers, does addition distribute over multiplication? That is, does $m + (n \cdot p) = (m + n) \cdot (m + p)$? If not, give an example which shows it does not.

6. Provide a justification for each step in the formal proof of the following theorem. Each justification should be one of the theorems Z1 through Z11, theorem 2.1, or "substitution of s_i into s_j" where s_i and s_j are prior statements in the proof.

   Theorem 2.2 For every integer m, $-(-m) = m$.

   (Or written symbolically, $\forall m \in \mathbf{Z}$, $-(-m) = m$.)

   Proof: For all $m \in \mathbf{Z}$

1. $m + (-m) = 0$	1.
2. $m + (-m) = (-m) + m$	2.
3. $(-m) + m = 0$	3.
4. $-(-m) = m$	4.

7. Provide a justification for each step in the formal proof of the following theorem. Each justification should be one of the theorems Z1 through Z11, theorem 2.1, theorem 2.2, or "substitution of s_i into s_j" where s_i and s_j are prior statements in the proof.

   Theorem 2.3 For integers m and n, $m(-n) = -(mn)$.

   (Or written symbolically, $\forall m, n \in \mathbf{Z}$, $m(-n) = -(mn)$.)

   Proof: For all $m, n \in \mathbf{Z}$

1. $m0 = 0$	1.
2. $n + (-n) = 0$	2.
3. $m(n + (-n)) = 0$	3.
4. $m(n + (-n)) = mn + m(-n)$	4.
5. $mn + m(-n) = 0$	5.
6. $m(-n) = -(mn)$	6.

8. Provide a justification for each step in the formal proof of the following theorem. Each justification should be one of the theorems Z1 through Z11, theorem 2.1, theorem 2.2, theorem 2.3, or "substitution of s_i into s_j" where s_i and s_j are prior statements in the proof.

   Theorem 2.4 For integers m and n, $(-m)(-n) = mn$.

   (Or written symbolically, $\forall m, n \in \mathbf{Z}$, $m(-n) = -(mn)$.)

Proof: For all $m, n \in \mathbf{Z}$

1. $(-m)0 = 0$	1.
2. $n + (-n) = 0$	2.
3. $(-m)(n + (-n)) = 0$	3.
4. $(-m)(n + (-n)) = (-m)n + (-m)(-n)$	4.
5. $(-m)n + (-m)(-n) = 0$	5.
6. $(-m)(-n) = -(-m)n$	6.
7. $-(-m) = m$	7.
8. $(-m)(-n) = mn$	8.

9. Provide a justification for each step in the formal proof of the following theorem. Each justification should be one of the theorems Z1 through Z11, theorem 2.1, theorem 2.2, theorem 2.3, theorem 2.4, "substitution of s_i into s_j" where s_i and s_j are prior statements in the proof, or the definition of subtraction for the integers. Recall that the definition is as follows: Let m and n be integers, the integer $m - n = x$ is the integer x which makes the statement $m = n + x$ true.

Theorem 2.5 If m and n are integers, then $m - n = m + (-n)$. (Or written symbolically, $\forall m, n \in \mathbf{Z}$, $m - n = m + (-n)$.)

Proof: For all $m, n \in \mathbf{Z}$

1. $m + [(-n) + n] = [m + (-n)] + n$	1.
2. $[m + (-n)] + n = n + [m + (-n)]$	2.
3. $m + [(-n) + n] = n + [m + (-n)]$	3.
4. $(-n) + n = n + (-n)$	4.
5. $n + (-n) = 0$	5.
6. $n + (-n) = (-n) + n$	6.
7. $(-n) + n = 0$	7.
8. $m + 0 = n + [m + (-n)]$	8.
9. $m + 0 = m$	9.
10. $m = n + [m + (-n)]$	10.
11. $m - n = m + (-n)$	11.

10. Provide a justification for each step in the following formal proof of the **Cancellation Property of Addition.**

Theorem 2.6 If k, m and n are integers and $k + m = k + n$, then $m = n$.

Proof: Let $k, m, n \in \mathbf{Z}$

1. $k + m = k + n$	1. Premise
2. $k + (-k) = 0$	2.
3. $k + (-k) = (-k) + k$	3.
4. $(-k) + k = 0$	4.

5. $(-k) + (k + m) = (-k) + (k + m)$ 5.
6. $(-k) + (k + m) = (-k) + (k + n)$ 6.
7. $(-k) + (k + m) = ((-k) + k) + m$ 7.
8. $((-k) + k) + m = (-k) + (k + n)$ 8.
9. $(-k) + (k + n) = ((-k) + k) + n$ 9.
10. $((-k) + k) + m = ((-k) + k) + n$ 10.
11. $0 + m = 0 + n$ 11.
12. $m = n$ 12.

11. Provide a justification for each step in the following formal proof.

Theorem 2.7 For all integers m, $(-1) \cdot m = -m$.

Proof: Let $m \in \mathbf{Z}$

1. $m(1 + (-1)) = m \cdot 1 + m \cdot (-1)$ 1.
2. $1 + (-1) = 0$ 2.
3. $m \cdot 0 = m \cdot 1 + m \cdot (-1)$ 3.
4. $m \cdot 0 = 0$ 4.
5. $m \cdot 1 + m \cdot (-1) = 0$ 5.
6. $m \cdot 1 = m$ 6.
7. $m + m \cdot (-1) = 0$ 7.
8. $m \cdot (-1) = (-1) \cdot m$ 8.
9. $m + (-1) \cdot m = 0$ 9.
10. $m + (-m) = 0$ 10.
11. $m + (-1) \cdot m = m + (-m)$ 11.
12. $(-1) \cdot m = -m$ 12.

2.2 Mathematical Proofs

In mathematics, a proof is a logically valid deduction of a theorem from the premises of the theorem, the axioms, or previously proven theorems. The truth of any statement which appears in a proof must be traceable back to the original axioms. In section 2.2.1, we present the basic proof techniques for proving conditional statements which include a direct proof, a proof by contraposition, and a proof by contradiction. In section 2.2.2, we discuss proof by cases and we show how to prove a biconditional statement by proving two conditional statements by any of the techniques cited above. Next, we consider proof by contradiction for any general statement—which includes statements that are not conditional statements. Then, we indicate how to prove statements which contain quantifiers. In section 2.2.3, we present some well-known mathematical conjectures and illustrate how to prove and disprove conjectures. Finally, we examine properties of the system of rational numbers and the system of real numbers.

2.2.1 Techniques for Proving the Conditional Statement P⇒Q

Most mathematical theorems are of the form "If P, then Q."—that is, in the form of the conditional statement $P \Rightarrow Q$. Additional theorems which are stated in the biconditional form "P if and only if Q." are proven by proving both of the conditional statements $P \Rightarrow Q$ and $Q \Rightarrow P$. At this point, we present three basic types of proofs and simple examples of proofs which involve the natural numbers and integers. First, we need the following definitions.

DEFINITIONS Even and Odd Integers

An integer n is **even**, if there exists an integer k such that $n = 2k$.

An integer n is **odd**, if there exists an integer k such that $n = 2k + 1$.

By definition, the set of **even integers** is

$$E = \{2k \mid k \in \mathbf{Z}\} = \{\ldots, -4, -2, 0, 2, 4, \ldots\}$$

and the set of **odd integers** is

$$O = \{2k + 1 \mid k \in \mathbf{Z}\} = \{\ldots, -3, -1, 1, 3, \ldots\}.$$

From the definitions of the set of integers, even integers, and odd integers, we see that every integer is either even or odd and no integer is both even and odd. Also we note that the negation of the statement "n is an even integer" is "n is not an even integer" which is logically equivalent to "n is an odd integer." Likewise, the negation of "n is an odd integer" is "n is an even integer."

Direct Proof

In a **direct proof** of the conditional statement $P \Rightarrow Q$, we assume P and then use axioms, definitions, rules of logical inference, previously proven theorems, and computations to infer Q. Hence, a direct proof has the following form.

Direct Proof of P⇒Q

Proof: Assume P. ... Therefore, Q. Hence, $P \Rightarrow Q$.

EXAMPLE 1 A Formal Direct Proof

Give a formal direct proof of the following theorem.

Theorem 2.8 If n is an even integer, then n^2 is an even integer.

SOLUTION

$<<$ Comments enclosed in $<< \; >>$ are not part of a proof and are included as information for the reader of the text. Our plan for this proof is to start by assuming the hypothesis—namely, that n is an even integer. Then use the definition of an even integer to assert that we can write $n = 2k$ for some integer k. Next, we calculate n^2 and perform the necessary computations to show that $n^2 = 2m$ for some integer m. Hence, proving n^2 is an even integer by definition. A formal proof consists of a list of true statements and associated justifications. Proofs of all kinds vary from writer to writer. Here is our formal direct proof of Theorem 2.8. $>>$

Proof:

Statement	Justification
1. n is an even integer	1. Premise
2. There exists a $k \in \mathbf{Z}$ such that $n = 2k$	2. Definition of an even integer
3. $n^2 = (n)(n)$	3. Definition of n^2
4. $n^2 = (2k)(2k)$ where $k \in \mathbf{Z}$	4. Substitution of 2 into 3
5. $2k = (2)(k)$	5. Definition of multiplication
6. $n^2 = ((2)(k))(2k)$	6. Substitution of 5 into 4
7. $((2)(k))(2k) = 2((k)(2k))$	7. Theorem Z6 (Associative Property of Multiplication)
8. $n^2 = 2((k)(2k))$	8. Substitution of 7 into 6
9. Since $2 \in \mathbf{Z}$ and $k \in \mathbf{Z}$, $(2k) \in \mathbf{Z}$	9. Theorem Z2 (Closure Property of Multiplication)
10. Since $k \in \mathbf{Z}$ and $(2k) \in \mathbf{Z}$, $k(2k) \in \mathbf{Z}$	10. Theorem Z2 (Closure Property of Multiplication)
11. Let $m = k(2k) \in \mathbf{Z}$	11. Notation
12. $n^2 = 2m$ where $m \in \mathbf{Z}$	12. Substitution of 11 into 8
13. n^2 is an even integer	13. Definition of an even integer

The advantage of a formal proof is that it makes you think about each step in the proof and a justification for the step. The disadvantage of a formal proof is that it is tedious and lengthy. An informal proof, on the other hand, is written in paragraph form and some justifications are omitted intentionally. The justifications which are omitted depend upon the reading audience. The following example presents an informal proof of Theorem 2.8.

EXAMPLE 2 An Informal Direct Proof

Give an informal direct proof of the following theorem.

Theorem 2.8 If n is an even integer, then n^2 is an even integer.

SOLUTION

<< In example 1, we presented a formal proof of this theorem, so our task is simply to rewrite that proof in paragraph form. >>

Proof: Let n be an even integer. By definition of even, there exists a $k \in \mathbf{Z}$ such that $n = 2k$. Squaring and using the associative property of multiplication, we find

$$n^2 = (2k)(2k) = 2(k(2k)) = 2m, \quad \text{where} \quad m = k(2k).$$

Since 2 and k are integers and since the integers are closed under multiplication, $m = k(2k)$ is an integer. Consequently, since $n^2 = 2m$ where m is an integer, it follows from the definition of an even integer that n^2 is an even integer.

Proof by Contraposition

One of the more useful logical equivalences is the that the conditional statement $P \Rightarrow Q$ is logically equivalent to its contrapositive $\neg Q \Rightarrow \neg P$. When you cannot easily provide a direct proof of the statement $P \Rightarrow Q$, it is often a good idea to try to find a proof of the contrapositive statement $\neg Q \Rightarrow \neg P$—either a direct proof or some other type proof. Since a statement and its contrapositive are logically equivalent, by proving the contrapositive you have proven the statement also. Consequently, a proof by contraposition has the form.

Contrapositive Proof of P⇒Q

Proof: Assume $\neg Q$. ... Therefore, $\neg P$. Hence, $\neg Q \Rightarrow \neg P$. And consequently, $P \Rightarrow Q$.

EXAMPLE 3 A Proof by Contraposition

Prove the following theorem by the method of contraposition.

Theorem 2.9 Let n be an integer. If n^2 is an even integer, then n is an even integer.

SOLUTION

<< We will prove this theorem by proving its contrapositive—namely, the statement "If n is not an even integer, then n^2 is not an even integer." More precisely, since the phrase "m is not an even integer" is logically equivalent to "m is an odd integer," we will provide a direct proof of the contrapositive statement "If n is an odd integer, then n^2 is an odd integer.">>

Proof: We will prove this theorem by contraposition. Let n be an odd integer. By definition of odd, there exists a $k \in \mathbf{Z}$ such that $n = 2k + 1$. Squaring and using the distributive properties, the associative property of multiplication, and the multiplicative identity property, we find

$$\begin{aligned} n^2 &= (2k+1)(2k+1) = (2k+1)(2k) + (2k+1)(1) \\ &= (2k)(2k) + 2k + 2k + 1 = 2(k(2k) + 2k) + 1 = 2p + 1, \end{aligned}$$

where $p = k(2k) + 2k$. Since 2 and k are integers and since the integers are closed under multiplication and addition, p is an integer. Consequently, since $n^2 = 2p + 1$ where p is an integer, it follows from the definition of an odd integer that n^2 is an odd integer. Hence, we have proven "If n is an odd integer, then n^2 is an odd integer;" and, therefore, by contraposition we have proven "If n^2 is an even integer, then n is an even integer."

<< Try to develop a direct proof of theorem 2.9. >>

Given two integers exactly one of the following statements is true and the other two are false:

(1) Both integers are odd.

(2) Both integers are even.

(3) One integer is odd and the other integer is even.

Therefore, we can conclude that the negation of the statement "both integers are odd" is the statement "both integers are even, or one integer is odd and the other integer is even" and vice-versa. Likewise, the negation of the statement "both integers are even or both integers are odd" is the statement "one integer is odd and the other integer is even." We will use this fact in the next example.

EXAMPLE 4 A Proof by Contraposition

Prove the following theorem by the method of contraposition.

Theorem 2.10 Let m and n be integers. If $m + n$ is an even integer, then either m and n are both even integers or m and n are both odd integers.

SOLUTION

Proof: We will prove this theorem by contraposition. Assume it is not the case that "m and n are both even integers or m and n are both odd integers." By our discussion above it follows that one integer is odd and the other is even. We assume that m is odd and n is even. Thus, we assume there exists an integer k such that $m = 2k + 1$ and there exists an integer ℓ such that $n = 2\ell$. Substituting, using the associative property of addition, the commutative property of addition, and a distributive property, we find

$$\begin{aligned} m + n &= (2k + 1) + 2\ell = 2k + (1 + 2\ell) = 2k + (2\ell + 1) \\ &= (2k + 2\ell) + 1 = 2(k + \ell) + 1 = 2p + 1 \end{aligned}$$

where $p = k + \ell$. Since k and ℓ are integers and since the integers are closed under addition, p is an integer. Consequently, since $m+n = 2p+1$ where p is an integer, it follows from the definition of an odd integer that $m+n$ is an odd integer. Hence, we have proven "If one integer is odd and the other integer is even, then $m + n$ is an odd integer." and, therefore, by contraposition we have proven "If $m + n$ is an even integer, then either m and n are both even integers or m and n are both odd integers."

$<<$ **NOTE**: Instead of assuming m is odd and n is even, we could have assumed that m is even and n is odd and obtained the same result. $>>$

Proof of P⇒Q by Contradiction

In a proof by contradiction, we prove that a statement is true by showing that it cannot possibly be false. That is, we assume a statement is false and then produce some contradiction. We conclude that the statement cannot be false; and, therefore, it must be true. Recall that the statement $P \Rightarrow Q$ is logically equivalent to the statement $(\neg P) \vee Q$. To prove by contradiction that the statement $P \Rightarrow Q$ is true, we assume $P \Rightarrow Q$ is false, which is equivalent to assuming $\neg(P \Rightarrow Q)$ is true. Since $P \Rightarrow Q \equiv (\neg P) \vee Q$, $\neg(P \Rightarrow Q) \equiv \neg((\neg P) \vee Q)$. By a De Morgan law and double negation, we obtain

$$\neg(P \Rightarrow Q) \equiv \neg((\neg P) \vee Q) \equiv \neg(\neg P) \wedge (\neg Q) \equiv P \wedge (\neg Q)$$

Consequently, a proof of $P \Rightarrow Q$ by contradiction has the form.

Proof of P⇒Q by Contradiction

Proof: Assume $P \wedge (\neg Q)$. ... Therefore, C, where C is a contradiction. Hence, $P \Rightarrow Q$.

Most often the contradiction C which is reached has the form $R \wedge (\neg R)$ where R is any statement. The advantage of a proof by contradiction is that the statement R may be any statement whatsoever. However, when actually developing the proof, the disadvantage is not knowing in advance what statement R to select to produce the desired contradiction. The following example illustrates proving a theorem by the technique of contradiction.

EXAMPLE 5 A Proof by Contradiction

Prove the following theorem by the method of contradiction.

Theorem 2.11 Let n be an integer. If n^2 is an odd integer, then n is an odd integer.

SOLUTION

Proof: Let P denote the statement "n^2 is an odd integer" and let Q denote the statement "n is an odd integer." We will prove this theorem by contradiction. Therefore, we assume P and $\neg Q$. The negation of the statement Q is "n is an even integer." Thus, there is an integer k such that $n = 2k$. Squaring and using the associative property of multiplication, we find $n^2 = (2k)(2k) = 2(k(2k)) = 2p$ where $p = k(2k)$. Since the integers are closed under multiplication, p is an integer; and, consequently, n^2 is an even integer. The statement "n^2 is an even integer" is the statement $\neg P$. Thus, we assumed the statements P and $\neg Q$ are true and we deduced that the statement $\neg P$ is true. Hence, we reached the contradiction $P \wedge (\neg P)$. Consequently, we have proven theorem 2.11 by contradiction.

<< Observe in this case, the statement R in the contradiction $R \wedge (\neg R)$ is the hypothesis P. >>

Although a direct proof, a proof by contraposition, and a proof by contradiction are all acceptable forms of proof, most mathematicians order of preference for a proof is a direct proof, followed by a proof by contraposition, followed by a proof by contradiction. In the following example, the same statement is proven using each type of proof.

EXAMPLE 6 A Direct Proof, a Proof by Contraposition, and a Proof by Contradiction

Prove the following theorem by (a) a direct proof, (b) a proof by contraposition, and (c) a proof by contradiction.

Theorem 2.12 If n is an odd integer, then $5n - 3$ is an even integer.

SOLUTION

(a) **A Direct Proof**

Let n be an odd integer. By definition of odd, there exists an integer k such that $n = 2k + 1$. Substituting for n, we find

$$5n - 3 = 5(2k + 1) - 3 = 10k + 5 - 3 = 10k + 2 = 2(5k + 1) = 2p$$

where $p = 5k + 1$. Since $p = 5k + 1$ is an integer, $5n - 3$ is an even integer.

(b) **A Proof by Contraposition**

Assume $5n - 3$ is not an even integer—that is, assume $5n - 3$ is an odd integer. Thus, there exists an integer k such that $5n - 3 = 2k + 1$. Since

$$\begin{aligned} n &= 5n - 4n = 5n - 3 - 4n + 3 = (5n - 3) - 4n + 3 \\ &= 2k + 1 - 4n + 3 = 2k - 4n + 4 = 2(k - 2n + 2) \end{aligned}$$

and since $k - 2n + 2$ is an integer, n is even. Hence, if $5n - 3$ is not an even integer, n is not an odd integer. By contraposition, if n is odd, then $5n - 3$ is even.

(c) **A Proof by Contradiction**

Assume n is an odd integer and $5n - 3$ is an odd integer. Because n is odd, there exists an integer k such that $n = 2k + 1$. Substituting, we find

$$5n - 3 = 5(2k + 1) - 3 = 10k + 5 - 3 = 10k + 2 = 2(5k + 1)$$

Since $5k + 1$ is an integer, $5n - 3$ is an even integer, which contradicts our assumption that $5n - 3$ is an odd integer.

<< Observe in this case, the statement R in the contradiction $R \wedge (\neg R)$ is the conclusion Q. >>

EXERCISES 2.2.1

In exercises 1-8, prove each theorem by a direct proof.

1. If m is an even integer, then $m + 1$ is an odd integer.
2. If m is an odd integer, then $m + 1$ is an even integer.
3. If m is an odd integer, then $m^2 + 1$ is an even integer.
4. For any integer m, $2m - 1$ is an odd integer.
5. If m and n are both even integers, then $m + n$ is an even integer.
6. If m and n are both odd integers, then $m + n$ is an even integer.
7. If m is an even integer and n is an odd integer, then $m + n$ is an odd integer.
8. If m is an even integer and n is an odd integer, then mn is an even integer.

In exercises 9-14, prove each theorem by a proof by contraposition.

9. If m is an odd integer, then $m + 1$ is an even integer.
10. If m is an even integer, then $m + 2$ is an even integer.
11. If mn is an even integer, then either m or n is an even integer.
12. If mn is an odd integer, then m and n are both odd integers.
13. If $m + n$ is an even integer, then either m and n are both odd integers or m and n are both even integers.
14. If $m + n$ and mn are both even integers, then m and n are both even integers.

In exercises 15-17, prove each theorem by a proof by contradiction.

15. If mn is an odd integer, then both m and n are odd integers.
16. If m is an even integer, then $m + 1$ is an odd integer.
17. Let m and n be integers. If $m - n = m + (-n)$ is an odd integer, then $m + n$ is an odd integer.

2.2.2 Additional Proof Techniques

In this section, we present a few more elementary techniques for proving theorems. First, we discuss a proof by cases which can be used when the hypothesis can be subdivided into a number of distinct cases and the conclusion proven separately for each case. Next, we show how to prove a biconditional statement. Then, we consider a proof by contradiction for any general statement. Finally, we show how to prove statements which contain quantifiers.

Proof by Cases or Proof by Exhaustion

The following is a proof of the logical equivalence

$$((P \vee Q) \Rightarrow R) \equiv (P \Rightarrow R) \wedge (Q \Rightarrow R).$$

Proof:

$((P \vee Q) \Rightarrow R) \equiv \neg(P \vee Q) \vee R$	Definition of implication
$\equiv ((\neg P) \wedge (\neg Q)) \vee R$	A De Morgan law
$\equiv ((\neg P) \vee R) \wedge ((\neg Q)) \vee R)$	A distributive law
$\equiv (P \Rightarrow R) \wedge (Q \Rightarrow R)$	Definition of implication (twice)

In a **proof by cases**, which is also called a **proof by exhaustion**, one proves $(P \vee Q) \Rightarrow R$ by proving $P \Rightarrow R$ by any technique and proving $Q \Rightarrow R$ by any technique. Thus, a proof by cases or a proof by exhaustion has the following basic form.

A Proof of $(P \vee Q) \Rightarrow R$ by Cases or Exhaustion

Proof: Prove $P \Rightarrow R$ by any technique. Prove $Q \Rightarrow R$ by any technique. Therefore, $(P \vee Q) \Rightarrow R$, since $((P \vee Q) \Rightarrow R) \equiv (P \Rightarrow R) \wedge (Q \Rightarrow R)$.

For example, to prove a theorem regarding the set of integers, we might subdivide the proof into two cases—one in which we prove the conclusion for the set of even integers and another in which we prove the conclusion for the set of odd integers. To prove a theorem for the set of real numbers, we might divide the proof into two or three cases. In the two case proof, we could divide the set of real numbers into the set of rational numbers and the set of irrational numbers and then prove the desired conclusion for the set of rational numbers and also prove the conclusion for the set of irrational numbers. In a three case proof regarding the real numbers, we might divide the real numbers into three sets: (1) the set of positive real numbers, (2) the set $\{0\}$, and (3) the set of negative real numbers and then prove the conclusion for cases (1), (2), and (3). In a proof by exhaustion, extreme care must be taken to be certain that every possible case has been considered. Of course, a proof by exhaustion is most appealing when the total number of possibilities is relatively small. In general, if the hypotheses can be divided into n distinct and exhaustive cases $P_1, P_2, \ldots, P_n$, then to prove $(P_1 \vee P_2 \vee \ldots \vee P_n) \Rightarrow R$ one proves $P_1 \Rightarrow R$, and $P_2 \Rightarrow R$, ..., and $P_n \Rightarrow R$.

EXAMPLE 1 A Proof by Cases

Prove the following theorem by cases.

Theorem 2.13 If n is an integer, then $n^2 + n$ is an even integer.

SOLUTION

Proof: We prove this theorem by considering two cases.

Case 1. Assume n is an even integer. Hence, there exists an integer k such that $n = 2k$. Computing, we find

$$n^2 + n = (2k)^2 + 2k = 4k^2 + 2k = 2(2k^2 + k) = 2p,$$

where $p = 2k^2 + k$. Since 2 and k are integers and since the integers are closed under multiplication and addition, p is an integer; and, consequently, $n^2 + n$ is an even integer.

Case 2. Assume n is an odd integer. Then there exists an integer m such that $n = 2m + 1$. Computing, we find

$$\begin{aligned} n^2 + n &= (2m+1)^2 + 2m + 1 = 4m^2 + 4m + 1 + 2m + 1 \\ &= 4m^2 + 6m + 2 = 2(2m^2 + 3m + 1) = 2q, \end{aligned}$$

where $q = 2m^2 + 3m + 1$. Since $1, 2, 3$ and m are integers and since the integers are closed under multiplication and addition, q is an integer; and, consequently, $n^2 + n$ is an even integer.

Proof of the Biconditional Statement P ⇔ Q

In section 1.3, we showed that the biconditional statement $P \Leftrightarrow Q$ is logically equivalent to $(P \Rightarrow Q) \wedge (Q \Rightarrow P)$. Hence, the proof of a biconditional statement has the following form.

A Proof of the Biconditional Statement $P \Leftrightarrow Q$

Proof: Prove $P \Rightarrow Q$ by any technique. Prove $Q \Rightarrow P$ by any technique. Therefore, $P \Leftrightarrow Q$, since $P \Leftrightarrow Q \equiv (P \Rightarrow Q) \wedge (Q \Rightarrow P)$.

EXAMPLE 2 A Proof of a Biconditional Statement

Prove the following theorem.

Theorem 2.14 An integer n is even if and only if n^2 is an even integer.

SOLUTION

Proof: In example 2 of section 2.2.1, we presented a direct proof of theorem 2.8: If n is an even integer, then n^2 is an even integer. << If we had not proven theorem 2.8 previously, we would write its proof here. >>

In example 3 of section 2.2.1, we gave a proof by contraposition of theorem 2.9.: If n^2 is an even integer, then n is an even integer. << If we had not proven theorem 2.9 earlier, we would present its proof here. >>

<< Let P be the statement "n is an even integer" and let Q be the statement "n^2 is an even integer." Since we have proven both $P \Rightarrow Q$ and $Q \Rightarrow P$, we have proven the biconditional statement $P \Leftrightarrow Q$, which is theorem 2.14. >>

EXAMPLE 3 A Proof of a Biconditional Statement

Prove the following theorem.

Theorem 2.15 Let m and n be integers. The integer mn is even if and only if m is even or n is even.

SOLUTION

Proof: First, we prove if mn is even, then m is even or n is even. Suppose mn is even. Then there exists an integer k such that (1) $mn = 2k$. If m is even, we are done. So suppose, to the contrary, m is odd. Then (2) $m = 2j+1$ for some integer j. Substituting (2) into (1), we find $(2j+1)n = 2k$, and so (3) $2jn + n = 2k$. Subtracting $2jn$ from both sides of equation (3), we get (4) $n = 2k - 2jn = 2(k - jn)$. Since j, k, and n are integers and the integers are closed under multiplication and subtraction, it follows from equation (4) that n is an even integer. Thus, if mn is even, then m is even or n is even.

Next, we prove if m is even or n is even, then mn is even. Since by hypothesis m or n is even, we may assume for definiteness that m is even. Thus, there exists an integer k such that $m = 2k$. Calculating, we find $mn = (2k)n = 2(kn)$. Consequently, mn is even. Hence, if m is even or n is even, mn is even.

<< Since we have proven "if mn is even, then m is even or n is even" and since we have also proven "if m is even or n is even, mn is even," we have proven theorem 2.15. >>

Proof by Contradiction

In section 2.2.1 we showed how to prove the statement $P \Rightarrow Q$ by contradiction. A proof by contradiction of the statement P (which may or may not be a conditional statement) is based on the logical equivalence $P \equiv ((\neg P) \Rightarrow C)$ where C is a contradiction. To prove this equivalence, we recall the conditional statement $(A \Rightarrow B) \equiv ((\neg A) \vee B)$, so $((\neg P) \Rightarrow C) \equiv (\neg(\neg P) \vee C) \equiv (P \vee C)$ by double negation. Since C is a contradiction, C is a false statement; and, therefore, $(P \vee C) \equiv P$. Hence, $P \equiv ((\neg P) \Rightarrow C)$ where C is a contradiction, and the proof of the statement P by contradiction has the form:

A Proof of the Statement P by Contradiction

Proof: Assume $\neg P$. ... Therefore, C where C is a contradiction. Hence, P.

Often the contradiction C will be of the form $R \vee (\neg R)$ where R is any statement. A proof by contradiction of the statement P is also called an **indirect proof** or a ***reductio ad absurdum* proof**.

EXAMPLE 4 A Proof of the Statement P by Contradiction

Prove the following statement by contradiction.

The integer 512 cannot be written as the sum of one odd integer and two even integers.

SOLUTION

Proof: Assume to the contrary that (1) $512 = x + y + z$ where x is an odd integer and y and z are both even integers. Since x is odd, (2) $x = 2k + 1$ for some integer k. And since y and z are both even, there exist integers ℓ and m such that (3) $y = 2\ell$ and (4) $z = 2m$. Substituting (2), (3), and (4) into (1) and rearranging algebraically, we find

$$(5) \qquad 512 = x + y + z = (2k + 1) + 2\ell + 2m = 2(k + \ell + m) + 1$$

Since $k + \ell + m$ is an integer, it follows from equation (5) that 512 is an odd integer which is a contradiction. Thus, by contradiction we have proved the statement: "The integer 512 cannot be written as the sum of one odd integer and two even integers."

At this point, we need the following definitions.

DEFINITIONS Divides, Prime, and Composite

Let a and b be natural numbers. The number a **divides** b if and only if there exists a natural number c such that $ac = b$. If a divides b, then we also say a **is a factor of** b and b **is divisible by** a.

A **prime number** (or, simply, a **prime**) is a natural number greater than one which is divisible only by itself and one.

A **composite number** (or, simply, a **composite**) is a natural number greater than one which is not a prime number.

The natural number 1 has special status among the natural numbers, it is not a prime number and it is not a composite number. All other natural numbers are either prime or composite but not both. Observe that 1 divides all natural numbers a, since $1a = a$ and that the only natural number which is a factor of 1 is 1 itself. It follows from the definition of composite number that the natural number n is composite if and only if there exist natural numbers a and b such that $n = ab$ where $1 < a < n$ and $1 < b < n$.

We state the Fundamental Theorem of Arithmetic which is also known as the Unique Factorization Theorem without proof. We prove this theorem later in section 6.2.

The Fundamental Theorem of Arithmetic

Every natural number greater than one is a prime or can be written uniquely as a product of primes except for the order in which the prime factors are written.

Observe that 11 is a prime and the composite $12 = 2 \cdot 2 \cdot 3 = 2 \cdot 3 \cdot 2$. Thus, the prime 2 is a factor of 12 exactly twice and the prime 3 is a factor of 12 exactly once and 12 can be written uniquely as $12 = 2 \cdot 2 \cdot 3$ except for the fact that the order of the prime factors on the right-hand side of the last equation may be changed.

About 350 B.C., Euclid proved: "The number of prime numbers is infinite." He published this result in his treatise, the *Elements*. His proof is by contradiction and goes as follows. Suppose on the contrary that the number of prime numbers is finite and let the list of primes be $p_1, p_2, \ldots, p_k$. Consider (1) $q = p_1 \cdot p_2 \cdot \ldots \cdot p_k + 1$. Since $p_1, p_2, \ldots, p_k$ are natural numbers, all of which are greater than one, q is a natural number greater than one.

Therefore, by the Fundamental Theorem of Arithmetic, either q is a prime or q can be written uniquely as a product of prime numbers. In either case, q has a prime factor. Since $p_1, p_2, \ldots, p_k$ is the complete list of prime numbers, some p_j where $1 \leq j \leq k$ is a factor of q and it is also a factor of the product $p_1 \cdot p_2 \cdot \ldots \cdot p_k$. We can rewrite equation (1) as (2) $q - (p_1 \cdot p_2 \cdot \ldots \cdot p_k) = 1$. Since for some j where $1 \leq j \leq k$, p_j is a factor of both q and $p_1 \cdot p_2 \cdot \ldots \cdot p_k$, p_j is a factor of the left hand side of equation (2); and, consequently, the prime p_j is a factor of 1, which is a contradiction. Hence, the assumption that there are only a finite number of prime numbers is false. Therefore, the number of prime numbers is infinite.

As of May 28, 2004, the largest known prime was the Mersenne prime $2^{24036583} - 1$. Written in standard decimal form, this number is 7,235,733 digits long.

Proofs of Statements Containing Quantifiers

Although quantifiers may not be stated explicitly, most mathematical theorems are actually quantified statements. Furthermore, most of the time the universe of discourse is not stated explicitly but usually understood from the context. In what follows, we will let U denote the universe of discourse. The quantified statement $(\forall x \in U)(P(x))$ is true if and only if the truth set of the statement $P(x)$ is the entire universe, U. Hence, to prove $(\forall x \in U)(P(x))$, we let x be an arbitrary element of U and deduce that $P(x)$ is true. Consequently, a proof of the universal statement $(\forall x \in U)(P(x))$ has the following form.

A Proof of the Universal Statement $(\forall x \in U)(P(x))$

Proof: Let $x \in U$. ... $P(x)$ is true. Hence, $(\forall x \in U)(P(x))$ is true.

The following example illustrates this method of proof.

EXAMPLE 5 A Proof of $(\forall x \in U)(P(x))$

Prove the theorem: "For all odd integers n, n^2 is an odd integer."

SOLUTION

Proof: Let n be an arbitrary odd integer. Since n is odd, there exists an integer k such that $n = 2k + 1$. Squaring this equation and rearranging, we find

$$n^2 = (2k+1)^2 = 4k^2 + 4k + 1 = 2(2k^2 + 2k) + 1$$

Since 2 and k are integers and since the set of integers is closed under addition and multiplication, $2k^2 + 2k$ is an integer and, by definition, n^2 is an odd integer.

Notice that the statement "For all odd integers n, n^2 is an odd integer." is logically equivalent to the conditional statement "If n is an odd integer, then n^2 is an odd integer." Consequently, to prove $(\forall x \in U)(P(x) \Rightarrow Q(x))$, we let x be an arbitrary element of the universal set U, we assume $P(x)$ is true, and we deduce that $Q(x)$ is true. That is, we show for arbitrary $x \in U$, $P(x) \Rightarrow Q(x)$ is a true statement.

The quantified statement $(\exists x \in U)(P(x))$ is true provided the truth set of $P(x)$ is not the empty set. There are two ways we can prove the **existence statement** $(\exists x \in U)(P(x))$. We can present a **constructive proof** by exhibiting or explaining how to construct an element x in U having the property $P(x)$. Or, we can present a **nonconstructive proof** by making a valid argument that some element x in U makes the statement $P(x)$ true. In a nonconstructive proof, we do not exhibit a specific element in the universal set which makes the statement $P(x)$ true.

For example, to prove the theorem "There exists an even prime." all we need do is to note that the natural number 2 is even and it is divisible by only 1 and itself only. Thus, by exhibiting the even, prime, natural number 2, we have given a constructive proof of the statement: "There exists an even prime."

Nonconstructive proofs often depend upon other theorems. For instance, to prove the theorem "There is a real solution of the equation $x^3 - 2x + 5 = 0$." we make use of both the Fundamental Theorem of Algebra and the Conjugate Root Theorem. The Fundamental Theorem of Algebra states: "If $p(x)$ is a polynomial of degree $n \geq 1$ with complex coefficients, then there exist n solutions of the equation $p(x) = 0$." And, the Conjugate Root Theorem states: "If $p(x)$ is a polynomial of degree $n \geq 2$ with real coefficients, then if $a + bi$ where $b \neq 0$ is a solution of the equation $p(x) = 0$, then $a - bi$ is a solution also." Thus, the Conjugate Root Theorem states that complex roots of polynomials with real coefficients of degree greater than or equal to two, occur in pairs (complex conjugate pairs). We will not prove these theorems in this text, but they are "well-known" theorems. A nonconstructive proof of the statement "There is a real solution of the equation $x^3 - 2x + 5 = 0$." proceeds as follows: Let $p(x) = x^3 - 2x + 5$. Observe that the polynomial $p(x)$ is of degree 3 and has real coefficients. By the Fundamental Theorem of Algebra the equation $p(x) = x^3 - 2x + 5 = 0$ has exactly three solutions. By the Conjugate Root Theorem $p(x) = 0$ has an even number of complex roots—either 2 or 0. Hence, the equation $p(x) = 0$ must have 1 or 3 real roots.

Sometimes there is one and only one element $x \in U$ which makes a statement $P(x)$ true. That is, sometimes there is a unique element $x \in U$ such that $P(x)$ is true. A theorem which states that there is one and only one element in a specific set which satisfies a certain statement is called a **uniqueness theorem**. A uniqueness theorem is most often proven by contradiction. Thus, one assumes that there are two elements x and y in the universe U such that $P(x)$ and $P(y)$ are true and $x \neq y$. Then by a valid argument one arrives at the contradiction $x = y$—that is, x and y are the same element of U. In the following example, we prove the uniqueness of the additive identity 0 of the set of integers **Z**.

EXAMPLE 6 A Proof of a Uniqueness Theorem

The existence of an additive identity theorem for the set of integers stated in section 2.1 was

Z10 $\quad (\exists 0 \in Z)(\forall m \in \mathbf{Z}),\ m + 0 = m$

Prove that the additive identity 0 for the set of integers **Z** is unique.

SOLUTION

Proof: Assume on the contrary, there exist two distinct additive identities $0, 0' \in \mathbf{Z}$ which satisfy (1) $m + 0 = m \quad \forall m \in \mathbf{Z}$ and (2) $m + 0' = m \quad \forall m \in \mathbf{Z}$. Since equation (1) is true for every $m \in \mathbf{Z}$, in (1) let $m = 0'$, then (1) becomes (3) $0' + 0 = 0'$. Since equation (2) is true for every $m \in \mathbf{Z}$, in (2) let $m = 0$, then (2) becomes (4) $0 + 0' = 0$. By $Z3$, the commutative theorem of addition for the integers, (5) $0' + 0 = 0 + 0'$. Hence, from (3), (4), and (5)

$$0' = 0' + 0 = 0 + 0' = 0.$$

In the exercises you will be asked to prove the additive inverse of an integer is unique and the multiplicative identity for the set of integers is unique. These proofs should be similar in form to the proof of this example.

Sometimes a uniqueness proof can be presented in such a way that it does not appear to be a proof by contradiction. For example, one proof of the uniqueness theorem "The natural number 2 is the only even prime." is as follows. The natural number 2 is the only even prime, because all other even natural numbers are divisible by 2 and, therefore, are not prime numbers.

Earlier, we proved there are an infinite number of primes. Therefore, for any natural number M, no matter how large, there are an infinite number of primes larger than M. Nonetheless in the next theorem, we show that we can find arbitrarily long sequences of consecutive natural numbers where there are no primes. Two natural numbers are **consecutive**, if their difference is one. Thus, 3 and 4 are consecutive natural numbers. Unknown consecutive natural numbers may be written as x and $x+1$, or $x-1$ and x. Likewise, a sequence of n consecutive natural numbers may be represented by $x, x+1, \ldots, x+(n-1)$. We wish to prove that for every natural number n, there exists a sequence of n consecutive natural numbers such that none of the numbers is a prime. The first several sequences of consecutive composite (non-prime) natural numbers with length two or more are $< 8, 9, 10 >$, $< 14, 15, 16 >$, $< 20, 21, 22 >$, $< 24, 25, 26, 27, 28 >$, $< 32, 33, 34, 35, 36 >$, $< 38, 39, 40 >$ ..... Thus, a few sequences of two consecutive composite natural numbers are $< 8,9 >$, $< 9,10 >$, $< 14,15 >$, $< 15,16 >$, $< 20,21 > \ldots$. And a few sequences of three consecutive composite numbers are $< 8,9,10 >$, $< 14,15,16 >$, $< 20,21,22 >$, $< 24,25,26 >$, $< 25,26,27 >$, $< 26,27,28 >, \ldots$. Notice in the following sequences of two consecutive composite numbers the first number is divisible by 2 and the second number is divisible by 3: $< 8,9 >$, $< 14,15 >$, $< 20,21 > \ldots$. Also notice in the following sequences of three consecutive composite numbers the first number is divisible by 2, the second number is divisible by 3, and the third number is divisible by 4: $< 14,15,16 >$, $< 26,27,28 >$, $< 34,35,36 >$, $< 38,39,40 >, \ldots$. Consequently, for each natural number n, we would like to find, if possible, a number x such that the n consecutive natural numbers $x, x+1, \ldots, x+(n-1)$ are all composite; and, furthermore, x is divisible by 2, $x+1$ is divisible by 3, ..., $x+i$ is divisible by $i+2$, ..., and $x+(n-1)$ is divisible by $n+1$. Since $2, 3, \ldots, (n+1)$ all divide $(n+1)!$, our first guess for x is $\hat{x} = (n+1)!$ Clearly, 2 divides $\hat{x}$. Checking $\hat{x}+1 = (n+1)!+1$, we find it is not divisible by 3. We make our second guess for x by adding 2 to $\hat{x}$, so that $x = (n+1)!+2$ will be divisible by 2. Rewriting $x+1 = [(n+1)!+2]+1$ as $x+1 = (n+1)!+3 = 3\{[2\cdot 4\cdot 5\cdots(n+1)]+1\}$, we see that $x+1$ is divisible by 3. And rewriting $x+2 = [(n+1)!+2]+2$ as $x+2 = (n+1)!+4 = 4\{[2\cdot 3\cdot 5\cdots(n+1)]+1\}$, we see that $x+2$ is divisible by 4. In general, for $0 \le i \le n-1$, we find

$$\begin{aligned} x+i &= [(n+1)!+2]+i = (n+1)!+(2+i) \\ &= (2+i)\{[2\cdots(1+i)(3+i)\cdots(n+1)]+1\} \end{aligned}$$

So, $x+i$ is divisible by $2+i$ for $0 \le i \le n-1$.

The previous paragraph indicates the trial-and-error process and the "scratch work," which one must perform before writing the informal proof which appears in the following example. Notice that very little of the original scratch work appears in the proof itself.

EXAMPLE 7 The Existence of n Consecutive Non-Primes

Prove the following theorem.

Theorem 2.16 For every natural number n, there exists a sequence of n consecutive natural numbers which contain no primes.

SOLUTION

Proof: Let n be an arbitrary natural number and let $x = (n+1)!+2$. Consider the n consecutive natural numbers $x+i$ where $0 \le i \le n-1$. The number $x+i$ is divisible by $2+i$ as the following computation shows. For $0 \le i \le n-1$,

$$\begin{aligned} x+i &= [(n+1)!+2]+i = (2\cdot 3\cdots(n+1))+(2+i) \\ &= (2+i)\{[2\cdots(1+i)(3+i)\cdots(n+1)]+1\} \end{aligned}$$

A **corollary** is a theorem which follows so obviously from the proof of another theorem that no proof, or almost no proof, is necessary. Since for all natural numbers n and for all natural numbers k, the n consecutive natural numbers $x_k + i = k(n+1)! + 2$ where $0 \leq i \leq n-1$ are divisible by $2+i$, we obtain the following corollary to theorem 2.16.

Corollary 2.16.1 For every natural number n, there exist an infinite number of sequences of n consecutive natural numbers which contain no primes.

It follows from this corollary that if one searches sequentially through the natural numbers for prime numbers, one will often encounter arbitrarily long sequences of consecutive natural numbers which contain no prime.

EXERCISES 2.2.2

In exercises 1-5, prove each theorem by cases.

1. If n is a natural number, then $n(n+1)$ is even.

2. If n is a natural number, then $n^2 + 3n + 5$ is odd.

3. For every integer m, the integer $m^3 + m$ is even.

4. For every integer m, the integer $m^2 + 5m + 7$ is odd.

5. Let m be an odd integer, then $m = 4j + 1$ for some integer j or $m = 4k - 1$ for some integer k.

In exercises 6-10, prove the given biconditional statement.

6. The natural number n is an even if and only if the natural number $n+1$ is odd.

7. For every integer m, the integer m^3 is odd if and only if m is an odd integer.

8. For every integer m, the integer $5m^3$ is odd if and only if $3m^2$ is an odd integer.

9. For every integer m, the integer $7m - 4$ is odd if and only if $5m + 3$ is an even integer.

10. Let m and n be integers. The integer $m^2(n+1)$ is even if and only if m is even or n is odd.

In exercises 11-14, prove each theorem by contradiction.

11. The natural number 111 cannot be written as the sum of three even natural numbers.

12. The natural number 110 cannot be written as the sum of two even natural numbers and an odd natural number.

13. There do not exist integers m and n such that $4m + 6n = 9$.

14. There do not exist integers m and n such that $3m + 6n = 4$.

In exercises 15-19, prove each of the given existence theorems.

15. There exists a natural number n such that $n^2 + 2n = 15$.

16. There exist integers m and n such that $3m + 5n = 1$.

17. There exist integers m and n such that $6m + 21n = 3$.

18. There exists an odd integer such that the sum of its digits is even and the product of its digits are even.

19. There exist two distinct integer solutions of the equation $x^3 = x^2$.

In exercises 20-22, prove the given uniqueness theorem.

20. Recall the following existence of additive inverses theorem for the integers which was stated in section 2.2.

    Z11 $\quad (\forall m \in \mathbf{Z})(\exists\,(-m) \in \mathbf{Z}),\ m + (-m) = 0$

    Prove that for every integer m, its additive inverse $-m$ is unique.

21. The existence of a multiplicative identity theorem for the set of integers stated in section 2.2 was

    Z9 $\quad (\exists\,1 \in \mathbf{Z})(\forall m \in \mathbf{Z}),\ 1m = m$

    Prove that the multiplicative identity for the set of integers is unique.

22. There exists a unique real number x such that for every real number y, $xy + x - 4 = 4y$.

    (HINT: Prove the existence of a real number x first, and then prove its uniqueness.)

23. Prove the following existence and uniqueness theorem.

    For every real number x, there exists a unique real number y such that $x^2y = x - y$.

In exercises 24-31, prove the given theorem by any technique.

24. Every even integer is the sum of two odd integers.

25. Every integer can be expressed as the sum of two unequal integers.

26. The sum of four consecutive integers is even.

27. For every two integers m and n, there exists an integer k such that $m + k = n$.

28. For every integer m, the integer $m^2 + m$ is even.

29. No odd integer can be written as the sum of three even integers.

30. For all odd integers m and n, the number 2 divides $m^2 + 3n^2$.

31. For any integer m, when m^2 is divided by 4, the remainder is either 0 or 1.

2.2.3 Conjectures, Proofs, and Disproofs

Recall that a theorem is a statement that has been proved true by a valid argument. While, in mathematics, a **conjecture** is a statement that has been proposed to be true. If, and when, a conjecture is proved true, it becomes a theorem. However, a conjecture may be disproved by producing a single **counterexample**—an example which proves that the conjecture is a false statement.

Since $2^2 - 1 = 4 - 1 = 3$ is a prime, since $2^3 - 1 = 8 - 1 = 7$ is a prime, since $2^5 - 1 = 32 - 1 = 31$ is a prime, and since $2^7 - 1 = 128 - 1 = 127$ is a prime, we might make the conjecture that "For p a prime, $2^p - 1$ is a prime." To disprove this statement, we must present a single counterexample. In this case, we note that $p = 11$ is a prime and compute $2^{11} - 1 = 2048 - 1 = 2047 = (23)(89)$, which shows that $2^{11} - 1$ is a composite. Thus, we have disproved the stated conjecture. However, by disproving the conjecture, we have also proven the theorem: "There is a prime p such that $2^p - 1$ is not a prime."

We now present several well known conjectures. Some have been proved to be true and are now theorems, some have been disproved by counterexamples, and still others remain "open" conjectures—conjectures for which the truth value remains unknown.

Fermat's Last Theorem In the USA, the word "arithmetic" refers to the computational algorithms and procedures associated with natural numbers, integers, and real numbers. Arithmetic is also commonly viewed as including solving problems which involve ratios, proportions, decimal fractions, and percents. To the ancient Greeks *arithmetica* was the study of the mathematical properties of natural numbers. In the USA, the *arithmetica* of the ancient Greeks is known as number theory.

Pythagoras (c. 540 B.C.) was born on the island of Samos, a Greek island in the Aegean Sea. One of his teachers was Thales. In order to acquaint himself with the knowledge of his time, Pythagoras traveled widely in Greece, Egypt, Babylonia, and India early in his life. Later, he founded a secret order at Crotona in southern Italy. Members of the Pythagorean Brotherhood devoted themselves to the study of mathematics and philosophy. Pythagoras associated magic and mysticism with the study of arithmetic and geometry. Pythagoras also stated and gave one or more proofs of the Pythagorean Theorem:

> Let a, b, and c be the lengths of the sides of a triangle with c being the length of the longest side. The triangle is a right triangle if and only if $a^2 + b^2 = c^2$.

Although this theorem bears Pythagoras's name, it was known centuries prior to his time. The Chinese stated the theorem in their text, *Arithmetic in Nine Sections*, which was written before 1000 B.C., and the Egyptians mention the use of the principle by their surveyors in 2000 B.C. A triple of natural numbers (a, b, c) is called a **Pythagorean triple**, if $a^2 + b^2 = c^2$. The Pythagoreans found a procedure for finding all such triples and there are an infinite number of them. A clay tablet dating from between 1900 B.C. and 1600 B.C. indicates that the Babylonians had studied such triples much earlier. The tablet contains fifteen Pythagorean triples. Based on the size of the numbers in the triples, it is reasonable to assume the Babylonians had a procedure for generating the triples. However, the procedure itself does not appear on the tablet. In the *Elements* (c. 350 B.C.), Euclid provides a general procedure for constructing all Pythagorean triples. The technique is as follows. If m and n are natural numbers with $m > n$, and if $a = m^2 - n^2$, $b = 2mn$, and $c = m^2 + n^2$, then $a^2 + b^2 = c^2$. When $m = 2$ and $n = 1$, we obtain the familiar, and smallest Pythagorean triple, $(3, 4, 5)$.

Once a theorem has been proved, mathematicians often try to modify or generalize the theorem in some manner. The "trick" is in knowing how to do so. Fermat's last theorem is one modification of the Pythagorean Theorem. Pierre Fermat (1601-1665) was the son of a wealthy French leather merchant. Fermat received his law degree from the university at Orléans and in 1631, he became a lawyer and government official at Toulouse. He lived in Toulouse for the rest of his life. Because he held public office, he was permitted to change his name to Pierre de Fermat. Fermat's vocation was the law, but his avid avocation was mathematics. Fermat is best known as the father of Number Theory and one of the founders of Analytic Geometry, Calculus, and Probability. However, he did not publish his results. Instead, he corresponded with famous mathematicians throughout Europe including Etienne Pascal and his son Blaise, René Descartes, Marin Mersenne, Gilles Persone de Roberval, and John Wallis. After Fermat's death in 1665, his friends became concerned that his life's work would be lost, since he never published. Therefore, they convinced his son, Samuel, to undertake the task of collecting, organizing, and publishing Pierre's correspondence, mathematical notes, and comments he had written in books. It is in this manner that Fermat's last theorem came to be known and published. In a translation of the Greek mathematician Diophantus's *Arithmetica*, Samuel found a marginal note stating what has come to be known as Fermat's last theorem:

> For n a natural number greater than 2, the equation $a^n + b^n = c^n$ has no natural number solutions a, b, c.

Also in the margin Fermat had written, "I have discovered a truly remarkable proof which this margin is too small to contain." This note was probably written about 1630 when Fermat first studied Diophantus's *Arithmetica*. At some point, Fermat may have discovered his remarkable proof was wrong, since he never sent the problem to any other mathematicians as a challenge problem. Fermat did know how to prove the theorem for the cases $n = 3$ and $n = 4$, and he did send those problems to others as a challenge. A very large number of false proofs of Fermat's last theorem have been published—between 1908 and 1912 alone, over 1000 such proofs appeared in print. In 1955, Yutaka Taniyama (1927-1958) posed some questions regarding elliptic curves. Additional investigations by André Weil (1906-1998) and G. Shimura resulted in the Shimura-Taniyama-Weil Conjecture. In 1986, a connection between this conjecture and Fermat's last theorem was discovered. In June 1993, the British mathematician Andrew Wiles (1953-) "proved" the Shimura-Taniyama-Weil Conjecture for a particular class of examples which included Fermat's last theorem as a corollary. During the reviewing process some difficulties with Wiles' proof emerged; but in October 1994, Wiles with the help of Richard Taylor completed a simpler revised proof. Thus, after more than 300 years Fermat's last theorem, which until 1994 should have been called a conjecture, is indeed a theorem.

Euler's Sum of Powers Conjecture In 1769, the renowned Swiss mathematician Léonard Euler (1707-1783) generalized Fermat's last theorem by conjecturing

> "For every natural number $n > 2$, a sum of at least n, nth powers of natural numbers is necessary for the sum to be the nth power of a natural number."

In other words, Euler conjectured:

> "For $n > 2$ for there to be natural numbers $a_1, a_2, \ldots, a_m$, and b which satisfy the equation $a_1^n + a_2^n + \cdots + a_m^n = b^n$, the natural number m must be greater than or equal to n."

Approximately 200 years later in 1967, L. J. Lander and T. R. Parkin used a systematic computer search to disprove this conjecture by finding the following counterexample for $n = 5$: $27^5 + 84^5 + 110^5 + 133^5 = 144^5$. In 1988,

Noam Elkies constructed the following counterexample for $n = 4$: $2682440^4 + 15365639^4 + 187960^4 = 20615673^4$. To date, no counterexamples to Euler's Sum of Powers Conjecture for $n > 5$ have been discovered.

Goldbach's Conjecture In 1725, the Prussian mathematician and historian Christian Goldbach (1690-1764) became a professor of the newly opened St. Petersburg Academy. In 1728, he went to Moscow to tutor Peter II (1715-1730), Tsar of Russia (1727-1730). In a letter written to Euler on June 7, 1742, Goldbach proposed what is called the "weak" Goldbach conjecture:

> "Every odd natural number greater than five can be written as a sum of three primes."

In his letter of reply, Euler stated the following "stronger" conjecture which is now known as Goldbach's Conjecture:

> "Every even natural number greater than 2 is the sum of two primes."

Euler further stated in his letter: "that every even number is the sum of two primes, I consider an entirely certain theorem in spite of that I am not able to demonstrate it." The strong version of the conjecture implies the weak version, since if n is an odd natural number greater than five, then $n - 3$ is even and if the stronger version is assumed to be true, then $n-3 = p+q$ where p and q are primes. And consequently, the odd natural number $n = p+q+3$ is the sum of three primes. To date, no one has proved or disproved Goldbach's conjecture. However, by October 2003, T. Oliveira e Salva had shown the conjecture to be true for all even natural numbers less than 6×10^{16}.

It is understood that the two primes mentioned in Goldbach's conjecture need not be distinct. Moreover, two primes (p, q) such that $p + q = 2k$ for some natural number k greater than or equal to two are called **Goldbach pairs**. The first few Goldbach pairs are $(2, 2)$ since $2 + 2 = 4$, $(3, 3)$ since $3 + 3 = 6$, $(3, 5)$ since $3 + 5 = 8$, $(3, 7)$ and $(5, 5)$ since $3 + 7 = 5 + 5 = 10$, $(5, 7)$ since $5 + 7 = 12$, and $(3, 11)$ and $(7, 7)$ since $3 + 11 = 7 + 7 = 14$. Most mathematicians believe Goldbach's conjecture is true, since the number of Goldbach pairs increases rapidly as the size of the even natural number, n, increases. For example, for $n = 10$ the number of pairs is 2, for $n = 10^4$ the number of pairs is 127, and for $n = 10^8$ the number of pairs is 291,400.

Twin Prime Conjecture In 1849, the French mathematician Alphonse de Polignac made the general conjecture:

> "For every natural number k, there are infinitely many prime pairs which are a distance $2k$ apart."

In the special case $k = 1$, we obtain the **Twin Prime Conjecture**:

> "There are an infinite number of primes p such that $p + 2$ is a prime also."

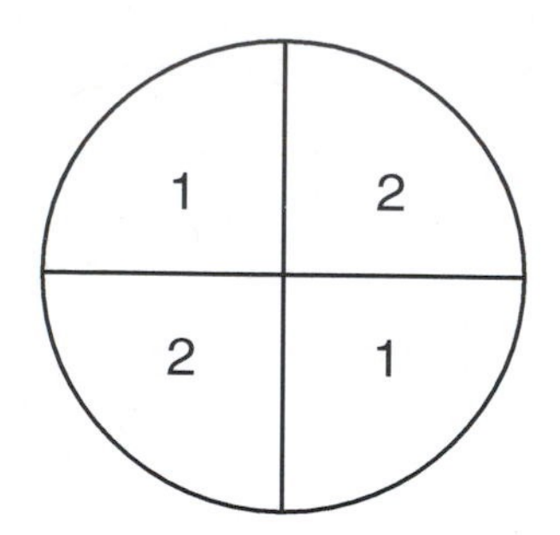

FIGURE 2.1: A Four Region Map which Requires Only Two Colors

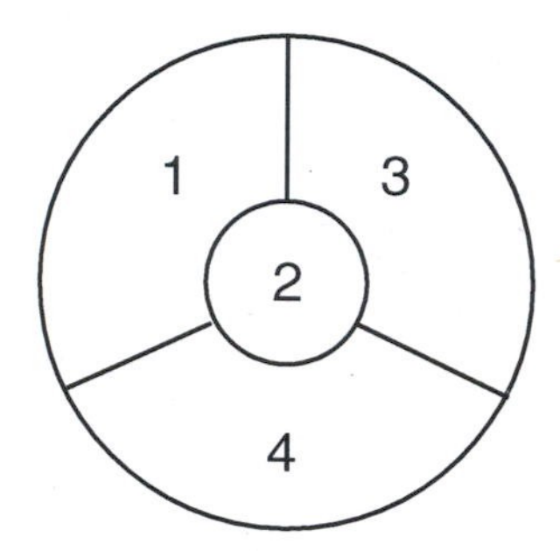

FIGURE 2.2: A Four Region Map which Requires Four Colors

Thus, **twin primes** are pairs of primes of the form $(p,\ p+2)$. The first few twin primes are $(3,5)$, $(5,7)$, $(11,13)$, $(17,19)$, $(29,31)\ldots$. Except for the twin prime pair $(3,5)$, all other twin prime pairs are of the form $(6n-1, 6n+1)$. Although there is "overwhelming evidence" that the twin prime conjecture is true, there is no valid proof, yet.

The Four Color Problem The four color problem asks if every planar map can be colored with four or fewer colors in such a way that no two adjacent regions are the same color. Two regions are adjacent, if they share a common border, not just a point. Figure 2.1 shows a map with four regions which can be colored using just two colors. While figure 2.2 shows a map with four regions which cannot be colored with fewer than four colors.

While coloring a map of the counties of England in 1852, Francis Guthrie observed that four colors appeared to be sufficient to color any map, but he was unable to prove it. Guthrie posed the problem to his former professor August De Morgan. De Morgan posed the problem to other mathematicians frequently, and in 1860 he published anonymously the first known discussion of the problem. In the July 17, 1879, issue of *Nature*, the British lawyer Alfred Kempe announced he had proved the four color problem. His proof was published later in 1879 in the *American Journal of Mathematics*. Eleven years later in 1890, Percy Heawood showed there was an error in Kempe's proof by giving an example of a map for which Kempe's proof technique did not work. However, Heawood was able to use Kempe's proof technique to prove that every planar map can be colored with five or fewer colors. The four color problem was finally proven by Kenneth Appel and Wolfgang Haken in 1977. Their proof was by cases and reduced the totality of all possible maps to 1936 configurations (which was later reduced to 1476 configurations). Since coloring verification could not be performed by hand, each configuration had to be verified by computer. Thus, the four color problem was the first major theorem proved in mathematics for which the use of a computer was absolutely essential.

EXAMPLE Proving and Disproving Statements

Prove or disprove each of the following statements. That is, for a statement which is true, provide a proof; and for a statement which is false, provide a counterexample.

1. For a, b, and c natural numbers, if a divides $b + c$, then a divides b or a divides c.

2. For a, b, and c natural numbers, if a divides b and a divides c, then a divides $b + c$.

SOLUTION

1. Let $P(a, b, c)$ be the statement "a divides $b+c$," let $Q(a, b)$ be the statement "a divides b," and let $R(a, c)$ be the statement "a divides c." Written symbolically, the given statement is $P(a, b, c) \Rightarrow (Q(a, b) \vee R(a, c))$. It appears this statement may be false. To disprove the statement, we must find natural numbers a, b, and c such that $P(a, b, c)$ and $\neg(Q(a, b) \vee R(a, c)) \equiv (\neg Q(a, b) \wedge \neg R(a, c))$ are true. Thus, to disprove $P \Rightarrow (Q \vee R)$, we must find natural numbers a, b, and c such that a divides $b + c$, a does not divide b, and a does not divide c. Since the sum of two odd numbers is even, and since 2 does not divide an odd number, we select $a = 2$, $b = 3$, and $c = 5$. This selection for a, b, and c disproves the given statement, because $a = 2$ divides $8 = 3 + 5 = b + c$, $a = 2$ does not divide $3 = b$, and $a = 2$ does not divide $5 = c$.

2. Let statements P, Q, and R be as stated above. Written symbolically, the given statement is $(Q(a, b) \wedge R(a, c)) \Rightarrow P(a, b, c)$. If $Q(a, b)$ is true, then by definition, there exists a natural number k such that $ak = b$. And, if $R(a, c)$ is true, then there exists a natural number ℓ such that $a\ell = c$. By substitution and the distributive property

$$b + c = ak + a\ell = a(k + \ell)$$

where $k+\ell$ is a natural number. Hence, a divides $b+c$—that is, $P(a, b, c)$ is true.

<< In this instance, we have presented only the "scratch work" from which we can construct a well written informal proof. What appears above is not an informal proof! You should write your own well written proof to complete this exercise. >>

EXERCISES 2.2.3

Either prove or disprove each of the following statements. That is, for a statement which is true, provide a proof; and for a statement which is false, provide a counterexample.

1. For a, b, and c natural numbers, if a is odd and $a+b=c$, then b is even and c is odd.

2. For a, b, and c natural numbers, if a divides b and b divides a, then $a=b$.

3. For a, b, and c integers, if a divides b and b divides a, then $a=b$.

4. For a, b, and c natural numbers, if a divides b and b divides c, then a divides c.

5. For a, b, and c natural numbers, if a divides bc, then a divides b or a divides c.

6. For a, b, and c natural numbers, if a divides b or a divides c, then a divides bc.

7. For a, b, c, and d natural numbers, if a divides $b-c$ and a divides $c-d$, then a divides $b-d$.

8. The sum of three consecutive integers is odd.

9. The sum of three consecutive integers is even.

10. The sum of four consecutive integers is odd.

2.2.4 The System of Rational Numbers and the System of Real Numbers

In 1937, while excavating in central Czechoslovakia, Karl Absolom discovered a prehistoric wolf bone dating from about 30,000 B.C. Fifty-five notches, grouped into sets of five, are cut into the bone. The first twenty-five notches are separated from the second twenty-five by a double length notch. It is reasonable to assume that the notches were made by some prehistoric man who was counting something such as the number of animals he had killed or the number of days since an important event. So at least 30,000 years ago, man had the concept of a one-to-one correspondence between elements in two different sets and the idea of a base for a system of numbers.

Spoken language existed long before written language. And the advent of writing appears to coincide with the transition from hunter-gatherer societies to agrarian societies. Most probably, writing was invented at different times in many different places—Mesopotamia, Egypt, China, and Central America to name a few.

By 3000 B.C. the Egyptians had developed their hieroglyphic writing. Hieroglyphs are pictures of objects which stand for a word, a concept, or a sound. There are over 700 known Egyptian hieroglyphic symbols. As long as hieroglyphics were only being carved into stone, there was no need for symbols which could be written quickly. Later, the Egyptians made a paper like material from the papyrus reed—a plant which grows along the Nile river. By peeling away the outer stem of the papyrus, cutting the soft pith inside into strips, laying the strips side by side, pounding them with a mallet, and drying them, the Egyptians were able to make a renewable, inexpensive material on which to write. The tip of the papyrus reed was used as a pen to write on the papyrus.

Most of our knowledge of Egyptian mathematics comes to us from writings on papyrus. At present, the best source we have for information about Egyptian arithmetic is the Rhind papyrus. A. Henry Rhind was a Scottish Egyptologist who purchased the text in Luxor, Egypt, in 1858 and sold it to the British Museum, where it is on display. The scroll is about 18 feet long and 1 foot wide. This papyrus is also known as the Ahmes papyrus in honor of the scribe who wrote it. The Ahmes papyrus was written about 1650 B.C.; but, according to Ahmes it was copied from an older document written between 2000 and 1800 B.C. The solution of eighty-one of the eighty-seven problems on the Ahmes papyrus require the use of fractions.

The Egyptians took a practical approach to mathematics. Their trade required that they could multiply, divide, and deal with fractions. However, the Egyptian number system was not well suited for performing arithmetic calculations. Therefore, by necessity, they devised methods for multiplication and division which involved addition only.

In our system of fractions, any integer may appear in the numerator. However, in the Egyptian system of fractions, with the exception of the fractions $\frac{2}{3}$ and $\frac{3}{4}$, the only number which may appear in the numerator for computational purposes is 1. A fraction of the form $\frac{1}{n}$ where n is a natural number is called an **Egyptian fraction** or a **unit fraction**. Egyptians were able to solve problems which would require calculations with a fraction such as $\frac{2}{5}$ by expressing the fraction as the sum of two or more Egyptian fractions. Thus, they would decompose $\frac{2}{5}$ into Egyptian fractions by writing $\frac{2}{5} = \frac{1}{3} + \frac{1}{15}$. Strangely enough, they would never repeat the same fraction in a decomposition—that is, they would not write $\frac{2}{5} = \frac{1}{5} + \frac{1}{5}$. Since any fraction can be written as a finite sum of Egyptian fractions, the Egyptians were able to solve many problems which required the use of fractions.

Prior to 2776 B.C. at the latest, the Egyptians had invented a calendar in which a year was 365 days long. They needed a calendar to aid in the prediction of the flooding of the Nile. Since the Nile flooded shortly after the star Sirius, the brightest star in the sky, first appeared in the heavens after the period when it was too close to the sun to be seen, the first day of the new year in the Egyptian calendar coincided with the rising of Sirius out of the sun, which occurs in our month of July. Eventually, the year was divided into 12 months of 30 days each with a 5 day period at the end of the year. The Egyptian calendar is the basis for the Julian and Gregorian calendars.

Mesopotamia is a fertile plain located between the Tigris and Euphrates rivers. Between 4500 and 4000 B.C. it was settled by the Ubaidians. About 3300 B.C. the Sumerians came to Mesopotamia. Initially, they used a stylus to carve symbols on soft clay tablets which they had fashioned from the clay in the region. Later, they developed an abstract form of writing on hand sized clay tablets using cuneiform (wedge-shaped) symbols. The symbols were wedge-shaped due to the instrument they used for writing. The clay tablets were dried in the hot sun and thousands of them still exist to this day. The counting system of the Sumerians was sexagesimal—that is, the base they used for counting was base 60. It is not known why they chose this base; however, 60 is divisible by the ten numbers 2, 3, 4, 5, 6, 10, 12, 15, 20, and 30 while our base 10 system has only two divisors—2 and 5. About 2000 B.C. the Babylonians invaded Mesopotamia and defeated the Sumerians. By 1900 B.C. the Babylonians had established their new capital at Babylon. Babylonian mathematics was somewhat more advanced than Egyptian mathematics. Egyptian mathematics was of a more practical nature, while Babylonian mathematics was, in some aspects, more abstract. Babylonian fractions were more general; however, not all fractions were permitted. The Babylonians had no division algorithm, instead they based their work with fractions on the identity $\frac{a}{b} = a \times (\frac{1}{b})$. That is, they calculated a divided by b by multiplying a by the reciprocal of b, $\frac{1}{b}$. Extensive Babylonian tables of reciprocals of numbers up to several billion are still existent.

The Babylonian year was 360 days long which gave rise to dividing the circle into 360 degrees, which we still do today. Their zodiac had twelve equal sectors of 30 degrees each. The Babylonian day was 12 hours long and each hour was 60 minutes in length. Hence, the Babylonian hour and minute were twice as long as ours. For practical applications, the Babylonians approximated π by 3; but from tablets unearthed in 1936 it is evident they knew $3\frac{1}{8}$ was a better approximation of π.

The System of Rational Numbers Given a mathematical system S which satisfies properties P but which does not satisfy a particular property P^*, mathematicians often try to construct a larger system S^* which contains S, which still satisfies properties P, and which also satisfies property P^*.

Since the set of natural numbers, **N**, is closed under the operations of addition and multiplication for every pair of natural numbers m and n, the equation $x = m + n$ always has a solution x in the set of natural numbers and the equation $y = mn$ always has a solution y in the set of natural numbers. However, the set of natural numbers is not closed under the operation of subtraction—that is, for m and n natural numbers the equation $w = m - n$ does not always have a solution w in the set of natural numbers. Earlier, we enlarged the set of natural numbers to the set of integers, **Z**, by adding to the set of natural numbers the number 0 and for every $m \in \mathbf{N}$ the number $(-m)$. In this manner, we obtained the set of integers which includes the set of natural numbers and which is closed under the operations of addition, multiplication and subtraction.

The set of integers is not closed under the operation of division, since for all integers m and n the equation $mx = n$ does not always have an integer solution x. Clearly, the equation $3x = 4$ does not have an integer solution. We wish to enlarge the set of integers to obtain a new set of numbers in such a way that if m and n are in the new set, then the solution x of the equation $mx = n$ is always in the new set and the new set is closed under addition, multiplication, and subtraction as well.

The operation of division is defined in terms of multiplication. If m and n are integers, then m divides n if there is a number x such that $m \cdot x = n$. Observe that division by the integer 0 is not defined, since for $n \neq 0$ there is no integer x such that $0 \cdot x = n$; and since for $n = 0$ every integer x satisfies $0 \cdot x = 0$. Thus, we define a rational number and the set of rational numbers as follows.

DEFINITIONS Rational Number and the Set of Rational Numbers

If m and n are integers and if $m \neq 0$, then by a **rational number** we mean the number x which satisfies the equation $m \cdot x = n$. We write the rational number x as $\frac{n}{m}$.

The **set of rational numbers** is the set

$$\mathbf{Q} = \{\frac{n}{m} \mid n, m \in \mathbf{Z} \text{ and } m \neq 0\}$$

The rational numbers $\frac{1}{2}, \frac{2}{4}, \frac{3}{6} \ldots$ are all different representations for the same rational number—the rational number x which satisfies the equation $2 \cdot x = 1$. Thus, we encounter for the first time in the system of rational numbers, a system in which equality is not taken for granted. We give the following definition of equality for two rational numbers.

DEFINITION Equality of Rational Numbers

The rational numbers $\frac{n}{m}$ and $\frac{p}{q}$ are **equal**, written $\frac{n}{m} = \frac{p}{q}$, if and only if $n \cdot q = m \cdot p$.

By this definition $\frac{2}{4} = \frac{3}{6}$, since $2 \cdot 6 = 4 \cdot 3 = 12$. Observe that equality for two rational numbers is a consequence of the properties for integers. Let n be an integer. By associating the integer n with the rational number $\frac{n}{1}$, we see that every integer can be represented as a rational number. That is, the set of integers is contained in the set of rational numbers.

Our definition of equality for rational numbers allows us to derive the following **cancellation law for rational numbers**.

Cancellation Law for Rational Numbers

If m, n, and p are integers and $m \neq 0$ and $p \neq 0$, then

$$\frac{n \cdot p}{m \cdot p} = \frac{n}{m}$$

The cancellation law follows easily from the commutative and associative properties of multiplication for integers and from the definition of equality for rational numbers, since

$$(n \cdot p) \cdot m = m \cdot (n \cdot p) = m \cdot (p \cdot n) = (m \cdot p) \cdot n$$

In practice, we write

$$\frac{n \cdot \not{p}}{m \cdot \not{p}} = \frac{n}{m}$$

A rational number $\frac{n}{m}$ is in **reduced form** or **simplest form** if n and m have no factors in common except for the number one.

We define multiplication and addition of rational numbers in terms of multiplication and addition of integers. In this section only, we will denote multiplication of rational numbers by $\odot$ and addition of rational numbers by $\oplus$ to emphasize that these operations are defined in terms of multiplication of integers denoted by $\cdot$ and addition of integers denoted by $+$. Thus, we define multiplication and addition for rational numbers as follows.

DEFINITION Multiplication and Addition for Rational Numbers

If $\frac{m}{n}$ and $\frac{p}{q}$ are rational numbers, then **multiplication**, $\odot$, is defined by the equation

$$\frac{m}{n} \odot \frac{p}{q} = \frac{m \cdot p}{n \cdot q}$$

If $\frac{m}{n}$ and $\frac{p}{q}$ are rational numbers, then **addition**, $\oplus$, is defined by the equation

$$\frac{m}{n} \oplus \frac{p}{q} = \frac{m \cdot q + n \cdot p}{n \cdot q}$$

Using these definitions the following basic theorems for rational numbers can be proven.

BASIC THEOREMS FOR THE RATIONAL NUMBERS

Closure Properties for Addition and Multiplication

Q1 $\forall r, s \in \mathbf{Q},\ r \oplus s \in \mathbf{Q}$
Q2 $\forall r, s \in \mathbf{Q},\ r \odot s \in \mathbf{Q}$

Commutative Properties for Addition and Multiplication

Q3 $\forall r, s \in \mathbf{Q},\ r \oplus s = s \oplus r$
Q4 $\forall r, s \in \mathbf{Q},\ r \odot s = s \odot r$

Associative Properties for Addition and Multiplication

Q5 $\forall r, s, t \in \mathbf{Q},\ r \oplus (s \oplus t) = (r \oplus s) \oplus t$
Q6 $\forall r, s, t \in \mathbf{Q},\ r \odot (s \odot t) = (r \odot s) \odot t$

Distributive Properties of Multiplication over Addition

Q7 $\forall r, s, t \in \mathbf{Q},\ r \odot (s \oplus t) = (r \odot s) \oplus (r \odot t)$
Q8 $\forall r, s, t \in \mathbf{Q},\ (r \oplus s) \odot t = (r \odot t) \oplus (s \odot t)$

Existence of a Multiplicative Identity Property

Q9 $(\exists \frac{1}{1} \in \mathbf{Q})(\forall r \in \mathbf{Q}),\ \frac{1}{1} \odot r = r$

Existence of an Additive Identity Property

Q10 $(\exists \frac{0}{1} \in \mathbf{Q})\ (\forall r \in \mathbf{Q}),\ r \oplus \frac{0}{1} = r$

Existence of Additive Inverses Property

Q11 $\quad (\forall\, r \in \mathbf{Q})\ (\exists\ (-r) \in \mathbf{Q}),\ r \oplus (-r) = 0$

Existence of Multiplicative Inverses Property

Q12 $\quad (\forall\, r \in \mathbf{Q})\ [(r \neq 0) \Rightarrow (\exists\ r^{-1} \in \mathbf{Q}),\ r \odot r^{-1} = \frac{1}{1}]$

Any set $\mathbf{Q}$ with operations $\oplus$ and $\odot$ defined on it which satisfies properties $Q1$ through $Q12$ is called a **field**. Thus, the set of rational numbers is our first example of a field.

The closure property for addition, $Q1$, holds in the rational numbers, because of the definition of addition $\oplus$ for the rational numbers; because the closure property for addition in the integers, $Z1$, holds; and because the closure property for multiplication in the integers, $Z2$, holds. In general, for $I = 1, 2, \ldots 11$, property QI holds for the rational numbers, because the corresponding property ZI holds for the integers and because of the definitions of addition and multiplication for the rational numbers. Hence, the set of rational numbers with addition and multiplication defined as above satisfy all the properties which the integers satisfy—properties $Z1$ through $Z11$. The rational numbers also satisfy the new property $Q12$. That is, every nonzero rational number has a multiplicative inverse. Hence, the rational numbers are closed under the operation of division with the exception of division by 0.

Let $\frac{m}{n}$ be any rational number. By the definition of multiplication and the cancellation law, for any nonzero integer p,

$$\frac{p}{p} \odot \frac{m}{n} = \frac{p \cdot m}{p \cdot n} = \frac{m}{n}.$$

Thus, $\frac{p}{p} = \frac{1}{1}$ in reduced form is the multiplicative identity for the rational numbers. Of course, $\frac{1}{1}$ is the rational number representation of the multiplicative identity 1 of the integers. Also, by the definition of addition; since $n \cdot 0 = 0$ for any integer n; by the existence of an additive identity property for the integers, Z10; and by the cancellation law for rational numbers, we have for any nonzero integer p,

$$\frac{m}{n} \oplus \frac{0}{p} = \frac{m \cdot p + n \cdot 0}{n \cdot p} = \frac{m \cdot p + 0}{n \cdot p} = \frac{m \cdot p}{n \cdot p} = \frac{m}{n}$$

Hence, $\frac{0}{p} = \frac{0}{1}$ in reduced form is the additive identity for the rational numbers. Clearly, $\frac{0}{1}$ is the rational number representation of the additive identity 0 of the integers. The rational number $\frac{(-m)}{n}$ is the additive inverse of the rational number $\frac{m}{n}$, since

$$\frac{m}{n} \oplus \frac{(-m)}{n} = \frac{m \cdot n + n \cdot (-m)}{n \cdot n} = \frac{n \cdot m + n \cdot (-m)}{n \cdot n}$$

$$= \frac{n \cdot (m + (-m))}{n \cdot n} = \frac{m + (-m)}{n} = \frac{0}{n} = \frac{0}{1}.$$

The commutative property of addition, Q3, the commutative property of multiplication, Q4, and the associative property of multiplication, Q6, follow easily from the definitions of $\oplus$ and $\odot$, the commutative properties of addition and multiplication for the integers, and the associative property of multiplication for the integers. In the following two examples, we prove the associative property of addition for the rational numbers, Q5, and the existence of a multiplicative inverse property for the rational numbers, Q12.

EXAMPLE 1 Proof of the Associative Property of Addition for the Rational Numbers

Prove for every $r, s, t \in \mathbf{Q}$, $r \oplus (s \oplus t) = (r \oplus s) \oplus t$.

SOLUTION

Let $r = \frac{m}{n}$, $s = \frac{p}{q}$, and $t = \frac{u}{v}$ where m, n, p, q, u, and v are integers and $n \neq 0$, $p \neq 0$, and $v \neq 0$. By the definition of addition for the rational numbers and the distributive properties for integers,

$$(1) \qquad r \oplus (s \oplus t) = \frac{m}{n} \oplus (\frac{p}{q} \oplus \frac{u}{v}) = \frac{m}{n} \oplus (\frac{p \cdot v + q \cdot u}{q \cdot v})$$

$$= \frac{m \cdot (q \cdot v) + n \cdot (p \cdot v + q \cdot u)}{n \cdot (q \cdot v)}$$

$$= \frac{m \cdot q \cdot v + n \cdot p \cdot v + n \cdot q \cdot u}{n \cdot q \cdot v}$$

and

$$(2) \qquad (r \oplus s) \oplus t = (\frac{m}{n} \oplus \frac{p}{q}) \oplus \frac{u}{v} = \frac{m \cdot q + n \cdot p}{n \cdot q} \oplus \frac{u}{v}$$

$$= \frac{(m \cdot q + n \cdot p) \cdot v + (n \cdot q) \cdot u}{(n \cdot q) \cdot v}$$

$$= \frac{m \cdot q \cdot v + n \cdot p \cdot v + n \cdot q \cdot u}{n \cdot q \cdot v}$$

Because the right hand sides of equations (1) and (2) are identical, for all $r, s, t \in \mathbf{Q}$, $r \oplus (s \oplus t) = (r \oplus s) \oplus t$.

EXAMPLE 2 Proof of the Existence of a Multiplicative Inverse Property for the Set of Rational Numbers

Prove for every nonzero rational number r there exists a rational number r^{-1} such that $r \odot r^{-1} = \frac{1}{1}$.

SOLUTION

Let $r = \frac{m}{n}$ be any nonzero rational number. Since r is a rational number, $n \neq 0$, and since r is nonzero, $m \neq 0$. Consider the product

$$\frac{m}{n} \odot \frac{n}{m} = \frac{m \cdot n}{n \cdot m} = \frac{n \cdot m}{n \cdot m} = \frac{1}{1}$$

Since $r \odot \frac{n}{m} = \frac{1}{1}$, $r^{-1} = \frac{n}{m}$ is a multiplicative inverse of r.

<< Just as each rational number does not have a unique representation, the multiplicative inverse of a nonzero rational number does not have a unique representation, because for any nonzero integer p,

$$r^{-1} = \frac{n}{m} = \frac{n \cdot p}{m \cdot p}.$$

That is, for any nonzero integer p, $\frac{n \cdot p}{m \cdot p}$ is a multiplicative inverse of $r = \frac{m}{n}$. If we require a multiplicative inverse of a nonzero rational number to be represented in reduced form, then the multiplicative inverse has a unique representation. >>

From their study of geometry, the ancient Greeks realized there were numbers which were not rational numbers. Virtually nothing is known about the date of the discovery of the existence of **irrational numbers** (numbers which cannot be written as the ratio of two integers) or the identity of the individual who made the discovery. Nonetheless, sometime between the time of Pythagoras (about 550 B.C.) and the time of Theodorus (about 400 B.C.) it was determined that non-rational numbers existed. The proof of the following theorem dates from between 550 B.C. and 400 B.C.

Theorem 2.17 There exist numbers which are not rational numbers.

Proof: Let c be the hypotenuse of an isosceles right triangle with sides $a = b = 1$. Since the triangle is a right triangle, by the Pythagorean theorem

$$(1) \qquad c^2 = a^2 + b^2 = 1^2 + 1^2 = 2$$

Suppose c is a rational number which is written in reduced form. Specifically, assume $c = p/q$ where p and $q \neq 0$ are integers and p and q have no common factors. Squaring c and substituting into (1), we see p and q must satisfy the equation

$$(2) \qquad c^2 = \left(\frac{p}{q}\right)^2 = \frac{p^2}{q^2} = 2$$

Multiplying the last equation by $q^2 \neq 0$, yields (3) $p^2 = 2q^2$. Hence, p^2 is an even integer, and it follows from theorem 2.9 that p is an even integer. Thus, $p = 2r$ for some integer r. Substituting $p = 2r$ into equation (3) and squaring, we obtain

$$p^2 = (2r)^2 = 4r^2 = 2q^2.$$

Dividing the last equation by 2, we find $2r^2 = q^2$. Therefore, q^2 is an even integer which implies by theorem 2.9 that q is an even integer. Thus, we have shown p is an even integer and q is an even integer. Hence, 2 is a common factor of p and q. Consequently, by contradiction, the number c which satisfies the equation (1) $c^2 = 2$ is not a rational number.

The irrational number which satisfies the equation $c^2 = 2$ is denoted by $\sqrt{2}$ or $2^{\frac{1}{2}}$. About 400 B.C., Theodorus proved that the numbers $\sqrt{3}$, $\sqrt{5}$, $\sqrt{6}$, $\sqrt{7}$, $\sqrt{8}$, $\sqrt{10}$, $\sqrt{11}$, $\sqrt{12}$, $\sqrt{13}$, $\sqrt{14}$, $\sqrt{15}$, and $\sqrt{17}$ are irrational numbers also.

An axiomatic development of the set of real numbers from the set of rational numbers did not occur until the latter part of the nineteenth century. In 1869, the French mathematician Hugues Charles Robert Méray (1835-1911) published the first rigorous development of the set of real numbers. He considered sequences of rational numbers and defined convergence for such sequences. Some convergent sequences of rational numbers converged to rational numbers, while others converged to "fictitious" numbers (irrational numbers).

Meŕay published his results in an obscure journal; consequently, his work went unnoticed for several years. Two years after Méray's publication, two similar developments of the real numbers from sequences of rational numbers were published in two different journals by two colleagues at the University of Halle in Germany. In October 1871, Eduard Heinrich Heine (1821-1881) published his article "The Elements of Function Theory" in the prestigious *Journal of Pure and Applied Mathematics.* One month later, Georg Ferdinand Ludwig Philipp Cantor (1845-1918) published his article "Extensions of Theorems Regarding Trigonometric Series" which included his development of the real numbers in the well-respected journal *Annals of Mathematics.*

Julius Wilhelm Richard Dedekind (1831-1916) was born in Braunschweig, Germany, and lived there most of his life. He was the last doctoral student of Gauss. To develop the set of real numbers, Dedekind began with the set of rational numbers and their properties. Dedekind observed that each rational number r divided the set of rational numbers into two nonempty sets—A_1, the set of all rational numbers less than r and A_2, the set of all rational numbers greater than r. The rational number r could be in either A_1 or A_2 but not both. To indicate the relationship between the rational number r and the division of the set of rational numbers into the sets A_1 and A_2, Dedekind wrote $r = \text{cut}(A_1, A_2)$. On October 24, 1858, Dedekind noted that every point on a line divides the line into two sets and that every such division into two sets is produced by a unique point. To extend the set of rational numbers to the set of real numbers, and to extend the properties of the rational numbers to the set of real numbers, Dedekind divided the set of rational numbers into two sets such that every element of A_1 is less than every element of A_2 and he wrote $\text{cut}(A_1, A_2)$ to represent this division. To honor Dedekind, such cuts are now called **Dedekind cuts**. Some Dedekind cuts are produced by rational numbers, while other cuts are produced by numbers which are not rational. The latter cuts produced new numbers called **irrational numbers**. For example, the cut in which the set A_1 consists of all rational numbers x less than a where $a > 0$ and $a^2 = 2$ and in which the set A_2 consists of all rational numbers x greater that a where $a > 0$ and $a^2 = 2$ defines the unique irrational number, $\sqrt{2}$. Later in 1872, after Heine and Cantor had published their developments of the real numbers from the rational numbers, Dedekind published his results in an article titled, "Continuity and the Irrational Numbers."

The numbers π and e are two other irrational numbers with which you may be familiar already. The Welsh mathematician William Jones was the first to use the symbol π to represent the ratio of the circumference of a circle to its diameter in 1706. In 1737, Euler began using the symbol π for this purpose and it soon became the standard notation. John Heinrich Lambert (1728-1777) proved in 1761 that π is irrational. The symbol e represents the value of the base of "natural logarithms." Euler first used this symbolism in a letter he wrote to Goldbach in 1731. Also, credit is given to Euler for showing that e is irrational.

The set of real numbers consists of the set of rational numbers together with the set of irrational numbers. The real numbers satisfy properties Q1 through Q12. So the real numbers is our second example of a field.

Earlier, we proved that $\sqrt{2}$ is irrational. Since $\sqrt{2} = 2^{\frac{1}{2}}$, there are rational numbers x and y such that x^y is an irrational number—namely, the numbers $x = 2$ and $y = \frac{1}{2}$. Out of curiosity, we might want to know if there are irrational numbers x and y such that x^y is an rational number. The following example answers this question to the affirmative.

EXAMPLE 3 Existence of Irrational Numbers x And y Such That x^y Is Rational

Prove there exist irrational numbers x and y such that x^y is rational.

SOLUTION

Consider $\sqrt{2}^{\sqrt{2}}$. The real number $\sqrt{2}^{\sqrt{2}}$ is either rational or irrational. If $\sqrt{2}^{\sqrt{2}}$ is rational, then we obtain the desired result by taking $x = y = \sqrt{2}$. If, on the other hand, $\sqrt{2}^{\sqrt{2}}$ is irrational, then we consider the number

$$\left(\sqrt{2}^{\sqrt{2}}\right)^{\sqrt{2}} = \sqrt{2}^{2} = 2$$

which is rational. Hence, if $\sqrt{2}^{\sqrt{2}}$ is irrational, we take $x = \sqrt{2}^{\sqrt{2}}$ and $y = \sqrt{2}$ and obtain the desired result—namely, x^y is rational.

In example 3, we showed that either $\sqrt{2}^{\sqrt{2}}$ is rational or $\left(\sqrt{2}^{\sqrt{2}}\right)^{\sqrt{2}}$ is rational. But we do not know from the example itself whether only one of these numbers is rational or if both of the numbers are rational. It has been proved that $\sqrt{2}^{\sqrt{2}}$ is irrational, so it is the number $\left(\sqrt{2}^{\sqrt{2}}\right)^{\sqrt{2}}$ which is rational.

EXERCISES 2.2.4

1. Use the definition of equality to show the following pairs of rational numbers are equal.

   a. $\frac{14}{8}, \frac{21}{12}$ b. $\frac{6}{14}, \frac{15}{35}$ c. $\frac{6}{15}, \frac{14}{35}$

2. Perform the indicated operations and express the results in reduced form.

   a. $\frac{7}{12} + \frac{4}{15}$ b. $\frac{9}{4} - \frac{11}{3}$ c. $\left(\frac{15}{4}\right)\left(\frac{7}{8}\right)$ d. $\left(\frac{15}{4}\right) \div \left(\frac{7}{8}\right)$

3. Prove the commutative property of addition for the set of rational numbers. That is, prove Q3.

4. Prove the associative property of multiplication for the set of rational numbers. That is, prove Q6.

5. Prove the left distributive property of multiplication over addition for the set of rational numbers. That is, prove Q7.

6. Which of the following real numbers are rational? irrational?

   a. $\frac{\sqrt{3}}{2}$ b. $\frac{\sqrt{4}}{2}$ c. $\frac{3\pi}{4\pi}$

   d. $\frac{-5\sqrt{7}}{6\sqrt{7}}$ e. $\sqrt{12}$ f. $\sqrt{24}\sqrt{6}$

   g. $(2 - \sqrt{3}) - (4 - \sqrt{3})$ h. $\frac{e}{3}$ i. $\sqrt{\frac{4}{25}}$

   j. $(-4 + 3\sqrt{3})(-4 - 3\sqrt{3})$ k. $\frac{1 + \sqrt{2}}{\sqrt{2}}$ l. $\frac{2}{\sqrt{3}} \div \frac{\sqrt{3}}{4}$

7. Prove that there is no rational number p such that $p^2 = 3$. That is, prove $\sqrt{3}$ is irrational.

8. Prove that $\sqrt{6}$ is irrational.

9. Prove that $\sqrt{3} + \sqrt{5}$ is irrational.

10. Use set-builder notation to specify the sets A_1 and A_2 which define the Dedekind cuts for the real numbers a. $\frac{3}{2}$ b. $\sqrt{3}$

11. Let x and y be real numbers. Prove if $xy = 0$, then $x = 0$ or $y = 0$. (**HINT:** Consider proving the contrapositive by contradiction.)

12. Prove the **Cancellation Property of Multiplication for Real Numbers:** If x, y, and z are real numbers, if $x \neq 0$, and if $xy = xz$, then $y = z$.

In exercises 13-20 prove or disprove the given statement.

13. If x is irrational and y is rational, then $x + y$ is irrational.

14. If $x + y$ is irrational, then both x and y are irrational.

15. If $x + y$ is irrational, then either x or y is irrational.

16. If x is irrational and y is rational, then xy is irrational.

17. If x is irrational and $y \neq 0$ is rational, then xy is irrational.

18. If x and y are irrational, then $x + y$ is irrational.

19. For every rational number x there exist irrational numbers y and z such that $x = y + z$.

20. If x is a positive irrational number, then $\sqrt{x}$ is an irrational number.

2.3 Chapter Review

Definitions

A **definition** is an agreement to use a symbol, a word, or a short phrase to substitute for something else, usually for some expression that is too long to write easily or conveniently.

An **axiom** or **postulate** is a statement that is assumed to be true. The axioms of a deductive mathematical system are the statements from which all other statements of the system can be derived.

A **theorem** is a true statement that has been proven by a valid argument.

A **proof** is a logically valid deduction of a theorem from the premises of the theorem, the axioms, or previously proven theorems.

A **formal proof** of a theorem is a finite sequence of statements $S_1, S_2, \ldots, S_k$ such that each statement S is a premise of the theorem to be established, an axiom of the system, follows from one or more of the preceeding S's by a logical rule of inference, or follows from one or more of the preceeding S's by a previously proven theorem.

A theorem which states that there is one and only one element in a specific set which satisfies a certain statement is called a **uniqueness theorem**.

A **corollary** is a theorem which follows so obviously from the proof of another theorem that no proof, or almost no proof, is necessary.

A **conjecture** is a statement that has been proposed to be a true. If, and when, a conjecture is proved true, it becomes a theorem. A conjecture may be disproved by producing a single **counterexample**—an example which proves that the conjecture is a false statement.

A **deductive mathematical system** consists of the following elements:

1. An underlying language.
2. A deductive logic system.
3. A list of undefined terms.
4. A list of formally defined technical terms, called definitions.
5. A list of statements which are assumed to be true, called postulates or axioms.
6. A list of deduced statements, called theorems.

A mathematical system is **consistent** if and only if contradictory statements are not implied by the axioms of the system.

A set of axioms is **inconsistent** if it is possible to deduce from the axioms that some statement is both true and false.

Subtraction Let S be a set of numbers and let m and n be elements of S. The number $x = m - n$, read "m minus n" or "n subtracted from m," is the number x which makes the statement $m = n + x$ true, if there be such a number x.

An integer n is **even**, if there exists an integer k such that $n = 2k$.

An integer n is **odd**, if there exists an integer k such that $n = 2k + 1$.

Let a and b be natural numbers. The number a **divides** b if and only if there exists a natural number c such that $ac = b$. If a divides b, then we also say a **is a factor of** b and b **is divisible by** a.

A **prime number** (or, simply, a **prime**) is a natural number greater than one which is divisible only by itself and one.

A **composite number** (or, simply, a **composite**) is a natural number greater than one which is not a prime number.

A triple of natural numbers (a, b, c) is called a **Pythagorean triple**, if $a^2 + b^2 = c^2$.

Two primes (p, q) such that $p + q = 2k$ for some natural number k greater than or equal to two are called **Goldbach pairs**.

Twin primes are pairs of primes of the form $(p, p+2)$.

A fraction of the form $\frac{1}{n}$ where n is a natural number is called an **Egyptian fraction** or a **unit fraction**.

The **set of rational numbers** is the set

$$\mathbf{Q} = \{\frac{n}{m} \mid n, m \in \mathbf{Z} \text{ and } m \neq 0\}$$

The rational numbers $\frac{n}{m}$ and $\frac{p}{q}$ are **equal**, written $\frac{n}{m} = \frac{p}{q}$, if and only if $n \cdot q = m \cdot p$.

If $\frac{m}{n}$ and $\frac{p}{q}$ are rational numbers, then **multiplication**, $\odot$, is defined by the equation

$$\frac{m}{n} \odot \frac{p}{q} = \frac{m \cdot p}{n \cdot q}$$

If $\frac{m}{n}$ and $\frac{p}{q}$ are rational numbers, then **addition**, $\oplus$, is defined by the equation

$$\frac{m}{n} \oplus \frac{p}{q} = \frac{m \cdot q + n \cdot p}{n \cdot q}$$

Forms of Proofs

Direct Proof of P⇒Q

Proof: Assume P. ... Therefore, Q. Hence, $P \Rightarrow Q$.

Contrapositive Proof of P⇒Q

Proof: Assume $\neg Q$. ... Therefore, $\neg P$. Hence, $\neg Q \Rightarrow \neg P$. And consequently, $P \Rightarrow Q$.

Proof of P⇒Q by Contradiction

Proof: Assume $P \wedge (\neg Q)$. ... Therefore, C, where C is a contradiction. Hence, $P \Rightarrow Q$.

A Proof of $(P \vee Q) \Rightarrow R$ by Cases or Exhaustion

Proof: Prove $P \Rightarrow R$ by any technique. Prove $Q \Rightarrow R$ by any technique. Therefore, $(P \vee Q) \Rightarrow R$, since $((P \vee Q) \Rightarrow R) \equiv (P \Rightarrow R) \wedge (Q \Rightarrow R)$.

A Proof of the Biconditional Statement $P \Leftrightarrow Q$

Proof: Prove $P \Rightarrow Q$ by any technique. Prove $Q \Rightarrow P$ by any technique. Therefore, $P \Leftrightarrow Q$, since $P \Leftrightarrow Q \equiv (P \Rightarrow Q) \wedge (Q \Rightarrow P)$.

A Proof of the Statement P by Contradiction

Proof: Assume $\neg P$. ... Therefore, C where C is a contradiction. Hence, P.

A Proof of the Universal Statement $(\forall x \in U)(P(x))$

Proof: Let $x \in U$. ... $P(x)$ is true. Hence, $(\forall x \in U)(P(x))$ is true.

A Proof of the Existence Statement $(\exists x \in U)(P(x))$

There are two ways we can prove the statement $(\exists x \in U)(P(x))$.

We can present a **constructive proof** by exhibiting or explaining how to construct an element x in U having the property $P(x)$.

Or, we can present a **nonconstructive proof** by making a valid argument that some element x in U makes the statement $P(x)$ true.

Named Theorems and Conjectures Mentioned in This Chapter

Cancellation Property of Addition for Integers If k, m and n are integers and $k + m = k + n$, then $m = n$.

Cancellation Law for Rational Numbers If m, n, and p are integers and $m \neq 0$ and $p \neq 0$, then

$$\frac{n \cdot p}{m \cdot p} = \frac{n}{m}$$

The Fundamental Theorem of Arithmetic Every natural number greater than one is a prime or can be written uniquely as a product of primes except for the order in which the prime factors are written.

The Pythagorean Theorem Let a, b, and c be the lengths of the sides of a triangle with c being the length of the longest side. The triangle is a right triangle if and only if $a^2 + b^2 = c^2$.

Fermat's last theorem For n a natural number greater than 2, the equation $a^n + b^n = c^n$ has no natural number solutions a, b, c.

Euler's Sum of Powers Conjecture For every natural number $n > 2$, a sum of at least n, nth powers of natural numbers is necessary for the sum to be the nth power of a natural number.

Goldbach's Conjecture Every even natural number greater than 2 is the sum of two primes.

Twin Prime Conjecture There are an infinite number of primes p such that $p + 2$ is a prime also.

The Four Color Problem The four color problem asks if every planar map can be colored with four or fewer colors in such a way that no two adjacent regions are the same color. Two regions are adjacent, if they share a common border, not just a point.

Review Exercises

1. Provide a justification for each step in the formal proof of the following statement.

   For integers k, m, and n, $k(m - n) = km - (kn)$.

   Proof: For all $k, m, n \in \mathbf{Z}$

1. $k(m - n) = k(m - n)$	1. $=$ is reflexive
2. $m - n = m + (-n)$	2.
3. $k(m - n) = k(m + (-n))$	3.
4. $k(m + (-n)) = km + k(-n)$	4.
5. $k(m - n) = km + k(-n)$	5.
6. $k(-n) = -(kn)$	6.
7. $k(m - n) = km + (-(kn))$	7.
8. $km + (-(kn)) = km - (kn)$	8.
9. $k(m - n) = km - (kn)$	9.

2. Give a direct proof of the following theorems.

   a. If m is an odd integer, then $7m + 5$ is an even integer.

   b. If m is an integer and $5m - 4$ is an even integer, then m is even.

3. Prove each theorem by a proof by contraposition.

   a. If m is an integer and m^2 is odd, then m is odd.

   b. Let m and n be integers. If $m - n$ is odd, then either m is even and n is odd or m is odd and n is even.

4. Prove the following theorems by a proof by contradiction.

   a. If m is an odd integer, then $m + 3$ is an even integer.

   b. Let m and n be integers. If mn is even, then m or n is even.

5. Prove each theorem by cases.

   a. If m is an integer, then $m(m + 3)$ is an even integer.

   b. For every integer m, the integer $m^2 + m - 1$ is odd.

6. Prove the following biconditional statements.

   a. Prove that 3 divides $2n^2 + 1$ if and only if 3 does not divide n.

   b. Let p be a prime and let m and n be natural numbers. Prove that the prime p divides mn if and only if p divides m or p divides n.

7. Prove the given existence theorems.

   a. There exists a natural number n such that $2n > n^2$.

   b. There exist natural numbers m and n such that $4m - 1 = 7n$.

   c. There exist three distinct integers k, m, and n such that $k^m = m^n$.

   d. If $x > 2$ is a real number, then there exists a unique real number $y < 0$ such that $x + xy - 2y = 0$.

8. Prove or disprove the following statements.

   a. Let k, m, and n be natural numbers. If k divides n and if m divides n, then km divides n

   b. The sum of five consecutative integers is divisible by five.

   c. The difference of two irrational numbers is irrational.

   d. The quotient of two irrational numbers is irrational.

9. Which of the following pairs of rational numbers are equal?

   a. $\dfrac{85}{102}, \quad \dfrac{75}{90}$ b. $\dfrac{-273}{299}, \quad \dfrac{399}{-436}$ c. $\dfrac{4773}{9878}, \quad \dfrac{7353}{15219}$

Chapter 3

Set Theory

In mathematics, it is often the case that there is a long period in which ideas are conceived and developed by several different individuals and then at a later time a significant breakthrough is made by more than one individual. Such was the case with the development of calculus and non-Euclidean geometry. However, on the contrary, set theory is primarily the creation of one individual, Georg Ferdinand Ludwig Philipp Cantor (1845-1918). Cantor was born in St. Petersburg, Russia. His father, Georg Waldemar Cantor, was born in Copenhagen, but moved to St. Petersburg as a young man. Cantor's mother, Maria Anna Böhm, was Russian. In 1856, when young Georg was eleven years old, the family consisting of Georg, a brother, a sister, and his parents moved to Wiesbaden, Germany, and later to Frankfurt due to the poor health of his father. Georg's father wanted him to become an engineer, so he could make a good living. After receiving the proper technical training in high school at Darmstadt from 1860 to 1862, Georg entered the Polytechnic of Zurich in the fall of 1862. Later in 1862, Georg requested his father's permission to study mathematics instead of engineering and his father consented. After his father's death in June 1863, Georg transfered to the University of Berlin where he completed his doctorate in December, 1867. In the spring of 1869, Cantor joined the faculty at the University of Halle as a lecturer. In 1872, he was promoted to assistant professor and in 1879 to professor. Cantor spent the remainder of his life in Halle. Cantor's early publications were in the area of number theory. However, Heinrich Eduard Heine, a senior colleague of Cantor at Halle, challenged Cantor to prove that a function can be represented uniquely as a trigonometric series. Cantor succeeded in doing so in 1870. In 1872, Cantor published a paper on trigonometric series in which he defined the irrational numbers in terms of convergent sequences of rational numbers. Then Cantor began his lifelong work on set theory and the concept of transfinite numbers. In 1873, he proved the rational numbers are countable—that is, that there is a one-to-one correspondence between the rational numbers and the natural numbers. In 1874, Cantor showed that the real numbers are not countable. After Cantor initiated research in the area of set theory, others made significant contributions. Conjectures which Cantor made opened fertile areas of research for others and the paradoxes which arose later because of his work resulted in important study with respect to the foundations of mathematics.

3.1 Sets and Subsets

The concept of a set is fundamental to the study of mathematics. Using set notation in mathematics promotes precision and clarity in communicating mathematical ideas. The theory of sets also provides a means of simplifying and unifying a large number of subdisciplines of mathematics.

In the late nineteenth century, Cantor defined the term "set." This definition led to a paradox known as Russell's paradox. Giving a formal definition of a set ultimately leads to circularity of definition, since a set probably would be defined as a "collection," a collection would be defined as an "aggregate," and so forth until we eventually circle back to the word "collection" or "aggregate." The modern method of developing mathematical theories such as the theory of sets is the axiomatic approach. In this approach to set theory, the terms "set" and "is an element of" are undefined terms—just as "lines," "points," and "intersects" are undefined terms in geometry. Intuitively, a **set** consists of objects called **elements** or **members**. Furthermore, a set is well-defined—that is, given a specific object, it is possible to determine if the object belongs to a given set or not. We will use the notation "$x \in A$" to denote that "x is an element of the set A," "x is a member of A," or, simply, "x is in A." To indicate "x is not an element of the set A," we write symbolically "$x \notin A$." Usually sets are described in **roster notation**, in which the elements of the set are enclosed in curly braces, $\{\ \}$, and separated by commas, or in **set-builder notation**, in which the set is specified in the form $\{x \mid P(x)\}$ which is read "the set of all x such that $P(x)$ is true". For example, "the set of all natural numbers less that 6" is specified in roster notation as $\{1, 2, 3, 4, 5\}$ and in set-builder notation as $\{x \mid x \in \mathbf{N} \text{ and } x < 6\}$.

There is one and only one set which has no elements. It is called the **empty set** or **null set**. The empty set is denoted by the symbol $\emptyset$. The following is a formal definition of the empty set.

DEFINITION Empty Set
The **empty set** is the set $\emptyset = \{x \mid x \neq x\}$.

Let $P(x)$ denote the statement "$x \neq x$." Since for all possible objects x, the statement $P(x)$ is false, the set denoted by the symbol $\emptyset$ contains no elements.

Consider the sets $X = \{a, 1, \$\}$ and $Y = \{1, 2, a, b, \$\}$. Notice that all of the elements of X—namely, a, 1, and \$—are all elements of Y. To indicate that all elements of X are elements of Y, we say "X is a subset of Y" and symbolically we write $X \subseteq Y$. We could also write $Y \supseteq X$ which is read "Y is a superset of X." The symbolism $Y \supseteq X$ has the same meaning as the symbolism $X \subseteq Y$. The formal definition of a subset follows.

DEFINITION Subset

Let A and B be sets. A is a **subset** of B, written $A \subseteq B$, if and only if every element of A is an element of B. Symbolically,

$$A \subseteq B \Leftrightarrow (\forall x)[(x \in A) \Rightarrow (x \in B)].$$

To indicate "A is not a subset of B" we write $A \nsubseteq B$ or $B \nsupseteq A$. We now prove every set is a subset of itself.

Theorem 3.1 For any set A, $A \subseteq A$.

Proof: Let $x \in A$. Since $(x \in A) \Rightarrow (x \in A)$ is a true statement, we have $(\forall x)[(x \in A) \Rightarrow (x \in A)]$ is true, and therefore, by definition of subset $A \subseteq A$.

Also, we can prove that the empty set is a subset of all sets.

Theorem 3.2 For any set A, $\emptyset \subseteq A$.

Proof: Let A be any set and let x be any object. We consider the following conditional statement $(\forall x)[(x \in \emptyset) \Rightarrow (x \in A)]$. Since for all objects x the hypothesis of the conditional statement, $x \in \emptyset$, is false, the implication $(x \in \emptyset) \Rightarrow (x \in A)$ is true. Therefore, by definition of subset $\emptyset \subseteq A$.

Subsets satisfy the following transitivity property.

Theorem 3.3 Let A, B, and C be sets. If $A \subseteq B$ and $B \subseteq C$, then $A \subseteq C$.

Proof: Let $x \in A$. By hypothesis $A \subseteq B$, so

$$(1) \qquad (x \in A) \Rightarrow (x \in B)$$

Also by hypothesis $B \subseteq C$, so

$$(2) \qquad (x \in B) \Rightarrow (x \in C)$$

From (1) and (2) by the rule of transitive inference,

$$(3) \qquad (\forall x)[(x \in A) \Rightarrow (x \in C)]$$

Hence, $A \subseteq C$.

Intuitively, we want two sets A and B to be **equal**, written "$A = B$," if and only if they are identical sets—that is, if and only if they have exactly the same elements. Thus, we could define equality as follows:

$$A = B \Leftrightarrow (\forall x)[(x \in A) \Leftrightarrow (x \in B)].$$

However, since $(\forall x)[(x \in A) \Leftrightarrow (x \in B)]$ is logically equivalent to

$$(\forall x)[[(x \in A) \Rightarrow (x \in B)] \wedge [(x \in B) \Rightarrow (x \in A)]]$$

which is logically equivalent to

$$[(\forall x)[(x \in A) \Rightarrow (x \in B)]] \wedge [(\forall x)[(x \in B) \Rightarrow (x \in A)]]$$

which is $A \subseteq B$ and $B \subseteq A$, we define equality of sets as follows.

DEFINITION Equality of Sets

Let A and B be sets. Then $A = B \Leftrightarrow [(A \subseteq B) \wedge (B \subseteq A)]$.

This definition of equality permits us to prove two sets A and B are equal (identical) by showing (i) A is a subset of B and (ii) B is a subset of A. As an example, let $A = \{1, 2, 3\}$ and $B = \{3, 1, 2, 2, 1\}$. First, consider the elements of A. The element $1 \in A$ and we see $1 \in B$ also. The elements $2, 3 \in A$ and they are members of the set B also. Since every element in the set A is in the set B, $A \subseteq B$. Now consider the elements in the set B. The element $3 \in B$, and we find $3 \in A$ as well. The elements $1, 2 \in B$ and they are elements of A too. Hence, $B \subseteq A$. Since $A \subseteq B$ and $B \subseteq A$, $A = B$.

EXAMPLE 1 Proving Two Sets Equal

Let $A = \{n \mid n \in \mathbf{Z} \text{ and } |n| < 2\}$ and let $B = \{x \mid x \in \mathbf{R} \text{ and } x^3 - x = 0\}$. Prove $A = B$.

SOLUTION

(i) Let $n \in A$. Since n is an integer and $|n| < 2$, n is -1, 0, or 1. Substituting $x = -1$ into the expression $x^3 - x$, we find $(-1)^3 - (-1) = -1 + 1 = 0$, so $-1 \in B$. Substituting, $x = 0$ into $x^3 - x$, we see $(0)^3 - (0) = 0 - 0 = 0$, so $0 \in B$. And substituting $x = 1$ into $x^3 - x$, yields $(1)^3 - (1) = 1 - 1 = 0$, so $1 \in B$. Hence, $A \subseteq B$.

(ii) Let $x \in B$. Factoring, yields

$$x^3 - x = x(x^2 - 1) = x(x - 1)(x + 1) = 0.$$

Consequently, the elements of the set B are the integers 0, 1, and -1. Since $|0| = 0 < 2$, $0 \in A$; since $|1| = 1 < 2$, $1 \in A$; and since $|-1| = 1 < 2$, $-1 \in A$. Hence, $B \subseteq A$.

By (i) $A \subseteq B$ and by (ii) $B \subseteq A$; therefore, $A = B$.

DEFINITION Proper Subset

The set A is a **proper subset** of the set B, written $A \subset B$, if and only if A is a subset of B and $A \neq B$. Symbolically,

$$A \subset B \Leftrightarrow [(A \subseteq B) \wedge (A \neq B)].$$

Clearly, A is a proper subset of B provided every element of A is an element of B and there exists at least one element of B which is not an element of A. Since the natural numbers, $\mathbf{N}$, are a proper subset of the integers, $\mathbf{Z}$, are a proper subset of the rational numbers, $\mathbf{Q}$, are a proper subset of the real numbers, $\mathbf{R}$, we can write $\mathbf{N} \subset \mathbf{Z} \subset \mathbf{Q} \subset \mathbf{R}$.

In 1880, the British logician John Venn (1834-1923) published an article titled "On the Diagrammatic and Mechanical Representation of Propositions and Reasonings." In this paper, Venn introduced his diagrams for illustrating syllogistic logic. Venn did not conceive of the idea of representing logical arguments by diagrams—Gottfried Leibniz, to name one person, had used diagrams for this purpose approximately 200 years earlier. Nonetheless, Venn was dissatisfied with the diagrams being used by his contemporary logicians George Boole and Augustus De Morgan, so he devised his own diagrams which are now called **Venn diagrams**. A Venn diagram for $A = B$ is shown in Figure 3.1 and a Venn diagram for $A \subset B$ is shown in Figure 3.2. If $A \subseteq B$, it is necessary to consider both possibilities displayed in Figures 3.1 and 3.2.

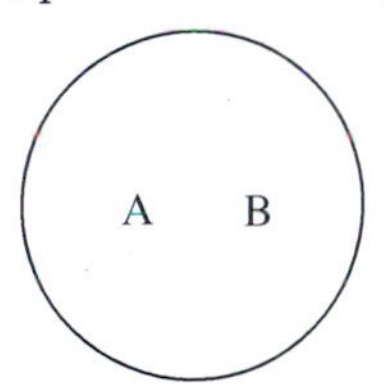

FIGURE 3.1: Venn Diagram for $A = B$

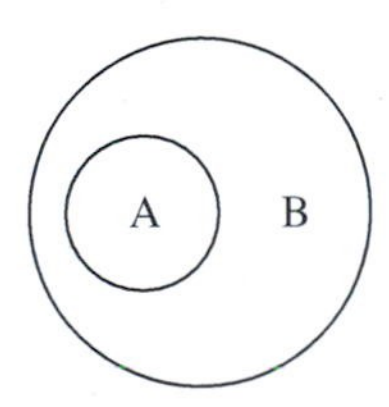

FIGURE 3.2: Venn Diagram for $A \subseteq B$

EXAMPLE 2 Listing All Subsets of a Given Set

List all subsets of the following sets.

a. $\emptyset$ b. $\{1\}$ c. $\{a, b\}$ d. $\{x, y, z\}$

SOLUTION

a. The only subset of the empty set is the empty set itself—that is, $\emptyset \subseteq \emptyset$.

b. The two subsets of the set $\{1\}$ are $\emptyset$ and $\{1\}$.

c. The four subsets of the set $\{a, b\}$ are $\emptyset$, $\{a\}$, $\{b\}$, and $\{a, b\}$

d. The eight subsets of the set $\{x, y, z\}$ are the sets $\emptyset$, $\{x\}$, $\{y\}$, $\{z\}$, $\{x, y\}$, $\{x, z\}$, $\{y, z\}$, and $\{x, y, z\}$

In example 2, observe that the number of subsets of the set with no elements—the empty set, $\emptyset$—is $2^0 = 1$; the number of subsets of a set with one element is $2^1 = 2$; the number of subsets of a set with two elements is $2^2 = 4$; and the number of subsets of a set with three elements is $2^3 = 8$. Let $A = \{a_1, a_2, \ldots, a_n\}$ be a set with $n \geq 1$ elements. In creating a subset X of A there are n decisions to be made—namely, whether $a_i \in X$ or $a_i \notin X$ for $i = 1, 2, \ldots, n$. Since there are n decisions to be made and two possible choices for each decision, there are a total of 2^n possible subsets X of A.

The set of all subsets of a finite set A is denoted by $\mathcal{P}(A)$ and is called the **power set** of A. The name power set comes from the fact that if the set A has n elements, then its power set $\mathcal{P}(A)$ has 2^n elements. From example 2, it is clear that $\mathcal{P}(\emptyset) = \{\emptyset\}$, $\mathcal{P}(\{1\}) = \{\emptyset, \{1\}\}$, $\mathcal{P}(\{a, b\}) = \{\emptyset, \{a\}, \{b\}, \{a, b\}\}$, and $\mathcal{P}(\{x, y, z\}) = \{\emptyset, \{x\}, \{y\}, \{z\}, \{x, y\}, \{x, z\}, \{y, z\}, \{x, y, z\}\}$. Notice that the power set is a set of sets.

EXERCISES 3.1

1. Which of the following sets of objects are well-defined?

   a. The set of all former Presidents of the United States.

   b. The set of all tall, former Presidents of the United States.

   c. The set of the ten best songs.

d. The set of all English names of the months of the year which begin with the letter J.

e. The set of all natural numbers which is less than -1.

2. Write the following sets using both roster notation and set-builder notation.

   a. The set of all letters in the word Mississippi.

   b. The set of odd natural numbers.

   c. The set of all prime numbers greater than 12 and less than 50.

   d. The set of all single digit integers.

   e. The set of all real numbers which satisfy the equation $x^2 + 1 = 0$.

3. Let $A = \{1, 2\}$ and $B = \{1, 2, 3\}$. Which of the following statements are true?

a. $1 \in A$	b. $1 \subset A$	c. $\{1\} \in A$	d. $\{1\} \subset A$
e. $\emptyset \in A$	f. $\emptyset \subseteq A$	g. $\emptyset \subset A$	h. $A \subseteq B$
i. $A \subset B$	j. $B \supset A$	k. $B \subset A$	l. $A = B$
m. $A \neq B$	n. $\emptyset \in \emptyset$	o. $\emptyset \subseteq \emptyset$	p. $\emptyset \subset \emptyset$
q. $\emptyset \in \{\emptyset, A\}$	r. $\emptyset \subset \{\emptyset, A\}$	s. $\{\emptyset\} \in \{\emptyset, A\}$	t. $\{\emptyset\} \subset \{\emptyset, A\}$

4. Pairs of sets A and B are given below. In each case, determine if $A = B$ or $A \neq B$.

   a. $A = \{1, 2, 3\}$ $\quad B = \{1, 2, 2, 3\}$

   b. $A = \{x \mid x \text{ is an even prime}\}$ $\quad B = \{n \mid n \in \mathbf{Z} \text{ and } |n| = 2\}$

   c. $A = \{x \mid x \text{ is an even prime}\}$ $\quad B = \{n \mid n \in \mathbf{N} \text{ and } |n| = 2\}$

   d. $A = \{n \mid n \in \mathbf{N} \text{ and } n \text{ is odd}\}$ $\quad B = \{n \mid n \in \mathbf{N} \text{ and } n^2 \text{ is odd}\}$

   e. $A = \{n \mid n \in \mathbf{N} \text{ and } n \text{ is odd}\}$ $\quad B = \{n \mid n \in \mathbf{Z} \text{ and } n^2 \text{ is odd}\}$

   f. $A = \{n \mid n \in \mathbf{Z} \text{ and } n^2 < 1\}$ $\quad B = \{n \mid n \in \mathbf{Z} \text{ and } n^2 < n\}$

5. Let $C = \{x \mid x \in \mathbf{R} \text{ and } x^2 - 1 = 0\}$ and $D = \{n \mid n \in \mathbf{Z} \text{ and } |n| = 1\}$. Prove $C = D$.

6. Let $E = \{x \mid x \in \mathbf{R} \text{ and } x^2 - 5x + 6 = 0\}$ and let $F = \{n \mid n \in \mathbf{N} \text{ and } 3 < n^2 < 10\}$. Prove $E = F$.

7. Draw three different Venn diagrams with sets A and B for which $A \not\subset B$.

8. a. Draw a Venn diagram in which $A \subset B$ and $x \notin B$.

   b. Prove if $A \subset B$ and $x \notin B$, then $x \notin A$.

   (**NOTE:** A Venn diagram does not constitute a proof. A Venn diagram is a visual aid.)

9. We showed if the set A has n elements, then the power set of A, $\mathcal{P}(A)$, has 2^n elements. How many proper sets does the set A have? How many nonempty subsets does the set A have? How many nonempty, proper subsets does A have?

10. Write the power set of the given sets. a. {Ann, Ben}

    b. {$, #, @} c. $\{\emptyset, \text{a}, \{\text{a}\}\}$ d. $\{0, \{0\}, \emptyset, \{\emptyset\}\}$

11. List the proper subsets of the following sets.

    a. {a} b. {1, 2} c. $\emptyset$ d. $\{\emptyset\}$

12. a. Prove if $A \subseteq B$ and $B \subset C$, then $A \subset C$.

    b. If $A \subset B$ and $B \subseteq C$, what do you think you can prove regarding the sets A and C?

13. Give examples of sets A, B, and C for which the following statements are true.

    a. $A \subset B$, $B \not\subset C$, and $A \subset C$. b. $A \subset B$, $B \not\subset C$, and $A \not\subset C$.

    c. $A \subseteq B$, $B \subseteq C$, and $C \subseteq A$. d. $A \not\subset B$, $B \not\subset C$, and $A \subset C$.

    e. $A \in B$, $B \notin C$, and $A \notin C$. f. $A \in B$, $A \subset C$, and $B \not\subset C$.

14. For each of the following statements determine if the statement is true or false. If the statement is true, prove it. If the statement is false, disprove it by giving a counterexample.

    a. If $a \in A$ and $A \in B$, then $a \in B$.

    b. If $a \in A$ and $A \not\subset B$, then $a \notin B$.

    c. If $A \subset B$ and $B \in C$, then $A \in C$.

    d. If $A \not\subset B$ and $B \subset C$, then $A \not\subset C$.

    e. If $A \not\subset B$ and $B \not\subset C$, then $A \not\subset C$.

    f. If $A \subseteq B$, then $\mathcal{P}(A) \subseteq \mathcal{P}(B)$.

    g. If $\mathcal{P}(A) \subseteq \mathcal{P}(B)$, then $A \subseteq B$.

3.2 Set Operations

Usually, at the beginning of a particular discussion a set called the **universe** is specified. The universe is selected in such a way that it contains all elements of all sets to be discussed. We will denote the universe by U. The universe need not be the same set for all discussions. For instance, in one discussion the universe might be the set of integers, in another discussion the universe might be the set of all triangles, or the set of points in the plane, etc. There are several elementary operations which may be performed on sets. In this section, we will examine two binary operations (operations involving two sets) and one unary operation (an operation involving a single set).

DEFINITION Union of Two Sets, $A \cup B$

Let A and B be two subsets of the universe U. The set A **union** B is the subset of U which contains all elements that belong either to A or to B or to both A and B. The set A union B is denoted by $A \cup B$, which is read "A union B." Symbolically,

$$A \cup B = \{x \mid x \in A \text{ or } x \in B\} = \{x \mid (x \in A) \vee (x \in B)\}.$$

For example, if the universe is the set of natural numbers, $A = \{1, 3, 5, 7\}$ and $B = \{2, 4, 6\}$, then $A \cup B = \{1, 2, 3, 4, 5, 6, 7\}$. Notice that $A \subset \mathbf{N}$, $B \subset \mathbf{N}$, and $A \cup B \subset \mathbf{N} = U$. The British logician Charles Dodgson (1832-1898) improved upon Venn diagrams by employing a circumscribed rectangle to represent the universal set. Dodgson was a distinguished mathematician and wrote three mathematical texts; however, he is probably best known under his pen name Lewis Carroll. Under that name he published two popular books *Alice's Adventure in Wonderland* and *Through the Looking Glass.* The shaded areas shown in Figure 3.3 represent the set $A \cup B$ for the three different Venn diagrams.

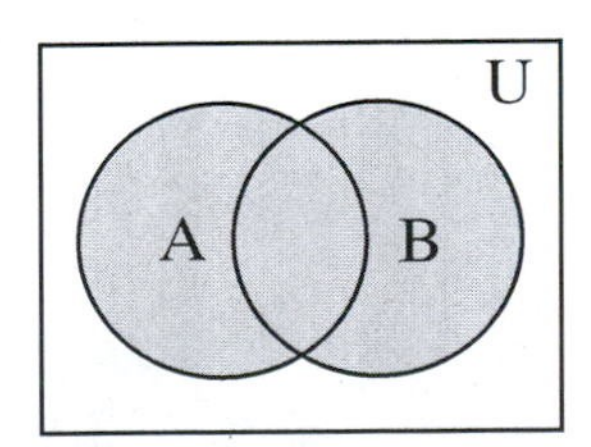

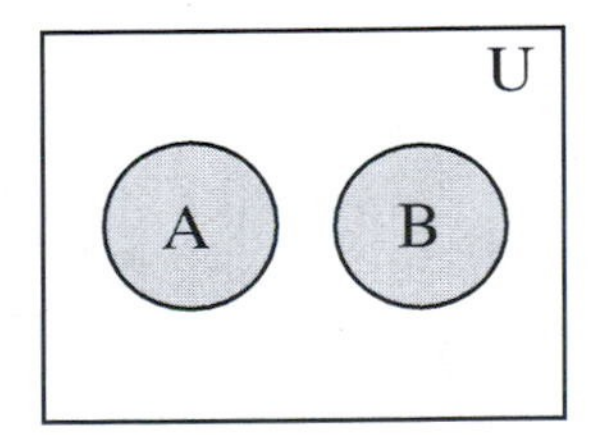

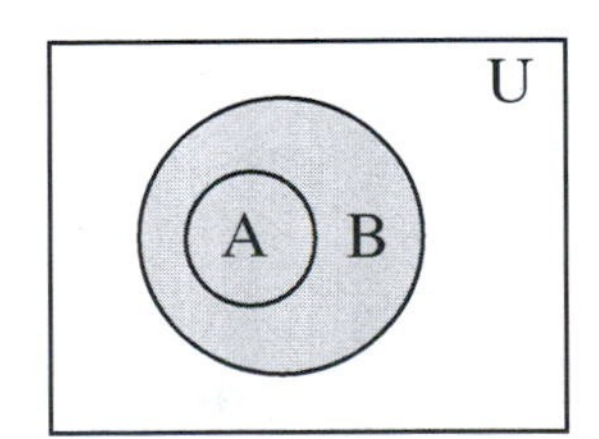

FIGURE 3.3: Three Venn Diagrams for $A \cup B$

DEFINITION Intersection of Two Sets, $A \cap B$

Let A and B be two subsets of the universe U. The set A **intersect** B is the subset of U which contains all elements that belong both A and B. The set A intersect B is denoted by $A \cap B$, which is read "A intersect B." Symbolically,

$$A \cap B = \{x \,|\, x \in A \text{ and } x \in B\} = \{x \,|\, (x \in A) \wedge (x \in B)\}.$$

For example, if the universe is the set of integers, $A = \{-2, -1, 0, 1, 2\}$ and $B = \{-4, -2, 0, 2\}$, then $A \cap B = \{-2, 0, 2\}$. The shaded areas shown in Figure 3.4 depict the set $A \cap B$ for the three different Venn diagrams. When $A \cap B = \emptyset$, the sets A and B are said to be **disjoint**. Notice in the center Venn diagram of Figure 3.4 that $A \cap B = \emptyset$. This diagram is the typical diagram for two disjoint sets.

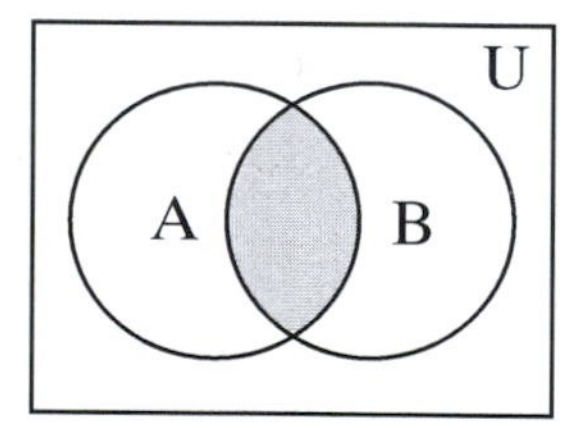

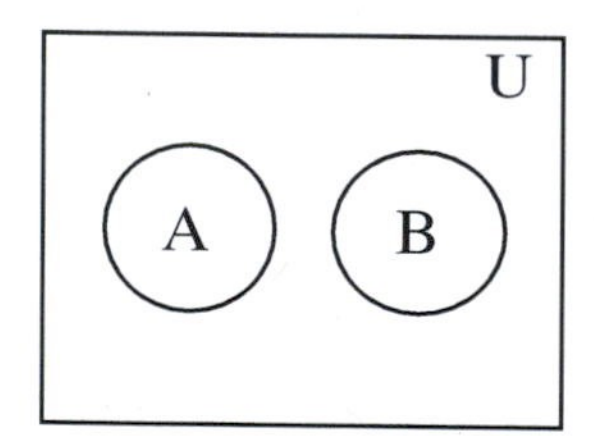

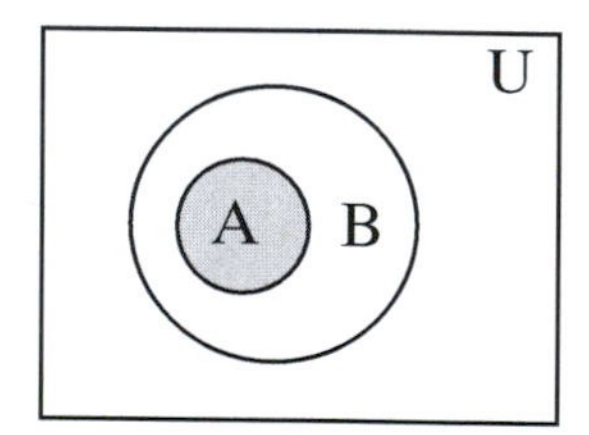

FIGURE 3.4: Three Venn Diagrams for $A \cap B$

It follows from the definition of union that $x \notin (A \cup B)$ is equivalent to

$$(1) \qquad \neg[x \in (A \cup B)] \equiv \neg[(x \in A) \vee (x \in b)] \equiv [\neg(x \in A)] \wedge [\neg(x \in B)]$$

by a De Morgan law from logic. Since $\neg(x \in A) \equiv x \notin A$ and since $\neg(x \in B) \equiv x \notin B$, it follows from (1) that $x \notin (A \cup B)$ is logically equivalent to $(x \notin A) \wedge (x \notin B)$. Likewise, $x \notin (A \cap B)$ is logically equivalent to $(x \notin A) \wedge (x \notin B)$.

DEFINITION Complement of the Set A, A'

The **complement** of the set A, written A', is the set of all elements in the universe, U, which are not in the set A. Hence,

$$A' = \{x \,|\, x \in U \text{ and } x \notin A\} = \{x \,|\, (x \in U) \wedge (x \notin A)\}.$$

For example, if the universe $U = \{1, 2, 3, 4, 5\}$ and $A = \{2, 4\}$, then $A' = \{1, 3, 5\}$. A Venn diagram for a set A and its complement A' is shown in Figure 3.5.

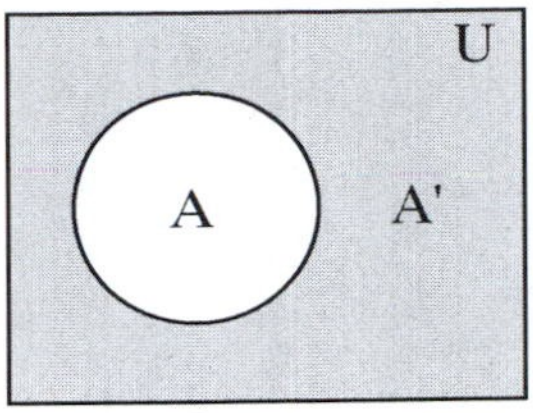

U
A
A'

FIGURE 3.5: A Venn Diagram of A and Its Complement A'

EXAMPLE 1 Performing Set Operations

Let $U = \{1, a, 2, b, \#, @\}$, $A = \{a, b, \#\}$, $B = \{1, a, 2, \#\}$, and $C = \{2, \#, @\}$. Find the following sets.

a. A' b. $(A')'$ c. $A \cup A'$ d. $A \cap A'$

e. $A \cup B$ f. $B \cup A$ g. $(A \cap B) \cap C$ h. $A \cap (B \cap C)$

i. $A \cup (B \cap C)$ j. $(A \cup B) \cap (A \cup C)$ k. $(A \cup B)'$ l. $A' \cap B'$

What do you notice about the answers of the pairs $e - f$, $g - h$, $i - j$, and $k - l$?

SOLUTION

a. By definition A' is the set of elements in the universe U which are not in the set A. Deleting the elements of the set A from the set U, we find

$$A' = \{1, 2, @\}.$$

b. By definition $(A')'$ is the set of elements in U which are not in the set A'. Deleting from U the elements of the set A', which we found in part a., we see

$$(A')' = \{a, b, \#\} = A.$$

In this example, we observe that the complement of the complement of the set A is the set A.

c. From the definition of A and the computation of the set A' performed in part a., we obtain

$$A \cup A' = \{a, b, \#, 1, 2, @\} = U.$$

In this instance, we notice $A \cup A' = U$.

d. From the definition of A and the result of part a.,

$$A \cap A' = \{a, b, \#\} \cap \{1, 2, @\} = \emptyset.$$

Thus, A and A' are disjoint.

e. $A \cup B = \{a, b, \#\} \cup \{1, a, 2, \#\} = \{a, b, \#, 1, 2\}$

f. $B \cup A = \{1, a, 2, \#\} \cup \{a, b, \#\} = \{1, a, 2, \#, b\}$

g. To calculate $(A \cap B) \cap C$, we calculate $A \cap B$ first (because of the placement of the parentheses) and then we calculate the intersection of the sets $A \cap B$ and C.

$$A \cap B = \{a, b, \#\} \cap \{1, a, 2, \#\} = \{a, \#\}$$

and

$$(A \cap B) \cap C = \{a, \#\} \cap \{2, \#, @\} = \{\#\}$$

h. To calculate $A \cap (B \cap C)$, we calculate $B \cap C$ first (because of the placement of the parentheses) and then we calculate the intersection of the sets A and $B \cap C$.

$$B \cap C = \{1, a, 2, \#\} \cap \{2, \#, @\} = \{2, \#\}$$

and

$$A \cap (B \cap C) = \{a, b, \#\} \cap \{2, \#\} = \{\#\}$$

i. From part h., $B \cap C = \{2, \#\}$, so

$$A \cup (B \cap C) = \{a, b, \#\} \cup \{2, \#\} = \{a, b, \#, 2\}$$

j. From part e., $A \cup B = \{a, b, \#, 1, 2\}$ and from the definitions of sets A and C,

$$A \cup C = \{a, b, \#\} \cup \{2, \#, @\} = \{a, b, \#, 2, @\}$$

The intersection of the sets $(A \cup B)$ and $(A \cup C)$ is

$$(A \cup B) \cap (A \cup C) = \{a, b, \#, 1, 2\} \cap \{a, b, \#, 2, @\} = \{a, b, \#, 2\}$$

k. From part e., $A \cup B = \{a, b, \#, 1, 2\}$. Deleting the elements of $A \cup B$ from U, we find $(A \cup B)' = \{@\}$.

l. From part a., $A' = \{1, 2, @\}$ and calculating the complement of B, we find $B' = \{b, @\}$. So

$$A' \cap B' = \{1, 2, @\} \cap \{b, @\} = \{@\}$$

The answers of the pairs $e - f$, $g - h$, $i - j$, and $k - l$ are equal. That is, for this example, $A \cup B = B \cup A$, $(A \cap B) \cap C = A \cap (B \cap C)$, $A \cup (B \cap C) = (A \cup B) \cap (A \cup C)$, and $(A \cup B)' = A' \cap B'$.

From example 1, it appears the operator $\cup$ may be a commutative operator—that is, for all subsets A and B of a universe U, it may be true that $A \cup B = B \cup A$. Hence, we formulate the following theorem and then prove it.

Theorem 3.4 Let A and B be subsets of the universal set U. Then $A \cup B = B \cup A$.

Proof: Let $x \in U$.

$x \in A \cup B \Leftrightarrow (x \in A) \vee (x \in B)$	Definition of $A \cup B$
$\Leftrightarrow (x \in B) \vee (x \in A)$	Commutative Law of Disjunction
$\Leftrightarrow x \in B \cup A$	Definition of $B \cup A$

Thus, $(\forall x \in U)[(x \in (A \cup B)) \Leftrightarrow (x \in (B \cup A))]$. Hence, $A \cup B = B \cup A$.

An equation, expression, or statement in set theory which is obtained by interchanging $\cup$ and $\cap$ and interchanging the sets $\emptyset$ and U is called the **dual** of the equation, expression, or statement accordingly. Hence, the dual of the statement $A \cup B = B \cup A$ is the statement $A \cap B = B \cap A$. Likewise, the dual of the statement $A \cup \emptyset = A$ is the statement $A \cap U = A$. The **principle of duality** for set theory says: "If T is a theorem which is written in terms of $\cup$, $\cap$, and $'$, then the dual statement T^D is also a theorem." Since we have proven $A \cup B = B \cup A$ is a theorem of set theory, by the principle of duality, $A \cap B = B \cap A$ is a theorem of set theory also. Using the principle of duality, by proving one theorem, we have proven two theorems.

The **algebra of sets** consists of a collection $\mathcal{A}$ of subsets $\emptyset$, U, A, B, $C, \ldots$ of some universe U. (Notice that $\emptyset$ and U must both be members of the collection $\mathcal{A}$.) On the collection $\mathcal{A}$ there is defined two relations, $\subseteq$ and $=$; two binary operations, $\cup$ and $\cap$; and one unary operation $'$. The collection $\mathcal{A}$ also has the properties that if A and B are arbitrary sets of $\mathcal{A}$, then $A \cup B \in \mathcal{A}$, $A \cap B \in \mathcal{A}$, and $A' \in \mathcal{A}$. For $\mathcal{A}$ an algebra of sets, the following statements and their duals are theorems.

Idempotent Laws

1. $A \cup A = A$	1^D. $A \cap A = A$

Identity Laws

2. $A \cup \emptyset = A$	2^D. $A \cap U = A$
3. $A \cup U = U$	3^D. $A \cap \emptyset = \emptyset$

Complement Laws

4. $A \cup A' = U$ $\quad$ 4^D. $A \cap A' = \emptyset$

5. $\emptyset' = U$ $\quad$ 5^D. $U' = \emptyset$

6. $(A')' = A$

Commutative Laws

7. $A \cup B = B \cup A$ $\quad$ 7^D. $A \cap B = B \cap A$

Associative Laws

8. $(A \cup B) \cup C = A \cup (B \cup C)$ $\quad$ 8^D. $(A \cap B) \cap C = A \cap (B \cap C)$

Distributive Laws

9. $A \cup (B \cap C) = (A \cup B) \cap (A \cup C)$ $\quad$ 9^D. $A \cap (B \cup C) = (A \cap B) \cup (A \cap C)$

De Morgan Laws

10. $(A \cup B)' = A' \cap B'$ $\quad$ 10^D. $(A \cap B)' = A' \cup B'$

The algebra of sets is another example of a mathematical system. Notice the similarity between the theorems (laws) of the algebra of sets and the theorems of mathematical logic when $\cup$ is associated with $\vee$, $\cap$ is associated with $\wedge$, and $'$ is associated with $\neg$. We now prove a few of the theorems stated above and leave the remainder to be proven as exercises.

EXAMPLE 2 Proof of the Idempotent Law $A \cup A = A$

Prove $A \cup A = A$.

SOLUTION

Proof: Let A be a subset in the algebra of sets $\mathcal{A}$ with universe U and let $x \in U$.

$$x \in A \cup A \Leftrightarrow (x \in A) \vee (x \in A) \qquad \text{Definition of } A \cup A$$

$$\Leftrightarrow (x \in A) \qquad \text{Idempotent Law of Disjunction}$$

Hence, $A \cup A = A$ and by the principle of duality the dual statement $A \cap A = A$ is a theorem also.

EXAMPLE 3 Proof of the Identity Law $A \cap \emptyset = \emptyset$

Prove $A \cap \emptyset = \emptyset$.

SOLUTION

Proof: Let A be any subset in the algebra of sets $\mathcal{A}$ and let $x \in U$, the universal set.

$x \in A \cap \emptyset \Leftrightarrow (x \in A) \wedge (x \in \emptyset)$	Definition of $A \cap \emptyset$
$\Leftrightarrow (x \in A) \wedge f$	$x \in \emptyset$ is a contradiction
$\Leftrightarrow f$	Contradiction Law of logic

This sequence of statements shows that the set $A \cap \emptyset$ has no elements. Hence, $A \cap \emptyset = \emptyset$ and by the principle of duality the dual statement $A \cup U = U$ is a theorem as well.

EXAMPLE 4 Proof of the Complement Law $A \cup A' = U$

Prove $A \cup A' = U$.

SOLUTION

Proof: Let A be a subset in the algebra of sets $\mathcal{A}$ and let $x \in U$, the universal set.

$x \in A \cup A'$

$\Leftrightarrow (x \in A) \vee (x \in A')$	Definition of $A \cup A'$
$\Leftrightarrow (x \in A) \vee [(x \in U) \wedge (x \notin A)]$	Definition of complement
$\Leftrightarrow [(x \in A) \vee (x \in U)] \wedge [(x \in A) \vee (x \notin A)]$	A De Morgan Law of logic
$\Leftrightarrow [(x \in A) \vee (x \in U)] \wedge [(x \in A) \vee (\neg(x \in A))]$	Definition of $\notin$
$\Leftrightarrow [(x \in A) \vee (x \in U)] \wedge t$	$P \vee (\neg P)$ is a tautology
$\Leftrightarrow [(x \in A) \vee (x \in U)]$	A Tautology Law
$\Leftrightarrow x \in A \cup U$	Definition of $A \cup U$

Thus, $A \cup A' = A \cup U$. But by the identity law 3. $A \cup U = U$, so $A \cup A' = A \cup U = U$. Furthermore, by the principle of duality the dual statement $A' \cap A = \emptyset$ is a theorem.

EXAMPLE 5 Proof of the Distributive Law
$A \cup (B \cap C) = (A \cup B) \cap (A \cup C)$

Prove $A \cup (B \cap C) = (A \cup B) \cap (A \cup C)$.

SOLUTION

Proof: Let A, B, $C \subseteq U$ and let $x \in U$.

$x \in A \cup (B \cap C)$

$\Leftrightarrow (x \in A) \vee (x \in (B \cap C))$	Definition of union
$\Leftrightarrow (x \in A) \vee [(x \in B) \wedge (x \in C)]$	Definition of intersection
$\Leftrightarrow [(x \in A) \vee (x \in B)] \wedge [(x \in A) \vee (x \in C)]$	A Distributive Law of logic
$\Leftrightarrow [x \in (A \cup B)] \wedge [x \in (A \cup C)]$	Definition of union
$\Leftrightarrow x \in (A \cup B) \cap (A \cup C)$	Definition of intersection

Hence, $A \cup (B \cap C) = (A \cup B) \cap (A \cup C)$ and by duality $A \cap (B \cup C) = (A \cap B) \cup (A \cap C)$

EXAMPLE 6 Proof of the De Morgan Law $(A \cap B)' = A' \cup B'$

Prove $(A \cap B)' = A' \cup B'$.

SOLUTION

Proof: Let A, $B \subseteq U$ and let $x \in U$.

$x \in (A \cap B)'$

$\Leftrightarrow (x \in U) \wedge (x \notin (A \cap B))$	Definition of complement
$\Leftrightarrow (x \in U) \wedge [\neg(x \in (A \cap B))]$	Definition of $\notin$
$\Leftrightarrow (x \in U) \wedge [\neg((x \in A) \wedge (x \in B))]$	Definition of intersection
$\Leftrightarrow (x \in U) \wedge [(\neg(x \in A)) \vee (\neg(x \in B))]$	A De Morgan Law of logic
$\Leftrightarrow (x \in U) \wedge [(x \notin A)) \vee (x \notin B)]$	Definition of $\notin$
$\Leftrightarrow [(x \in U) \wedge (x \notin A)] \vee [(x \in U) \wedge (x \notin B)]$	A Distributive Law of logic
$\Leftrightarrow (x \in A') \vee (x \in B')$	Definition of A' and B'
$\Leftrightarrow x \in A' \cup B'$	Definition of $A' \cup B'$

Hence, $(A \cap B)' = A' \cup B'$ and by duality $(A \cup B)' = A' \cap B'$.

EXERCISES 3.2

1. Let the universe be the set $U = \{1, 2, 4, a, e, \emptyset\}$. Let $A = \{a, e, \emptyset\}$, $B = \{1, 4, \emptyset\}$, and $C = \{2, 4\}$. Write the following sets using roster notation.

   a. A' b. B' c. $A \cup B$

   d. $A' \cup B'$ e. $A \cap B$ f. $A' \cap B'$

   g. $(A \cup B)'$ h. $(A \cap B)'$ i. $A \cap C$

   j. $(A \cap C)'$ k. $(A \cap B) \cup C$ l. $A \cap (B \cup C)$

   m. $(A \cap C') \cup (A \cap C)$ n. $(A \cup C) \cap (A \cap C)$

2. Let $U = \{1, 2, \ldots, 10\}$, $E = \{x \in U \mid x \text{ is even}\}$, $O = \{x \in U \mid x \text{ is odd}\}$, and $P = \{x \in U \mid x \text{ is a prime}\}$.

   a. Write E, O, and P using roster notation.

b. Write the following sets using roster notation.

(i) $E \cap P$ (ii) $E \cup P$ (iii) $E' \cap P$ (iv) $E \cap P'$

(v) $E \cap O$ (vi) $E \cup O$ (vii) $O \cap P$ (viii) $O \cup P$

(ix) $(O \cap P) \cup E$ (x) $E' \cap O'$

3. Let U be the set of all people, B be the set of all blonds, F be the set of all females, and T be the set of all people taller than five feet, six inches. The following sets are defined in set builder notation. Write these sets in terms of B; F; T; union, $\cup$; intersection, $\cap$; and complement, $'$.

   a. $\{x \mid x$ is not blond$\}$

   b. $\{x \mid x$ is a male$\}$

   c. $\{x \mid x$ is a blond female$\}$

   d. $\{x \mid x$ is a blond male$\}$

   e. $\{x \mid x$ is a female with dark hair$\}$

   f. $\{x \mid x$ is a dark haired male who is five feet, six inches tall or less$\}$

   g. $\{x \mid x$ is a blond female who is taller than five feet, six inches$\}$

4. Copy the following three Venn diagram configurations four times and shade the indicated sets in parts a-d.

   a. $A' \cap B$ b. $A \cap B'$ c. $A' \cup B'$ d. $A' \cap B'$

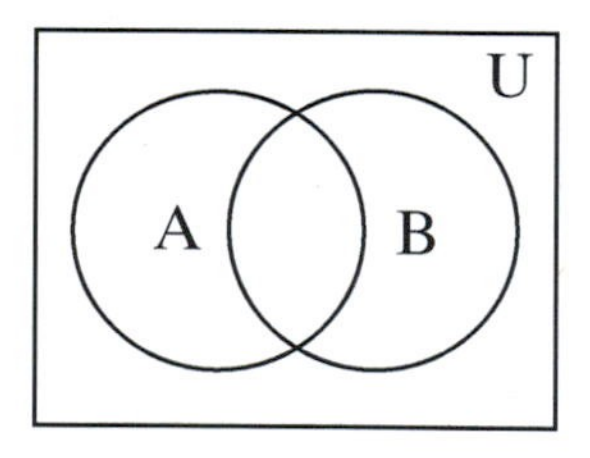

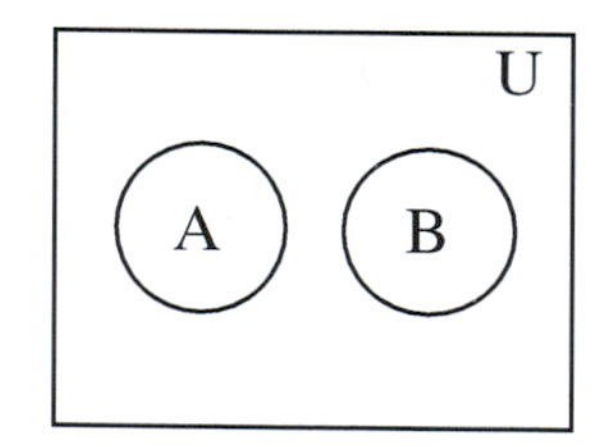

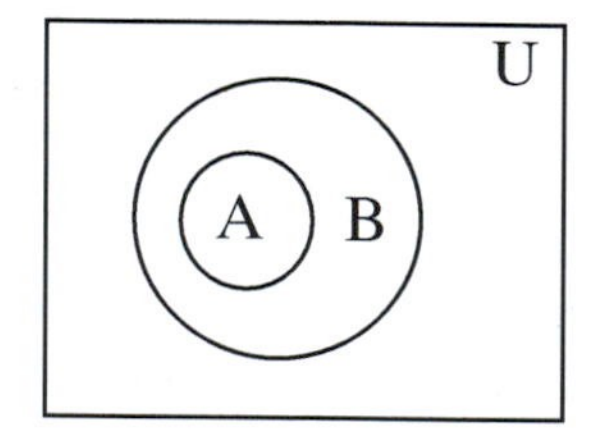

In exercises 5-8, let A, B, C be subsets of the universe U.

5. Prove the identity law: $A \cap U = A$.

6. Prove the complement law: $\emptyset' = U$.

7. Prove the complement law: $(A')' = A$.

8. Prove the associative law: $(A \cup B) \cup C = A \cup (B \cup C)$.

In exercises 9-15, let A and B be sets.

9. Prove $A \subseteq A \cup B$.

10. Prove $A \cap B \subseteq A$.

    (Observe that from exercises 9 and 10 we can deduce $A \cap B \subseteq A \cup B$.)

11. Prove $A \subseteq B \Leftrightarrow A \cup B = B$.

12. Prove $A \subseteq B \Leftrightarrow A \cap B = A$.

13. Prove $A \subseteq B \Leftrightarrow B' \subseteq A'$.

14. Prove $A \cap B = \emptyset \Leftrightarrow A \subseteq B'$.

15. Prove $A \cup B = A \cap B \Leftrightarrow A = B$.

16. Let A, B, and C be sets. Give a counterexample to each of the following statements.

    a. If $A \cup B \subseteq A \cup C$, then $B \subseteq C$.

    b. If $A \cap B \subseteq A \cap C$, then $B \subseteq C$.

17. Let A, B, and C be sets. Prove the following two theorems.

    a. If $B \subseteq C$, then $A \cup B \subseteq A \cup C$ for any set A.

    b. If $B \subseteq C$, then $A \cap B \subseteq A \cap C$ for any set A.

In exercises 18-21, let A, B, C, and D be sets.

18. Prove $(A \cap B) \cup C = A \cap (B \cup C) \Leftrightarrow C \subseteq A$.

19. Prove if $A \subseteq C$ and $B \subseteq D$, then $A \cup B \subseteq C \cup D$.

20. Prove if $A \subseteq C$ and $B \subseteq D$, then $A \cap B \subseteq C \cap D$.

21. Prove if $A \cup B \subseteq C \cup D$, $A \cap B = \emptyset$, and $C \subseteq A$, then $B \subseteq D$.

3.3 Additional Set Operations

In this section, we discuss two more binary operations—the difference of two sets and the Cartesian product of two sets.

DEFINITION The Difference of Two Sets, $A - B$

Let A and B be two subsets of the universe U. The **set difference of A and B**, written $A - B$ and read "A minus B," is the set of all elements in the set A that are not in the set B. Symbolically,

$$A - B = \{x \mid x \in A \text{ and } x \notin B\} = \{x \mid (x \in A) \wedge (x \notin B)\}.$$

For example, if the universe is the set of natural numbers, $A = \{3, 5, 7, 8\}$, and $B = \{1, 3, 7, 9\}$, then $A - B = \{5, 8\}$ and $B - A = \{1, 9\}$. Observe that in this example, $A - B \neq B - A$. Hence, the operator $-$ is not commutative. In addition, let $C = \{5, 9\}$. Then $(A - B) - C = \{5, 8\} - \{5, 9\} = \{8\}$. Furthermore, since $B - C = \{1, 3, 7\}$, $A - (B - C) = \{3, 5, 7, 8\} - \{1, 3, 7\} = \{5, 8\}$. Thus, for this example, $(A - B) - C \neq A - (B - C)$. Thus, the operator $-$ is not an associative operator.

It follows immediately from the definition of the complement of the set A, $A' = \{x \mid x \in U \text{ and } x \notin A\}$, and the definition of U minus A, $U - A = \{x \mid x \in U \text{ and } x \notin A\}$, that $U - A = A'$. It also follows easily from the definition of the difference of sets that

a. $A - U = \emptyset$ b. $A - \emptyset = A$ c. $\emptyset - A = \emptyset$ d. $A - A = \emptyset$ e. $A - B \subseteq A$

The shaded areas of Figure 3.6 represent the set difference $A - B$ for three Venn diagrams configurations.

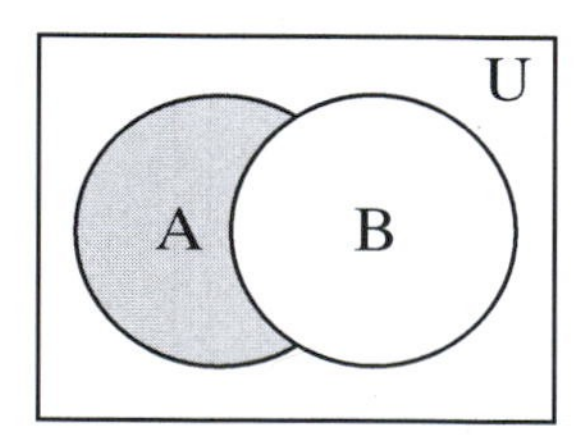

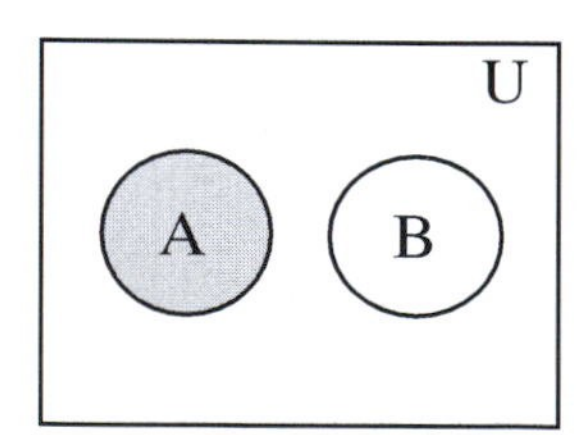

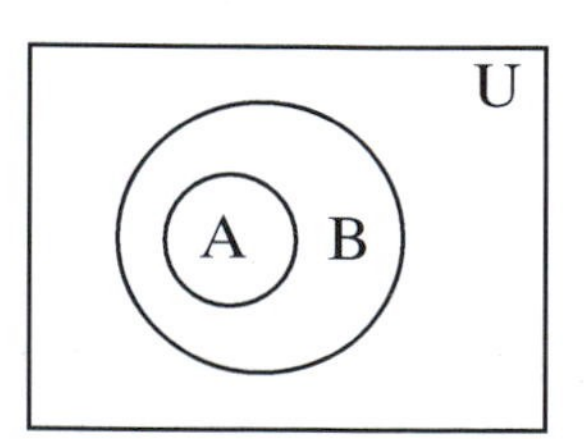

FIGURE 3.6: Three Venn Diagrams for $A - B$

EXAMPLE 1 Set Operations Involving Union, Intersection, and Complementation

Let $U = \{3, 5, \pi, e, i, \mu\}$, $A = \{3, \pi, i\}$, $B = \{5, \pi, e\}$, and $C = \{i, \mu\}$. Write the following sets using roster notation.

a. $A - B$ b. $A \cap B'$ c. $(A - B) - C$ d. $A - (B \cup C)$

e. $A - C$ f. $A - (A \cap C)$ g. $(A \cap B) - C$ h. $A \cap (B - C)$

What do you notice about the answers of pairs a-b, c-d, e-f, and g-h?

SOLUTION

a. $A - B = \{3, \pi, i\} - \{5, \pi, e\} = \{3, i\}$.

b. Since $B' = U - B = \{3, 5, \pi, e, i, \mu\} - \{5, \pi, e\} = \{3, i, \mu\}$, $A \cap B' = \{3, \pi, i\} \cap \{3, i, \mu\} = \{3, i\}$.

Thus, from parts a. and b. it follows for the given sets U, A, and B that $A - B = A \cap B'$.

c. From part a., $A - B = \{3, i\}$, so $(A - B) - C = \{3, i\} - \{i, \mu\} = \{3\}$.

d. $B \cup C = \{5, \pi, e\} \cup \{i, \mu\} = \{5, \pi, e, i, \mu\}$. So $A - (B \cup C) = \{3, \pi, i\} - \{5, \pi, e, i, \mu\} = \{3\}$.

Hence, from parts c. and d. for the given sets A, B, and C, we have $(A - B) - C = A - (B \cup C)$.

e. $A - C = \{3, \pi, i\} - \{i, \mu\} = \{3, \pi\}$.

f. $A \cap C = \{3, \pi, i\} \cap \{i, \mu\} = \{i\}$, so $A - (A \cap C) = \{3, \pi, i\} - \{i\} = \{3, \pi\}$.

Thus, from e. and f. for the given sets A and C, it follows that $A - C = A - (A \cap C)$.

g. Since $A \cap B = \{3, \pi, i\} \cap \{5, \pi, e\} = \{\pi\}$, $(A \cap B) - C = \{\pi\} - \{i, \mu\} = \{\pi\}$.

h. Because $B - C = \{5, \pi, e\} - \{i, \mu\} = \{5, \pi, e\}$, $A \cap (B - C) = \{3, \pi, i\} \cap \{5, \pi, e\} = \{\pi\}$.

Therefore, it follows from parts g. and h. that for the specified sets A, B, and C, $(A \cap B) - C = A \cap (B - C)$.

From example 1, parts a. and b., we conjecture that for A and B subsets of some universe U, it is true that $A - B = A \cap B'$. This leads us to state and prove the following theorem.

Theorem 3.5 Let A and B be subsets of the universal set U. Then $A - B = A \cap B'$.

Proof: Let $x \in U$.

$x \in A \cap B' \Leftrightarrow (x \in A) \wedge (x \in B')$	Definition of $A \cap B'$
$\Leftrightarrow (x \in A) \wedge [(x \in U) \wedge (x \notin B)]$	Definition of B'
$\Leftrightarrow [(x \in A) \wedge (x \in U)] \wedge (x \notin B)$	Associative Law of Conjunction
$\Leftrightarrow [x \in (A \cap U)] \wedge (x \notin B)$	Definition of $A \cap U$
$\Leftrightarrow (x \in A) \wedge (x \notin B)$	Substitution $A \cap U = A$
$\Leftrightarrow x \in A - B$	Definition of $A - B$

Thus, $(\forall x \in U)[(x \in (A \cap B')) \Leftrightarrow (x \in (A - B))]$. Hence, $A - B = A \cap B'$.

Theorem 3.5 shows how to write the set $A - B$ in terms of set intersection and set complementation. Thus, to prove theorems which involve the difference of two sets, we may write the difference of sets in terms of set intersection and complementation and then use the algebra of sets to prove the desired result. The proof of theorem 3.6 illustrates this technique of proof.

Theorem 3.6 Let A and B be subsets of universe U. Then $A - B = A - (A \cap B)$.

Proof:

$A - (A \cap B) = A \cap (A \cap B)'$	By Theorem 3.5
$= A \cap (A' \cup B')$	De Morgan Law 10^D
$= (A \cap A') \cup (A \cap B')$	Distributive Law 9^D
$= \emptyset \cup (A \cap B')$	Complement Law 4^D
$= (A \cap B') \cup \emptyset$	Commutative Law 7
$= A \cap B'$	Identity Law 2
$= A - B$	By Theorem 3.5

As we noted earlier, the set $\{a, b\}$ is equal to the set $\{b, a\}$, and it is immaterial which way we choose to write the set. However, the order in which elements appear and events occur is sometimes very important. Therefore, entities must be defined for which the order of occurrence of the elements is significant.

DEFINITION Ordered Pair

By an **ordered pair** we mean an entity consisting of two elements in a specific order. We denote the ordered pair with a as **first element** and b as **second element** by (a, b).

In 1921, the Polish mathematician Kazimieri Kuratowski (1896-1980) gave the following set theoretic definition of the ordered pair (a, b):

$$(a, b) = \{\{a\}, \{a, b\}\}$$

Using this definition, the following basic theorem regarding the equality of ordered pairs can be proved. (See exercise 21 of this section.)

Theorem 3.7 The ordered pair $(a, b) = (c, d)$ if and only if $a = c$ and $b = d$.

By theorem 3.7, the ordered pair $(2, -3) \neq (2, 4)$ because $-3 \neq 4$; the ordered pair $(1, 5) \neq (-2, 5)$ because $1 \neq -2$; and the ordered pair $(3, 6) \neq (6, 3)$ because $3 \neq 6$.

Generalizing the concept of an ordered pair, we define an **ordered triple** to be an entity consisting of three elements in a specific order such as (a, b, c) in which a is the first element, b is the second element, and c is the third element. Two order triples (a, b, c) and (x, y, z) are equal if and only if $a = x$ and $b = y$ and $c = z$. For a natural number $n \geq 2$, an **ordered n-tuple** is denoted by $(a_1, a_2, \ldots, a_n)$ and $(a_1, a_2, \ldots, a_n) = (b_1, b_2, \ldots, b_n)$ if and only if $a_i = b_i$ for $i = 1, 2, \ldots, n$.

DEFINITION Cartesian Product

Let A and B be sets. The **Cartesian product of A and B**, written as $A \times B$, is the set of all ordered pairs (a, b) such that $a \in A$ and $b \in B$. That is,

$$A \times B = \{(a, b) \mid a \in A \text{ and } b \in B\}.$$

The symbolism $A \times B$ is read "the Cartesian product of A and B" or often simply as "A cross B." The Cartesian product is named in honor of the French philosopher and mathematician René Descartes (1596-1650). In the process of developing analytic geometry, the synthesis of algebra and geometry, Descartes invented the Cartesian product and the Cartesian coordinate system. If $(a, b) \in A \times B$, then it must be true that $a \in A$ and $b \in B$. Whereas, if $(a, b) \notin A \times B$, then either $a \notin A$ or $b \notin B$.

As an example, let $A = \{1, a, \#\}$ and $B = \{\$, @\}$, then

$$A \times B = \{(1, \$), (1, @), (a, \$), (a, @), (\#, \$), (\#, @)\}$$

and

$$B \times A = \{(\$, 1), (\$, a), (\$, \#), (@, 1), (@, a), (@, \#)\}$$

As this example illustrates, in general, $A \times B \neq B \times A$. That is, the Cartesian product is not commutative.

Let A be a finite set with m elements and let B be a finite set with n elements, then the Cartesian products $A \times B$ and $B \times A$ both have mn elements which are ordered pairs. To visualize the Cartesian product $A \times B$ we draw a graph. First, we draw a horizontal line, choose m arbitrary points on the line, and label the points with the elements in set A. Next, we draw a vertical line, choose n arbitrary points on this line, and label these points with the elements in set B. Through each point on the horizontal line chosen to represent an element of A, we draw a vertical line; and through each point on the vertical line chosen to represent an element of B, we draw a horizontal line. The m vertical lines intersect the n horizontal lines in mn points. These points represent the mn ordered pairs of the set $A \times B$. A graph of $A \times B$ for $A = \{1, a, \#\}$ and $B = \{\$, @\}$ is displayed in Figure 3.7.

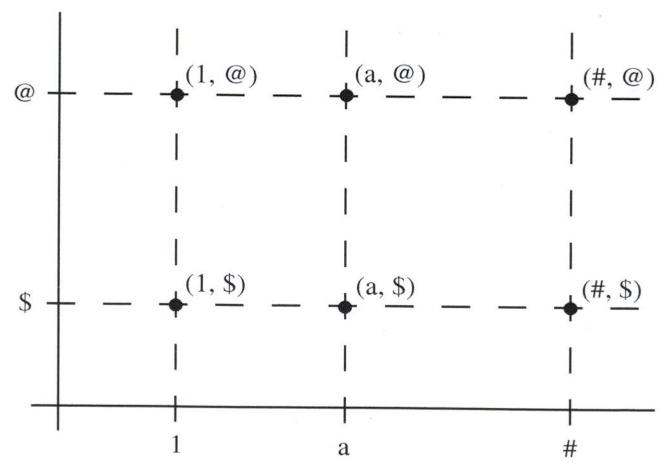

FIGURE 3.7: A Graph of $A \times B$

Now let $A = \{a\}$, $B = \{b\}$, and $C = \{c\}$. Then

$$A \times (B \times C) = \{a\} \times (\{b\} \times \{c\}) = \{a\} \times \{(b, c)\} = \{(a, (b, c))\}$$

and

$$(A \times B) \times C = (\{a\} \times \{b\}) \times \{c\} = \{(a, b)\} \times \{c\} = \{((a, b), c)\}$$

Consequently, in general, $A \times (B \times C) \neq (A \times B) \times C$. That is, the Cartesian product is not associative. The Cartesian product of three nonempty sets A, B, and C is defined as

$$A \times B \times C = \{(a, b, c) \mid a \in A \text{ and } b \in B \text{ and } c \in C\}$$

Thus, the Cartesian product of three nonempty sets A, B, and C is a set of ordered triples.

The following theorem states a relationship between the Cartesian product and the empty set.

Theorem 3.8 $A \times B = \emptyset$ if and only if $A = \emptyset$ or $B = \emptyset$.

Proof: We will prove this theorem by proving its contrapositive: $A \neq \emptyset$ and $B \neq \emptyset$ if and only if $A \times B \neq \emptyset$.

First, we assume $A \neq \emptyset$ and $B \neq \emptyset$. Thus, there exists an $a \in A$ and there exists a $b \in B$. Hence, $(a, b) \in A \times B$ and, consequently, $A \times B \neq \emptyset$.

Next, we assume $A \times B \neq \emptyset$. Thus, there exists an $(a, b) \in A \times B$ which implies there exists an $a \in A$ and there exists a $b \in B$. Therefore, $A \neq \emptyset$ and $B \neq \emptyset$.

The following example illustrates combining the Cartesian product operation with the operations of union and complementation.

EXAMPLE 2 Set Operations Involving the Cartesian Product

Let $A = \{1\}$, $B = \{2, 3\}$, $C = \{1, 2\}$, and $D = \{3\}$. Write the following sets using roster notation.

a. $A \times (B \cup C)$ b. $(A \times B) \cup (A \times C)$ c. $A \times (B - C)$

d. $(A \times B) - (A \times C)$ e. $(A \times B) \cup (C \times D)$ f. $(A \cup C) \times (B \cup D)$

What do you observe about the answers of pairs a-b, c-d, and e-f?

SOLUTION

a. $A \times (B \cup C) = \{1\} \times (\{2, 3\} \cup \{1, 2\}) = \{1\} \times \{2, 3, 1\} = \{(1, 2), (1, 3), (1, 1)\}$

b. $$\begin{aligned}(A \times B) \cup (A \times C) &= (\{1\} \times \{2, 3\}) \cup (\{1\} \times \{1, 2\}) \\ &= \{(1, 2), (1, 3)\} \cup \{(1, 1), (1, 2)\} \\ &= \{(1, 2), (1, 3), (1, 1)\}\end{aligned}$$

From parts a. and b. we see that $A \times (B \cup C) = (A \times B) \cup (A \times C)$ for the given sets A, B, and C.

c. $A \times (B - C) = \{1\} \times \{3\} = \{(1,3)\}$

d. $$\begin{aligned}(A \times B) - (A \times C) &= (\{1\} \times \{2,3\}) - (\{1\} \times \{1,2\}) \\ &= \{(1,2),(1,3)\} - \{(1,1),(1,2)\} = \{(1,3)\}\end{aligned}$$

It follows from parts c. and d. that $A \times (B - C) = (A \times B) - (A \times C)$ for the given sets A, B, and C.

e. $$\begin{aligned}(A \times B) \cup (C \times D) &= (\{1\} \times \{2,3\}) \cup (\{1,2\} \times \{3\}) \\ &= \{(1,2),(1,3)\} \cup \{(1,3),(2,3)\} \\ &= \{(1,2),(1,3),(2,3)\}\end{aligned}$$

f. $$\begin{aligned}(A \cup C) \times (B \cup D) &= (\{1\} \cup \{1,2\}) \times (\{2,3\} \cup \{3\}) \\ &= \{1,2\} \times \{2,3\} = \{(1,2),(1,3),(2,2),(2,3)\}\end{aligned}$$

From parts e. and f. it follows that for the specified sets A, B, C, and D, $(A \times B) \cup (A \times C) \subset (A \cup C) \times (B \cup D)$.

From parts a. and b. of example 2, we conjecture that for any sets A, B, and C it is true that $A \times (B \cup C) = (A \times B) \cup (A \times C)$. Hence, we formally state and prove the following theorem.

Theorem 3.9 For sets A, B, and C, $A \times (B \cup C) = (A \times B) \cup (A \times C)$.

Proof: $(a,d) \in [A \times (B \cup C)]$

$\Leftrightarrow [(a \in A) \wedge (d \in (B \cup C))]$	Def. of Cartesian Product
$\Leftrightarrow [(a \in A) \wedge [(d \in B) \vee (d \in C)]]$	Definition of Union
$\Leftrightarrow [(a \in A) \wedge (d \in B)] \vee [(a \in A) \wedge (d \in C)]$	Distributive Law Conjunction
$\Leftrightarrow [(a,d) \in (A \times B)] \vee [(a,d) \in (A \times C)]$	Def. of Cartesian Product
$\Leftrightarrow (a,d) \in [(A \times B) \cup (A \times C)]$	Definition of Union

Theorem 3.9 states that the Cartesian product distributes over union.

You are undoubtedly familiar with the geometric representation of the set of real numbers by the "real number line." The real number line is usually drawn as a horizontal straight line. A point is selected to represent 0 and a second, distinct point to the right of 0 is selected to represent 1. The choice of the locations of 0 and 1 determine the scale on the number line. It follows from the axioms of Euclidean geometry that to each point on the line there corresponds one and only one real number and, conversely, to each real number there corresponds one and only one point on the line. For $x, y \in \mathbf{R}$, if $x < y$, then the point corresponding to x lies to the left of the point corresponding to y. Positive real numbers correspond to points to the right of 0 and negative real numbers correspond to points to the left of 0. If $a, b, x \in \mathbf{R}$ and $a < b$, then x satisfies the inequality $a < x < b$ if and only if x corresponds to a point "between" the points corresponding to a and b. A graph of the real number line is displayed in Figure 3.8.

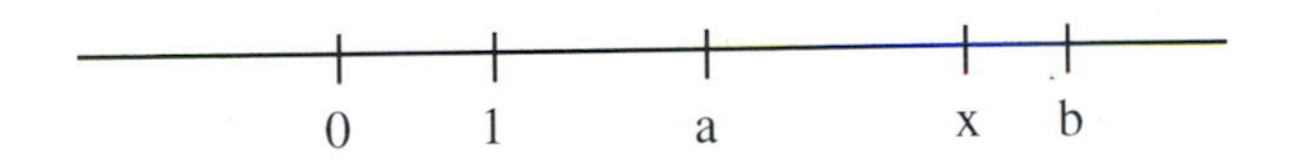

FIGURE 3.8: Set of Real Numbers Represented Geometrically on a Line

We can define subsets of the real numbers which are intervals using inequalities as follows.

DEFINITION Intervals of Real Numbers

Let $a, b \in \mathbf{R}$ and let $a < b$.

The **open interval** (a, b) is the subset of the real numbers

$$(a, b) = \{x \in \mathbf{R} \mid a < x < b\}$$

The **closed interval** $[a, b]$ is the subset of the real numbers

$$[a, b] = \{x \in \mathbf{R} \mid a \leq x \leq b\}$$

The **half-open (half-closed) intervals** are

$$[a, b) = \{x \in \mathbf{R} \mid a \leq x < b\}$$

and

$$(a, b] = \{x \in \mathbf{R} \mid a < x \leq b\}$$

DEFINITION Intervals of Real Numbers

Let $a, b \in \mathbf{R}$ and let $a < b$.

The **infinite intervals** are

$$[a, \infty) = \{x \in \mathbf{R} \mid a \leq x\}$$

$$(a, \infty) = \{x \in \mathbf{R} \mid a < x\}$$

$$(-\infty, b] = \{x \in \mathbf{R} \mid x \leq b\}$$

$$(-\infty, b) = \{x \in \mathbf{R} \mid x < b\}$$

$$(-\infty, \infty) = \mathbf{R}$$

A graph of some intervals is shown in Figure 3.9. The notation ∘ means the real number corresponding to the point is not in the interval; whereas, the notation • means the real number corresponding to the point is in the interval.

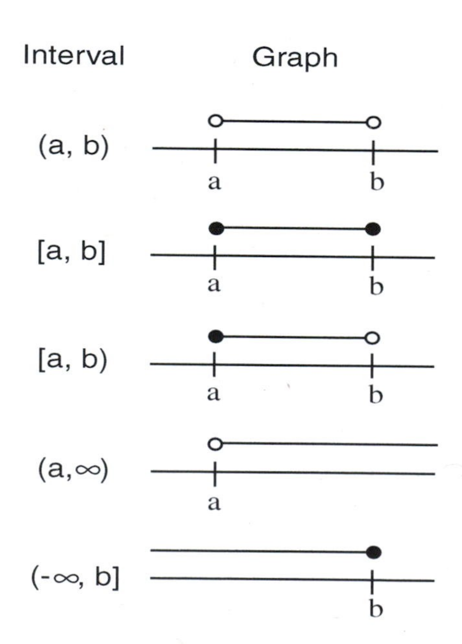

FIGURE 3.9: Graphs of Some Intervals

EXAMPLE 3 Set Operations on Intervals

Let $A = [-5, 2)$ and $B = [1, \infty)$. Graph each of the following sets and write the set using both set notation and interval notation.

a. $A \cup B$ b. $A \cap B$ c. A' d. B'

e. $A - B$ f. $B - A$ g. $A \times B$ h. $B \times A$

SOLUTION

In set notation, $A = [-5, 2) = \{x \in \mathbf{R} \mid -5 \leq x < 2\}$ and $B = [1, \infty) = \{x \in \mathbf{R} \mid 1 \leq x\}$.

a. We graph A and B on the same number line as shown in Figure 3.10. Since $A \cup B$ is the set of all elements in the set A or in the set B, the graph of $A \cup B$ is the set which appears at the bottom of Figure 3.10. By the definition of union, $A \cup B = \{x \in \mathbf{R} \mid (-5 \leq x < 2) \text{ or } (1 \leq x)\}$. From this expression and from the graph of the set $A \cup B$ in Figure 3.10, we see that the set notation for $A \cup B$ is $A \cup B = \{x \in \mathbf{R} \mid -5 \leq x\}$. Hence, in interval notation $A \cup B = [-5, \infty)$.

FIGURE 3.10: A Graph of A, B, and $A \cup B$

b. In Figure 3.11 we graph the sets A and B. Since $A \cap B$ is the set of all elements in the set A and in the set B, $A \cap B$ in Figure 3.11 is the region where the sets A and B "overlap." Hence, the graph of $A \cap B$ is the set which appears at the bottom of Figure 3.11. By the definition of intersection, $A \cap B = \{x \in \mathbf{R} \mid (-5 \leq x < 2) \text{ and } (1 \leq x)\}$. From this definition and Figure 3.11, we find that the set notation for $A \cap B$ is $A \cap B = \{x \in \mathbf{R} \mid 1 \leq x < 2\}$ and that the interval notation is $A \cap B = [1, 2)$.

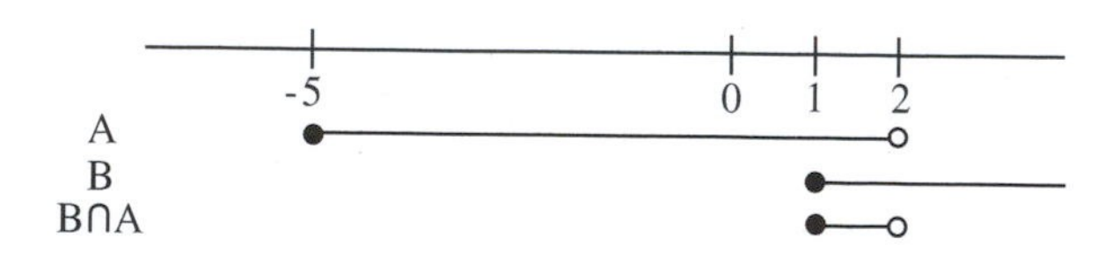

FIGURE 3.11: A Graph of A, B, and $A \cap B$

c. A graph of A and its complement A' is shown in Figure 3.12. It is clear from the graph that $A' = \{x \in \mathbf{R} \mid (x < -5) \text{ or } (2 \leq x)\}$ in set notation and $A' = (-\infty, -5) \cup [2, \infty)$ in the interval notation.

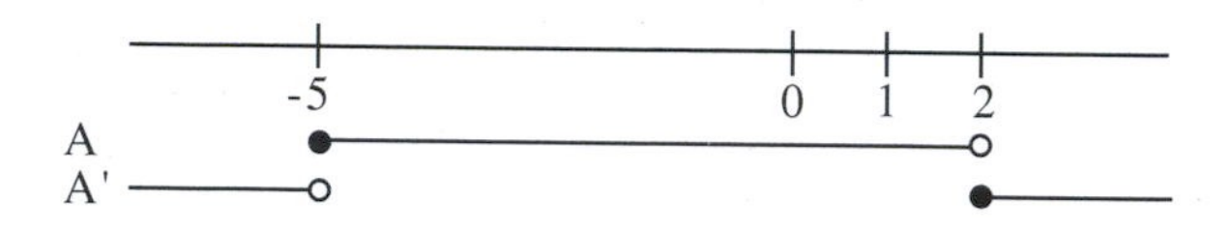

FIGURE 3.12: A Graph of A and A'

d. A graph of B and its complement B' is shown in Figure 3.13. From the graph, it follows that in set notation $B' = \{x \in \mathbf{R} \mid x < 1\}$ and in interval notation $B' = (-\infty, 1)$.

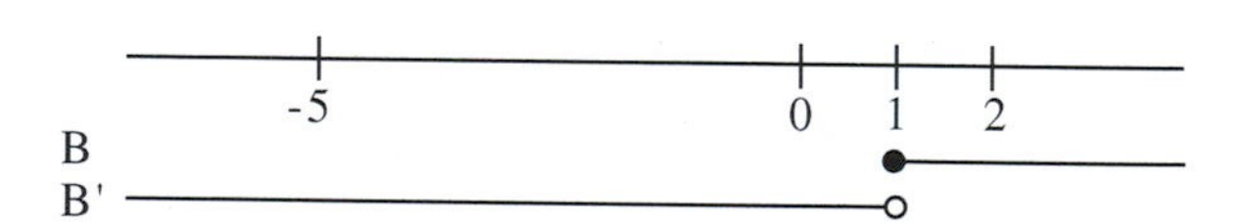

FIGURE 3.13: A Graph of B and B'

e. First, we graph A and B. Then since $A - B = A - (A \cap B)$, we remove from the set A the "overlap" of A and B, the set $A \cap B$, to obtain the graph of $A - B$ shown in Figure 3.14. From the graph, it is obvious that $A - B = \{x \in \mathbf{R} \mid -5 \leq x < 1\} = [-5, 1)$.

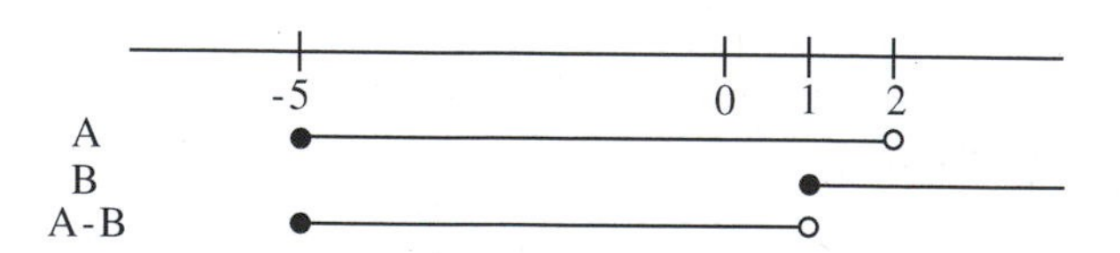

FIGURE 3.14: A Graph of A, B, and $A - B$

f. A graph of $B - A$ is displayed in Figure 3.15. It was obtained by removing from the set B the overlap of sets A and B. From this graph it is clear that $B - A = \{x \in \mathbf{R} \mid 2 \leq x\} = [2, \infty)$.

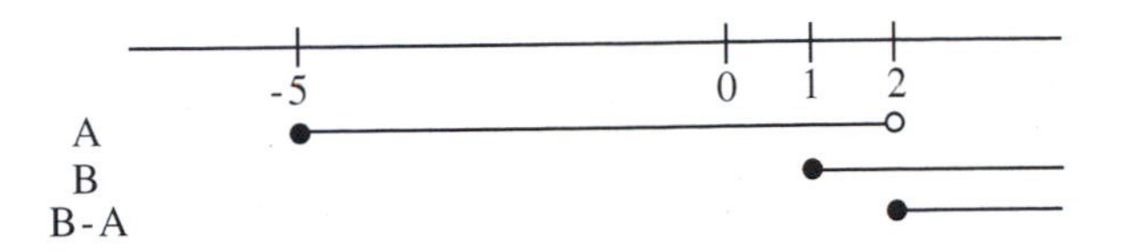

FIGURE 3.15: A Graph of A, B, and $B - A$

g. First, we draw the Cartesian plane $\mathbf{R} \times \mathbf{R}$. Then, we graph the set A along the horizontal axis and the set B along the vertical axis. The graph of $A \times B$ is the "infinite strip" shaded in Figure 3.16. In the graph the right boundary of $A \times B$ (a ray of the line $x = 2$) appears as a dashed line, because it does not belong to the set $A \times B$. The point $(2, 1)$ appears in the graph as $\circ$, because $(2, 1) \notin A \times B$; and the point $(-5, 1)$ appears in the graph as $\bullet$, because $(-5, 1)$ is in the set $A \times B$. In set notation, $A \times B = \{(x, y) \in \mathbf{R} \times \mathbf{R} \mid (-5 \leq x < 2) \text{ and } (1 \leq y)\}$ and in interval notation $A \times B = [-5, 2) \times [1, \infty)$.

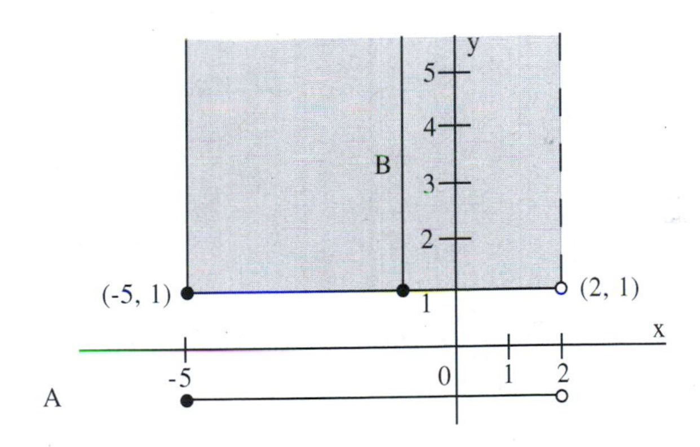

FIGURE 3.16: A Graph of A, B, and $A \times B$

h. The graph of $B \times A$ is the "infinite strip" shaded in Figure 3.17. In set notation $B \times A = \{(x, y) \in \mathbf{R} \times \mathbf{R} \mid (1 \leq x) \text{ and } (-5 \leq y < 2)\}$ and in interval notation $B \times A = [1, \infty) \times [-5, 2)$.

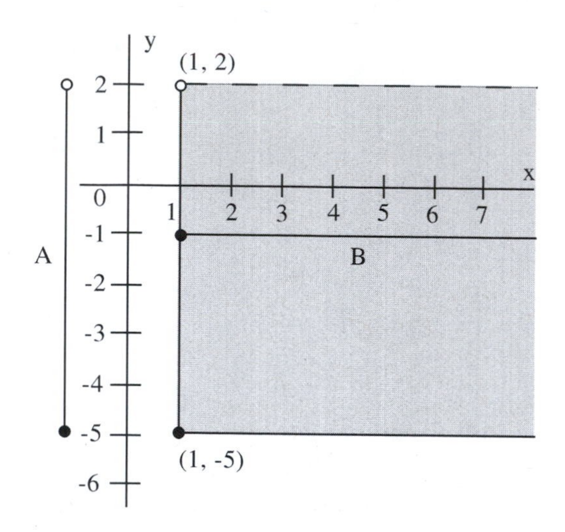

FIGURE 3.17: A Graph of A, B, and $B \times A$

EXERCISES 3.3

In exercises 1-7 let A, B, and C be subsets of the universe U. Prove each of the given statements.

1. $(A \cap B) - C = A \cap (B - C)$

2. $A = (A - B) \cup (A \cap B)$

3. $A - B$ and $A \cap B$ are disjoint sets.

4. If $A \subseteq B$, then $A - B = \emptyset$.

5. If $A \cap B = \emptyset$, then $A - B = A$ and $B - A = B$.

6. $A - B = B - A$ if and only if $A = B$.

7. $(A - B) - C = (A - C) - (B - C)$

8. A typical Venn diagram for three sets A, B, and C is shown below. Observe that the three sets divide the universal set U into eight regions.

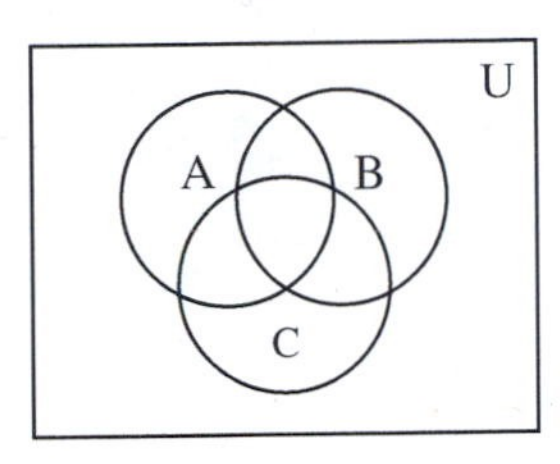

   a. Copy the Venn diagram twice. On one diagram, shade the region which corresponds to the set $A - (B \cap C)$. On the other diagram, shade the region which corresponds to the set $(A - B) \cup (A - C)$.

   b. What do you conjecture about the relationship between the sets $A - (B \cap C)$ and $(A - B) \cup (A - C)$?

   c. Once again copy the Venn diagram twice. On one diagram, shade the region which corresponds to the set $A - (B \cup C)$, and on the other diagram, shade the region which corresponds to the set $(A - B) \cap (A - C)$.

   d. What do you conjecture about the relationship between the sets $A - (B \cup C)$ and $(A - B) \cap (A - C)$?

   e. Prove that $A - (B \cap C) = (A - B) \cup (A - C)$.

   f. Prove that $A - (B \cup C) = (A - B) \cap (A - C)$.

9. The symmetric difference of two sets A and B is the set

$$A \triangle B = (A - B) \cup (B - A)$$

A Venn diagram representing the symmetric difference of A and B is shown below.

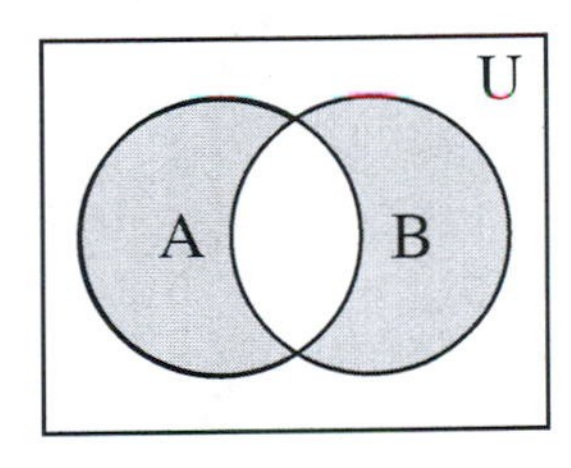

Prove the following properties of the symmetric difference.

   a. $A \triangle B = B \triangle A$
   (The operator $\triangle$ is commutative.)

   b. $(A \triangle B) \triangle C = A \triangle (B \triangle C)$
   (The operator $\triangle$ is associative.)

c. For all sets A, $A \triangle \emptyset = A$
(The empty set is the identity operator for $\triangle$.)

d. $A \cap (B \triangle C) = (A \cap B) \triangle (A \cap C)$
(Intersection distributes over the symmetric difference.)

e. If the sets A and B are specified, then the equation $A \triangle X = B$ has a solution X.
(**HINT:** What does $A \triangle (A \triangle B) = ?$)

10. For the given sets A and B list the ordered pairs of $A \times B$ and $B \times A$.

a. $A = \{-1, 0, 1\}$, $B = \{-i, i\}$

b. $A = B = \{2, 4, 6\}$

c. $A = \{1, 3\}$, $B = \{(2, 4), (6, 8)\}$

11. Graph $A \times B$ for parts a., b., and c. of exercise 10.

12. Given that the Cartesian product $A \times A$ has nine elements, $(a, b) \in A \times A$, and $(b, c) \in A \times A$, list the remaining seven elements of $A \times A$.

13. Let A and B be nonempty sets. Prove $A \times B = B \times A$ if and only if $A = B$.

14. Prove if A, B, and C are sets and $A \subseteq B$, then $A \times C \subseteq B \times C$.

15. Prove or disprove the following statements.

a. If $A \times C = B \times C$, then $A = B$.

b. If $A \times C = B \times C$ and $C \neq \emptyset$, then $A = B$.

In exercises 16-20, let A, B, C, and D be sets and prove the given statements.

16. $A \times (B - C) = (A \times B) - (A \times C)$
(The Cartesian product distributes over complementation.)

17. $A \times (B \cap C) = (A \times B) \cap (A \times C)$
(The Cartesian product distributes over intersection.)

18. $(A \times B) \cup (C \times D) \subseteq (A \cup C) \times (B \cup D)$

19. $(A \times B) \cap (C \times D) = (A \cap C) \times (B \cap D)$

20. $(A \times B) \cap (B \times A) = (A \cap B) \times (A \cap B)$

21. Use Kuratowski's definition $(a, b) = \{\{a\}, \{a, b\}\}$ to prove the following theorem: $(a, b) = (c, d)$ if and only if $a = c$ and $b = d$.

22. Let $A = \{x \mid 1 \leq x < 5\}$, $B = \{x \mid 3 < x \leq 8\}$, and $C = \{x \mid 5 < x < 8\}$. Write the following sets using both set notation and interval notation.

a. $A \cup B$ b. $A \cap B$ c. $A \cup C$ d. $A \cap C$ e. $A - B$

f. $B - A$ g. $A - C$ h. $A \times B$ i. $B \times A$ j. $C \times C$

23. Let $A = (-1, 2)$, $B = (-4, 1]$, and $C = [1, \infty)$. Write the following sets using both interval notation and set notation.

a. $A \cup B$ b. $A \cap B$ c. $B \cup C$ d. $B \cap C$

e. A' f. C' g. $A - B$ h. $B - C$

24. Graph the following Cartesian products.

a. $(1, 3] \times \mathbf{R}$ b. $\mathbf{R} \times \{-2\}$ c. $(-\infty, 2) \times [3, \infty)$ d. $\{2\} \times \{3\}$

e. $((1, 2) \cup (3, 4)) \times ([2, 3) \cup (4, 5])$

f. $((1, 2) \times [2, 3)) \cup ((3, 4) \times (4, 5])$

3.4 Generalized Set Union and Intersection

The set operations of union and intersection are binary operations. That is, at present $\cup$ and $\cap$ are defined only for two sets. However, by using parentheses, we have been able to calculate more complex expressions such as $(A \cup B) \cup C$, $(A \cap B) \cup (C \cap D)$, and so forth. Also, until now, we have denoted individual sets by different capital letters. In this section we will extend the definitions of union and intersection to arbitrary collections of sets.

Let A_1, A_2, and A_3 (instead of the usual A, B, and C) be subsets of the universe U. The associative law for set union is $(A_1 \cup A_2) \cup A_3 = A_1 \cup (A_2 \cup A_3)$. Essentially, the associative law says we may omit the parentheses and simply write $A_1 \cup A_2 \cup A_3$, since it is immaterial if we calculate $A_1 \cup A_2$ first and then calculate the union of the sets $A_1 \cup A_2$ and A_3 or if we calculate $A_2 \cup A_3$ first and then calculate the union of the sets A_1 and $A_2 \cup A_3$. The result of either calculation is the same. Likewise, the set $((A_1 \cup A_2) \cup A_3) \cup A_4$ may be written as $A_1 \cup A_2 \cup A_3 \cup A_4$ and, in general, for $n \in \mathbf{N}$ we may write $A_1 \cup A_2 \cup \cdots \cup A_n$ without parentheses and without ambiguity. Note that for $n \in \mathbf{N}$, $x \in A_1 \cup A_2 \cup \cdots \cup A_n$ if and only if $x \in A_i$ for some $i = 1, 2, \ldots n$. Thus, we define the union of a finite number of sets $A_1, A_2, \ldots, A_n$ as follows.

DEFINITION The Finite Union of Sets $A_1, A_2, \ldots, A_n$

Let $n \in \mathbf{N}$ and let $A_1, A_2, \ldots, A_n$ be subsets of the universe U, then

$$A_1 \cup A_2 \cup \cdots \cup A_n = \{x \in U \mid x \in A_i \text{ for some } i = 1, 2, \ldots, n\}$$

$$= \{x \in U \mid (\exists i \in \{1, 2, \ldots, n\})(x \in A_i).\}$$

It is convenient to condense the notation for the finite union by writing

$$\bigcup_{i=1}^{n} A_i = A_1 \cup A_2 \cup \cdots \cup A_n.$$

We extend the definition of the union of a finite number of sets to the infinite number of sets $A_1, A_2, \ldots$ in the following manner.

DEFINITION The Infinite Union of Sets $A_1, A_2, \ldots$

For each $n \in \mathbf{N}$, let $A_1, A_2, \ldots$ be subsets of the universe U, then

$$\bigcup_{i=1}^{\infty} A_i = A_1 \cup A_2 \cup \cdots = \{x \in U \mid (\exists i \in \mathbf{N})(x \in A_i)\}.$$

EXAMPLE 1 Computations of $\bigcup_{i=j}^{k} A_i$

For $i \in \mathbf{N}$, let $A_i = [i, i+1)$. Determine the following sets

a. $\bigcup_{i=1}^{5} A_i$ b. $\bigcup_{i=1}^{50} A_i$ c. $\bigcup_{i=10}^{37} A_i$ d. $\bigcup_{i=1}^{\infty} A_i$

SOLUTION

a. $\bigcup_{i=1}^{5} A_i = A_1 \cup A_2 \cup A_3 \cup A_4 \cup A_5 = [1, 2) \cup [2, 3) \cup [3, 4) \cup [4, 5) \cup [5, 6) = [1, 6)$

b. $\bigcup_{i=1}^{50} A_i = [1, 2) \cup [2, 3) \cdots \cup [50, 51) = [1, 51)$

c. $\bigcup_{i=10}^{37} A_i = [10, 11) \cup [11, 12) \cdots \cup [37, 38) = [10, 38)$

d. $\bigcup_{i=1}^{\infty} A_i = [1, 2) \cup [2, 3) \cdots = [1, \infty)$

In an analogous manner, we define the intersection of a finite number of sets and the intersection of an infinite number of sets as below.

DEFINITION The Finite Intersection of Sets $A_1, A_2, \ldots, A_n$

Let $n \in \mathbf{N}$ and let $A_1, A_2, \ldots, A_n$ be subsets of the universe U, then

$$\begin{aligned}\bigcap_{i=1}^{n} A_i &= A_1 \cap A_2 \cap \cdots \cap A_n \\ &= \{x \in U \mid x \in A_i \text{ for all } i = 1, 2, \ldots, n\} \\ &= \{x \in U \mid (\forall i \in \{1, 2, \ldots, n\})(x \in A_i)\}.\end{aligned}$$

DEFINITION The Infinite Intersection of Sets $A_1, A_2, \ldots$

For each $n \in \mathbf{N}$, let $A_1, A_2, \ldots$ be subsets of the universe U, then

$$\bigcap_{i=1}^{\infty} A_i = A_1 \cap A_2 \cap \cdots = \{x \in U \mid (\forall i \in \mathbf{N})(x \in A_i)\}.$$

EXAMPLE 2 Computations of $\bigcap_{i=j}^{k} A_i$

For $i \in \mathbf{N}$, let $A_i = [0, \frac{1}{i})$. Determine the following sets

a. $\bigcap_{i=1}^{5} A_i$ b. $\bigcap_{i=1}^{50} A_i$ c. $\bigcap_{i=10}^{37} A_i$ d. $\bigcap_{i=1}^{\infty} A_i$

SOLUTION

a. $\bigcap_{i=1}^{5} A_i = [0,1) \cap [0, \frac{1}{2}) \cap [0, \frac{1}{3}) \cap [0, \frac{1}{4}) \cap [0, \frac{1}{5}) = [0, \frac{1}{5})$

b. $\bigcap_{i=1}^{50} A_i = [0,1) \cap [0, \frac{1}{2}) \cap \cdots \cap [0, \frac{1}{50}) = [0, \frac{1}{50})$

c. $\bigcap_{i=10}^{37} A_i = [0, \frac{1}{10}) \cap [0, \frac{1}{11}) \cap \cdots \cap [0, \frac{1}{37}) = [0, \frac{1}{37})$

d. $\bigcap_{i=1}^{\infty} A_i = [0,1) \cap [0, \frac{1}{2}) \cap \cdots = \{0\}$, since for all $x \in (0,1)$ there exists an $i \in \mathbf{N}$ such that $\frac{1}{i} < x$.

A set of sets is often called a **family of sets**, or simply a **family**. We will denote a family of sets by a script capital letter. For instance, the family $\mathcal{A} = \{\{a\}, \{a,b\}, \{a,b,c\}, \{a,b,c,d\}\}$ is a family with four sets and the family $\mathcal{B} = \{(-a,a) \mid a \in [0,\infty)\}$ is an infinite family of open intervals. The union and intersection of a family of sets is defined as follows.

DEFINITION The Union and Intersection of a Family of Sets

Let $\mathcal{F}$ be a family of sets which are all subsets of a universe U. The **union over $\mathcal{F}$** is

$$\bigcup_{A \in \mathcal{F}} A = \{x \in U \mid (\exists A \in \mathcal{F})(x \in A)\}.$$

The **intersection over $\mathcal{F}$** is

$$\bigcap_{A \in \mathcal{F}} A = \{x \in U \mid (\forall A \in \mathcal{F})(x \in A)\}.$$

Thus, for $\mathcal{A} = \{\{a\}, \{a,b\}, \{a,b,c\}, \{a,b,c,d\}\}$

$$\bigcup_{A \in \mathcal{A}} A = \{a,b,c,d\} \quad \text{and} \quad \bigcap_{A \in \mathcal{A}} A = \{a\}$$

and for $\mathcal{B} = \{(-a, a) \mid a \in [0, \infty)\}$

$$\bigcup_{B \in \mathcal{B}} B = \mathbf{R} \quad \text{and} \quad \bigcap_{B \in \mathcal{B}} B = \{0\}$$

Theorem 3.10 Let $\mathcal{F}$ be a family of sets and let B be any member of the family $\mathcal{F}$. Then

$$\bigcap_{A \in \mathcal{F}} A \subseteq B \quad \text{and} \quad B \subseteq \bigcup_{A \in \mathcal{F}} A$$

Proof: Let $\mathcal{F}$ be a family of sets and let $B \in \mathcal{F}$. Suppose $x \in \bigcap_{A \in \mathcal{F}} A$. Then by the definition of intersection, $x \in A$ for every $A \in \mathcal{F}$. Since $B \in \mathcal{F}$, $x \in B$. Therefore, $\bigcap_{A \in \mathcal{F}} A \subseteq B$.

Let $\mathcal{F}$ be a family of sets, let $B \in \mathcal{F}$, and let $x \in B$. By the definition of union, $x \in B$ implies $x \in \bigcup_{A \in \mathcal{F}} A$, since B is a set in the family $\mathcal{F}$. Hence, every element of B is an element of the union $\bigcup_{A \in \mathcal{F}} A$. Consequently, $B \subseteq \bigcup_{A \in \mathcal{F}} A$.

It is tempting to conjecture that if $\mathcal{F}$ is a family of sets, then

$$\bigcap_{A \in \mathcal{F}} A \subseteq \bigcup_{A \in \mathcal{F}} A.$$

The next example shows this statement is false when the family of sets $\mathcal{F}$ is the empty set. However, the statement $\bigcap_{A \in \mathcal{F}} A \subseteq \bigcup_{A \in \mathcal{F}} A$ is true whenever the family of sets $\mathcal{F}$ is not the empty set.

EXAMPLE 3 Interesting Results for the Empty Family

Let $\mathcal{F}$ be the empty family of subsets of the natural numbers, $\mathbf{N}$. Show that

a. $\bigcap_{A \in \mathcal{F}} A = \mathbf{N}$ *b.* $\bigcup_{A \in \mathcal{F}} A = \emptyset$, and *c.* $\bigcap_{A \in \mathcal{F}} A \not\subseteq \bigcup_{A \in \mathcal{F}} A$

SOLUTION

a. The statement $(\forall A \in \mathcal{F})(x \in A)$ is equivalent to $(\forall A)[(A \in \mathcal{F}) \Rightarrow (x \in A)]$. Since $\mathcal{F}$ is assumed to be the empty set, the hypothesis of the last statement, $A \in \mathcal{F}$, is false. Consequently, the implication $[(A \in \mathcal{F}) \Rightarrow (x \in A)]$ is true for all $x \in \mathbf{N}$. Hence, $\bigcap_{A \in \mathcal{F}} A = \mathbf{N}$.

b. The statement $(\exists A \in \mathcal{F})(x \in A)$ is logically equivalent to the statement $(\exists A)[(A \in \mathcal{F}) \wedge (x \in A)]$. The negation of the last statement is

$$(\forall A)[\neg(A \in \mathcal{F}) \vee (\neg(x \in A))] \equiv (\forall A)[(A \in \mathcal{F}) \Rightarrow (x \notin A)].$$

Since $\mathcal{F} = \emptyset$, $A \in \mathcal{F}$ is false; and, therefore, the implication $(A \in \mathcal{F}) \Rightarrow (x \notin A)$ is true. Thus, it follows from the definition of union for a family of sets that for $\mathcal{F} = \emptyset$, $\bigcup_{A \in \mathcal{F}} A = \emptyset$.

c. Let $\mathcal{F}$ be the empty family of subsets of $\mathbf{N}$. Since by part a., $\bigcap_{A \in \mathcal{F}} A = \mathbf{N}$; since by part b., $\bigcup_{A \in \mathcal{F}} A = \emptyset$; and since $\emptyset \subset \mathbf{N}$, $\bigcup_{A \in \mathcal{F}} A \subset \bigcap_{A \in \mathcal{F}} A$. That is, for $\mathcal{F} = \emptyset$, $\bigcap_{A \in \mathcal{F}} A \not\subseteq \bigcup_{A \in \mathcal{F}} A$.

It is always possible to identify each set in a nonempty family of sets with an identification tag called an index. The following definition and examples should help clarify this concept.

DEFINITION An Indexed Family of Sets

Let $\mathcal{F}$ be a nonempty family of sets and let I be a nonempty set with the property that for each $i \in I$ there corresponds a set $A_i \in \mathcal{F}$. Then the family of sets $\mathcal{F} = \{A_i \mid i \in I\}$ is called an **indexed family of sets**. The set I is called the **indexing set** and each $i \in I$ is called an **index**.

In example 1, $A_i = [i, i+1)$ are the members of the family of sets $\mathcal{F}$, the natural numbers $\mathbf{N}$ is the indexing set I, and each natural number i is an index. The family $\mathcal{A} = \{\{a\}, \{a,b\}, \{a,b,c\}, \{a,b,c,d\}\}$ contains four sets and may be indexed by the indexing set $I = \{1,2,3,4\}$ where 1 is associated with the set $\{a\}$, 2 is associated with the set $\{a,b\}$, 3 is associated with the set $\{a,b,c\}$, and 4 is associated with the set $\{a,b,c,d\}$. Of course, associating 1 with $\{a,b\}$, 2 with $\{a,b,c,d\}$, 3 with $\{a\}$, and 4 with $\{a,b,c\}$ is another indexing of the family $\mathcal{A}$ by the indexing set I. In addition, the family $\mathcal{A}$ may be indexed by the set $J = \{a,b,c,d\}$ where a is associated with the set $\{a\}$, b is associated with the set $\{a,b\}$, c is associated with the set $\{a,b,c\}$, and d is associated with the set $\{a,b,c,d\}$. These observations illustrate that the indexing set for a family of sets is not unique and neither is the indexing itself. The family $\mathcal{B} = \{(-a,a) \mid a \in [0,\infty)\}$ may be indexed by the set $I = [0,\infty)$ by associating with each $a \in [0,\infty)$ the interval $(-a,a)$ in the family $\mathcal{B}$. Notice that $\mathcal{B}$ can be indexed by the set $[0,\infty)$ by associating with each $a \in [0,\infty)$ the interval $(-\frac{a}{2}, \frac{a}{2})$. Using this technique, we can clearly index $\mathcal{B}$ in an infinite number of ways.

Every nonempty family of sets $\mathcal{F}$ may be indexed. Simply choose the indexing set to be $\mathcal{F}$ itself and associate with each set A in the indexing set the set A in the family of sets. When $\mathcal{F}$ is a nonempty family of sets indexed by the set I, the following alternate definitions may be used for $\bigcup_{A \in \mathcal{F}} A$ and

$\bigcap_{A\in\mathcal{F}} A$ respectively

$$\bigcup_{i\in I} A_i = \{x \in U \mid (\exists i \in I)(x \in A_i)\}$$

and

$$\bigcap_{i\in I} A_i = \{x \in U \mid (\forall i \in I)(x \in A_i)\}$$

Many theorems regarding set operations for finitely many sets can be generalized to theorems regarding set operations for arbitrary families. For example, the distributive laws $A \cup (B \cap C) = (A \cup B) \cap (A \cup C)$ and $A \cap (B \cup C) = (A \cap B) \cup (A \cap C)$ and the De Morgan laws $(A \cup B)' = A' \cap B'$ and $(A \cap B)' = A' \cup B'$ may be generalized as stated in theorems 3.11 and 3.12.

Theorem 3.11 (Generalized Distributive Laws) Let U be the universe, let A be any subset of the universe, and let $\mathcal{F} = \{B_i \mid i \in I\}$ be an indexed family of subsets of U. Then

a. $A \cup (\bigcap_{i\in I} B_i) = \bigcap_{i\in I} (A \cup B_i)$ (union distributes over intersection)

b. $A \cap (\bigcup_{i\in I} B_i) = \bigcup_{i\in I} (A \cap B_i)$ (intersection distributes over union)

Proof: a. There are two cases to consider: $\mathcal{F} = \emptyset$ and $\mathcal{F} \neq \emptyset$.

Case 1. If $\mathcal{F} = \emptyset$, then $\bigcap_{i\in I} B_i = U$ and $\bigcap_{i\in I}(A \cup B_i) = U$. Hence,

$$A \cup (\bigcap_{i\in I} B_i) = A \cup U = U = \bigcap_{i\in I} (A \cup B_i).$$

Case 2. Suppose $\mathcal{F} \neq \emptyset$ and let $x \in U$.

$x \in A \cup (\bigcap_{i\in I} B_i)$

$\Leftrightarrow (x \in A) \vee (x \in \bigcap_{i\in I} B_i)$	Definition of union
$\Leftrightarrow (x \in A) \vee [(\forall i \in I)(x \in B_i)]$	Definition of $\bigcap_{i\in I} B_i$
$\Leftrightarrow (\forall i \in I)[(x \in A) \vee (x \in B_i)]$	A theorem for quantifiers
$\Leftrightarrow (\forall i \in I)(x \in A \cup B_i)$	Definition of union
$\Leftrightarrow x \in \bigcap_{i\in I}(A \cup B_i)$	Definition of $\bigcap_{i\in I}(A \cup B_i)$

b. The proof of part b. is an exercise.

Theorem 3.12 (Generalized De Morgan Laws) Let U be the universe and let $\mathcal{F} = \{A_i \mid i \in I\}$ be an indexed family of subsets of U. Then

$$a.\ (\bigcup_{i \in I} A_i)' = \bigcap_{i \in I} A_i' \qquad \text{and} \qquad b.\ (\bigcap_{i \in I} A_i)' = \bigcup_{i \in I} A_i'$$

Proof: a. The proof of part a. is an exercise.

b. There are two cases to consider: $\mathcal{F} = \emptyset$ and $\mathcal{F} \neq \emptyset$.

Case 1. If $\mathcal{F} = \emptyset$, then $\bigcap_{i \in I} A_i' = U$ and $\bigcup_{i \in I} A_i' = \emptyset$. Hence,

$$(\bigcup_{i \in I} A_i') = \emptyset' = U = \bigcap_{i \in I} A_i'.$$

Case 2. Suppose $\mathcal{F} \neq \emptyset$ and let $x \in U$.

$x \in (\bigcap_{i \in I} A_i)' \Leftrightarrow \neg(x \in \bigcap_{i \in I} A_i)$	Definition of complement
$\Leftrightarrow \neg[(\forall i \in I)(x \in A_i)]$	Definition of $\bigcap_{i \in I} A_i$
$\Leftrightarrow (\exists i \in I)(x \notin A_i)$	A negation theorem for quantifiers
$\Leftrightarrow (\exists i \in I)(x \in A_i')$	Definition of complement
$\Leftrightarrow x \in \bigcup_{i \in I} A_i'$	Definition of $\bigcup_{i \in I} A_i'$

EXERCISES 3.4

1. Let $\mathcal{F}$ be the family of sets $\mathcal{F} = \{\{3, 4, 6, 8\}, \{4, 5, 8, 9\}, \{7, 8, 11, 12\}\}$ and let $U = \{1, 2, \ldots, 12\}$. Determine

   a. $\bigcup_{A \in \mathcal{F}} A$ b. $\bigcap_{A \in \mathcal{F}} A$.

2. Let the universe be **N** and let $\mathcal{F}$ be the family of sets of natural numbers which contain 5. Find

   a. $\bigcup_{A \in \mathcal{F}} A$ b. $\bigcap_{A \in \mathcal{F}} A$.

3. Let $\mathcal{F}$ be the family $\mathcal{F} = \{\{2, 3, 4\}, \{3, 4, 5, 6\}, \{3, 4, 5\}\}$ and let $U = \{1, 2, 3, 4, 5, 6, 7\}$. Calculate the following sets.

   a. $\bigcup_{A \in \mathcal{F}} A$ b. $\bigcap_{A \in \mathcal{F}} A$ c. $(\bigcup_{A \in \mathcal{F}} A)'$

   d. $\bigcap_{A \in \mathcal{F}} A'$ e. $(\bigcap_{A \in \mathcal{F}} A)'$ f. $\bigcup_{A \in \mathcal{F}} A'$

   g. $\{3, 5\} \cap (\bigcup_{A \in \mathcal{F}} A)$ h. $\bigcup_{A \in \mathcal{F}}(\{3, 5\} \cap A)$

   i. $\{2, 4, 6\} \cup (\bigcap_{A \in \mathcal{F}} A)$ j. $\bigcap_{A \in \mathcal{F}}(\{2, 4, 6\} \cup A)$

4. Let $U = \{a, b, \{a\}, \{b\}, \{a, b\}\}$ and let $A_1 = \{a\}$, $A_2 = \{a, b\}$, $A_3 = \{a, b, \{a, b\}\}$, $A_4 = \{a, b, \{a\}\}$, and $A_5 = \{\{a\}, \{b\}\}$. Find $\bigcup_{i \in I} A_i$ and $\bigcap_{i \in I} A_i$ for the following index sets

   a. $I = \{3, 4, 5\}$ b. $I = \{1, 2, 4, 5\}$ c. $I = \{1, 2, 3, 4, 5\}$.

5. For $n \in \mathbf{N}$, let $A_n = \{1, 2, \ldots, n\}$. Determine

   a. $\bigcup_{i=1}^{5} A_i$ b. $\bigcup_{i=6}^{10} A_i$ c. $\bigcap_{i=1}^{5} A_i$

   d. $\bigcap_{i=6}^{10} A_i$ e. $\bigcup_{i=1}^{5} (\mathbf{N} - A_i)$ f. $\bigcap_{i=1}^{5} (\mathbf{N} - A_i)$

6. For $n \in \mathbf{N}$, let $A_n = \{nm \mid m \in \mathbf{Z}\}$. Thus, $A_3 = \{\ldots, -6, -3, 0, 3, 6, \ldots\}$. Find

   a. $A_3 \cap A_4$ b. $\displaystyle\bigcap_{n \in \mathbf{N}} A_n$ c. $A_3 \cup A_4$ d. $\displaystyle\bigcup_{n \in \mathbf{N}} A_n$

7. Let the universe be the set of real numbers. For each of the following sets A_n find (i) $\bigcup_{n \in \mathbf{N}} A_n$ and (ii) $\bigcap_{n \in \mathbf{N}} A_n$

   a. $A_n = [n, n+1]$ b. $A_n = (n, n+1)$ c. $A_n = (-n, n)$

   d. $A_n = (-\frac{1}{n}, 1 + \frac{1}{n})$ e. $A_n = (-\frac{1}{n}, 1)$ f. $A_n = [0, 1 + \frac{1}{n})$

8. Let the index set $I = [0, \infty)$. For each $r \in I$ define

$$S_r = \{(x, y) \in \mathbf{R} \times \mathbf{R} \mid |x| + |y| = r\}$$

   and

$$T_r = \{(x, y) \in \mathbf{R} \times \mathbf{R} \mid |x| + |y| > r\}.$$

   Determine

   a. $\displaystyle\bigcup_{r \in I} S_r$ b. $\displaystyle\bigcap_{r \in I} S_r$ c. $\displaystyle\bigcup_{r \in I} T_r$ d. $\displaystyle\bigcap_{r \in I} T_r$

9. Prove theorem 3.11.b. $A \cap (\bigcup_{i \in I} B_i) = \bigcup_{i \in I} (A \cap B_i)$

10. Prove theorem 3.12.a. $(\bigcup_{i \in I} A_i)' = \bigcap_{i \in I} A_i'$

11. a. Expand $(A_1 \cup A_2) \cap (B_1 \cup B_2 \cup B_3)$ into a union of intersections.

    b. Expand $(\bigcup_{i=1}^{m} A_i) \cap (\bigcup_{j=1}^{n} B_j)$ into a union of intersections.

    c. Expand $(\bigcup_{i \in I} A_i) \cap (\bigcup_{j \in J} B_j)$ where I and J are nonempty index sets into a union of intersections.

12. a. Expand $(A_1 \cap A_2) \cup (B_1 \cap B_2 \cap B_3)$ into an intersection of unions.

    b. Expand $(\bigcap_{i=1}^{m} A_i) \cup (\bigcap_{j=1}^{n} B_j)$ into an intersection of unions.

    c. Expand $(\bigcap_{i \in I} A_i) \cup (\bigcap_{j \in J} B_j)$ where I and J are nonempty index sets into an intersection of unions.

13. Let U be the universe, let A_i where $i \in I$ be a nonempty indexed family of subsets of U, and let B be any subset of U. Prove

    a. $B - \bigcup_{i \in I} A_i = \bigcap_{i \in I}(B - A_i)$ and b. $B - \bigcap_{i \in I} A_i = \bigcup_{i \in I}(B - A_i)$

3.5 Chapter Review

Definitions

A **set** consists of objects called **elements** or **members**.

Sets are described in **roster notation**, in which the elements of the set are enclosed in curly braces, $\{\ \}$, and separated by commas, or in **set-builder notation**, in which the set is specified in the form $\{x \mid P(x)\}$.

There is one and only one set which has no elements. It is called the **empty set** or **null set**. The **empty set** is the set $\emptyset = \{x \mid x \neq x\}$.

The set A is a **subset** of the set B, written $A \subseteq B$, if and only if every element of A is an element of B. That is, $A \subseteq B \Leftrightarrow (\forall x)[(x \in A) \Rightarrow (x \in B)]$.

Two sets A and B are **equal** if and only if $[(A \subseteq B) \wedge (B \subseteq A)]$.

The set A is a **proper subset** of the set B, written $A \subset B$, if and only if A is a subset of B and $A \neq B$. That is, $A \subset B \Leftrightarrow [(A \subseteq B) \wedge (A \neq B)]$.

The set of all subsets of a finite set A is denoted by $\mathcal{P}(A)$ and is called the **power set** of A.

Let A and B be two subsets of the universe U.

The set A **union** B is the subset of U which contains all elements that belong either to A or to B or to both A and B. Thus, $A \cup B = \{x \mid x \in A \text{ or } x \in B\} = \{x \mid (x \in A) \vee (x \in B)\}$.

The set A **intersect** B is the subset of U which contains all elements that belong both A and B. That is, $A \cap B = \{x \mid x \in A \text{ and } x \in B\} = \{x \mid (x \in A) \wedge (x \in B)\}$.

When $A \cap B = \emptyset$, the sets A and B are said to be **disjoint**.

The **complement** of the set A, written A', is the set of all elements in the universe, U, which are not in the set A. Hence, $A' = \{x \mid x \in U \text{ and } x \notin A\} = \{x \mid (x \in U) \wedge (x \notin A)\}$.

An equation, expression, or statement in set theory which is obtained by interchanging $\cup$ and $\cap$ and interchanging the sets $\emptyset$ and U is called the **dual** of the equation, expression, or statement accordingly.

Let A and B be two subsets of the universe U. The **set difference of A and B**, written $A - B$, is the set of all elements in the set A that are not in the set B. That is, $A - B = \{x \mid x \in A \text{ and } x \notin B\} = \{x \mid (x \in A) \wedge (x \notin B)\}$.

An ordered pair with a as **first element** and b as **second element** is denoted by (a, b).

The **Cartesian product of two sets A and B**, written as $A \times B$, is the set of all ordered pairs (a, b) such that $a \in A$ and $b \in B$. That is, $A \times B = \{(a, b) \mid a \in A \text{ and } b \in B\}$.

Let $a, b \in \mathbf{R}$ and let $a < b$.
The **open interval** (a, b) is the subset of the real numbers

$$(a, b) = \{x \in \mathbf{R} \mid a < x < b\}.$$

The **closed interval** $[a, b]$ is the subset of the real numbers

$$[a, b] = \{x \in \mathbf{R} \mid a \leq x \leq b\}.$$

The **half-open (half-closed) intervals** are

$$[a, b) = \{x \in \mathbf{R} \mid a \leq x < b\} \text{ and } (a, b] = \{x \in \mathbf{R} \mid a < x \leq b\}.$$

The **infinite intervals** are

$$[a, \infty) = \{x \in \mathbf{R} \mid a \leq x\}$$

$$(a, \infty) = \{x \in \mathbf{R} \mid a < x\}$$

$$(-\infty, b] = \{x \in \mathbf{R} \mid x \leq b\}$$

$$(-\infty, b) = \{x \in \mathbf{R} \mid x < b\}$$

$$(-\infty, \infty) = \mathbf{R}.$$

Let $n \in \mathbf{N}$ and let $A_1, A_2, \ldots, A_n$ be subsets of the universe U, then the **finite union of sets $A_1, A_2, \ldots, A_n$** is

$$\begin{aligned} A_1 \cup A_2 \cup \cdots \cup A_n &= \{x \in U \mid x \in A_i \text{ for some } i = 1, 2, \ldots, n\} \\ &= \{x \in U \mid (\exists i \in \{1, 2, \ldots, n\})(x \in A_i)\}. \end{aligned}$$

For each $n \in \mathbf{N}$, let $A_1, A_2, \ldots$ be subsets of the universe U, then the **infinite union of sets** $A_1, A_2, \ldots$ is

$$\bigcup_{i=1}^{\infty} A_i = A_1 \cup A_2 \cup \cdots = \{x \in U \mid (\exists i \in \mathbf{N})(x \in A_i)\}.$$

Let $n \in \mathbf{N}$ and let $A_1, A_2, \ldots, A_n$ be subsets of the universe U, then the **finite intersection of sets** $A_1, A_2, \ldots, A_n$ is

$$\begin{aligned}\bigcap_{i=1}^{n} A_i &= A_1 \cap A_2 \cap \cdots \cap A_n \\ &= \{x \in U \mid x \in A_i \text{ for all } i = 1, 2, \ldots, n\} \\ &= \{x \in U \mid (\forall i \in \{1, 2, \ldots, n\})(x \in A_i)\}.\end{aligned}$$

For each $n \in \mathbf{N}$, let $A_1, A_2, \ldots$ be subsets of the universe U, then the **infinite intersection of sets** $A_1, A_2, \ldots$ is

$$\bigcap_{i=1}^{\infty} A_i = A_1 \cap A_2 \cap \cdots = \{x \in U \mid (\forall i \in \mathbf{N})(x \in A_i)\}.$$

Let $\mathcal{F}$ be a family of sets which are all subsets of a universe U.
The **union over** $\mathcal{F}$ is

$$\bigcup_{A \in \mathcal{F}} A = \{x \in U \mid (\exists A \in \mathcal{F})(x \in A)\}.$$

The **intersection over** $\mathcal{F}$ is

$$\bigcap_{A \in \mathcal{F}} A = \{x \in U \mid (\forall A \in \mathcal{F})(x \in A)\}.$$

Let $\mathcal{F}$ be a nonempty family of sets and let I be a nonempty set with the property that for each $i \in I$ there corresponds a set $A_i \in \mathcal{F}$. Then the family of sets $\mathcal{F} = \{A_i \mid i \in I\}$ is called an **indexed family of sets**. The set I is called the **indexing set** and each $i \in I$ is called an **index**.

Useful Laws

The **principle of duality** for set theory says: "If T is a theorem which is written in terms of $\cup$, $\cap$, and $'$, then the dual statement T^D is also a theorem."

The **algebra of sets** consists of a collection $\mathcal{A}$ of subsets $\emptyset$, U, A, B, $C, \ldots$ of some universe U. On the collection $\mathcal{A}$ there is defined two relations, $\subseteq$ and $=$; two binary operations, $\cup$ and $\cap$; and one unary operation $'$. The collection $\mathcal{A}$ also has the properties that if A and B are arbitrary sets of $\mathcal{A}$, then $A \cup B \in \mathcal{A}$, $A \cap B \in \mathcal{A}$, and $A' \in \mathcal{A}$. For $\mathcal{A}$ an algebra of sets, the following statements and their duals are theorems.

Idempotent Laws

1. $A \cup A = A$ $\qquad$ 1^D. $A \cap A = A$

Identity Laws

2. $A \cup \emptyset = A$ $\qquad$ 2^D. $A \cap U = A$
3. $A \cup U = U$ $\qquad$ 3^D. $A \cap \emptyset = \emptyset$

Complement Laws

4. $A \cup A' = U$ $\qquad$ 4^D. $A \cap A' = \emptyset$
5. $\emptyset' = U$ $\qquad$ 5^D. $U' = \emptyset$
6. $(A')' = A$

Commutative Laws

7. $A \cup B = B \cup A$ $\qquad$ 7^D. $A \cap B = B \cap A$

Associative Laws

8. $(A \cup B) \cup C = A \cup (B \cup C)$ $\qquad$ 8^D. $(A \cap B) \cap C = A \cap (B \cap C)$

Distributive Laws

9. $A \cup (B \cap C) = (A \cup B) \cap (A \cup C)$ $\qquad$ 9^D. $A \cap (B \cup C) = (A \cap B) \cup (A \cap C)$

De Morgan Laws

10. $(A \cup B)' = A' \cap B'$ $\qquad$ 10^D. $(A \cap B)' = A' \cup B'$

The algebra of sets is an example of a mathematical system.

Generalized Distributive Laws Let U be the universe, let A be any subset of the universe, and let $\mathcal{F} = \{B_i \mid i \in I\}$ be an indexed family of subsets of U. Then

a. $A \cup (\bigcap_{i \in I} B_i) = \bigcap_{i \in I} (A \cup B_i)$ (union distributes over intersection)

b. $A \cap (\bigcup_{i \in I} B_i) = \bigcup_{i \in I} (A \cap B_i)$ (intersection distributes over union)

Generalized De Morgan Laws Let U be the universe and let $\mathcal{F} = \{A_i \mid i \in I\}$ be an indexed family of subsets of U. Then

a. $(\bigcup_{i \in I} A_i)' = \bigcap_{i \in I} A_i'$ and *b.* $(\bigcap_{i \in I} A_i)' = \bigcup_{i \in I} A_i'$

Review Exercises

1. Which of the following sets are well-defined?

   a. The set of all honest people.
   b. The set of natural numbers which are divisible by 5.
   c. The set of English names for the days in the middle of the week.
   d. The set of integer solutions of the equation $x^2 + 9 = 0$.

2. Write the following sets using roster notation.

   a. $A = \{n \in \mathbf{N} \mid -3 < n \leq 2\}$
   b. $B = \{n \in \mathbf{Z} \mid -3 < n \leq 2\}$
   c. $C = \{n \in \mathbf{Z} \mid n^2 \leq 9\}$
   d. $D = \{x \in \mathbf{R} \mid x^2 + 16 = 0\}$

3. Let $A = \{a, b, c\}$, and $B = \{\emptyset, a, b, c, d\}$. Which of the following statements are true?

   a. $a \in A$ b. $a \subset A$ c. $\{a\} \in A$ d. $\{a\} \subset A$

   e. $\emptyset \subset A$ f. $\{\emptyset\} \in A$ g. $\emptyset \subset B$ h. $\{\emptyset\} \in B$

   i. $A \subset B$ j. $A \subseteq B$ k. $B \subset A$ l. $A = B$

4. Let $X = \{1, a\}$.

   a. Write the power set of X, $\mathcal{P}(X)$.
   b. Write the proper subsets of X.
   c. Write the nonempty, proper subsets of X.

5. Let $U = \{1, 2, 3, 4, 5, 6, 7\}$, $A = \{3, 5\}$, $B = \{2, 6\}$, and $C = \{1, 3, 6\}$. Calculate

   a. $A \cup B$ b. $A \cap B$ c. $A \cup A'$ d. $B \cap B'$

   e. $(A \cup B) \cap C$ f. $(B \cap C) \cup C$ g. $B - C$ h. $C - B$

   i. $(A \cup B \cup C)'$ j. $A \times C$ k. $C \times A$ l. $A \times (B - C)$

6. Let $A = \{x \in \mathbf{R} \mid -2 < x \leq 4\}$, $B = [-1, 5)$, and $C = \{x \in \mathbf{R} \mid x > 3\}$. Write the following sets using both set notation and interval notation.

   a. $A \cup B$ b. $A \cap C$ c. $B \cup C$ d. $B \cap C$

   e. $A - B$ f. $B - A$ g. B' h. $(C - B)'$

7. Graph the following Cartesian products.

   a. $\mathbf{R} \times [-1, 2)$ b. $\{3\} \times \mathbf{R}$ c. $\{-1\} \times \{2\}$ d. $(-3, \infty) \times (-\infty, 2]$

8. For each of the following statements, determine if the statement is true or false. If the statement is true, prove it. If the statement is false, give a counterexample.

   a. If $A \subset B$ and $B \subseteq C$, then $A \subset C$.
   b. If $a \in A$ and $A \not\subseteq B$, then $a \notin B$.
   c. If $A \subseteq B$ and $c \notin B$, then $c \notin A$.
   d. If $A \not\subseteq B$ and $B \not\subseteq C$, then $A \not\subseteq C$.
   e. If $A \subseteq C$ and $B \subseteq C$, $(A \cup B) \subseteq C$.

9. Let $A_1 = \{a, b, c, d\}$, $A_2 = \{b, d\}$, and $A_3 = \{a, c\}$. Find

   a. $\bigcup_{i=1}^{3} A_i$ b. $\bigcap_{i=1}^{3} A_i$

10. Let $A_n = [2, 2 + \frac{3}{2n})$. Find

   a. $\bigcup_{i=1}^{5} A_i$ b. $\bigcap_{i=1}^{5} A_i$ c. $\bigcup_{i=1}^{\infty} A_i$ d. $\bigcap_{i=1}^{\infty} A_i$

Chapter 4

Relations

Relations and functions play a major role in many branches of mathematics and sciences. Historically, the concept of a function was introduced prior to the concept of a relation. However, a relation is more general entity than a function. Therefore, in this chapter we will consider relations and in the next chapter we will consider functions.

4.1 Relations

In everyday conversations, we often hear statements such as "Richard is the husband of Sue," "Kimberly is the sister of Jerry," "Alaska is larger than Arizona," and so forth. Each of these sentences includes a predicate ("is the husband of," "is the sister of," or "is larger than,") which expresses a "relation" between two objects. The word "relation" implies an association between two objects—people, states, numbers, concepts, etc.—based on some property of the objects. In elementary mathematics, you previously encountered the relations "is equal to," "is less than," "is congruent to," "is parallel to," and many more. Let $P(x, y)$ denote the open sentence in two variables: "x is the husband of y." If we specify that the variable x is an element of the set of all living male people, M, and y is an element of the set of all living female people, F, then given a particular ordered pair $(x, y) \in M \times F$ we can substitute $x = m$ and $y = f$ into the open sentence $P(x, y)$ and obtain the statement $P(m, f)$. This statement will be true or false but not both. Let H denote the predicate "is the husband of." When $P(m, f)$ is true, we will write $(m, f) \in H$, which is read "m is H-related to f." In this instance, we would say "m is the husband of f." On the other hand, when $P(m, f)$ is false, we will write $(m, f) \notin H$, which is read "m is not H-related to f." In general, two sets of ordered pairs are determined by each relation R—the set of ordered pairs which satisfies the relation, R, and the set of ordered pairs which does not satisfy the relation, R'.

We now state the mathematical definition of a relation formally.

DEFINITIONS Relation, Domain of a Relation, and Range of a Relation

Let A and B be sets.

A **relation from** A **to** B is any subset of $A \times B$. In particular, when $B = A$, a relation R from A to A is called a **relation on** A.

The **domain** of a relation R from A to B is the set

$$\text{Dom}(R) = \{x \in A \mid (\exists y \in B)((x, y) \in R)\}$$

The **range** of a relation R from A to B is the set

$$\text{Rng}(R) = \{y \in B \mid (\exists x \in A)((x, y) \in R)\}$$

The domain of a relation R from A to B is the set of all first coordinates of the ordered pairs in the set R and by definition of the domain of R is a subset of A—that is, $\text{Dom}(R) \subseteq A$. The range of a relation R from A to B is the set of all second coordinates of the ordered pairs in the set R and by definition of range $\text{Rng}(R) \subseteq B$.

Let R be any set of ordered pairs, let A be any set such that $\text{Dom}(R) \subseteq A$, and let B be any set such that $\text{Rng}(R) \subseteq B$. Then by definition R is a relation from A to B. Consequently, **every set of ordered pairs is a relation**. For example, let $R = \{(1, a), (b, 2), (1, 2)\}$. Then $\text{Dom}(R) = \{1, b\}$, $\text{Rng}(R) = \{a, 2\}$, and R is a relation from $A = \text{Dom}(R)$ to $B = \text{Rng}(R)$.

EXAMPLE 1 Determining Relations, Domains, and Ranges

Let $A = \{1, 2, 3\}$, and $B = \{2, 3, 4\}$. Let S be the relation from A to B defined by

$$S = \{(x, y) \in A \times B \mid x \text{ is less than } y\}$$

Let T be the relation from B to A defined by

$$T = \{(x, y) \in B \times A \mid x \text{ is less than } y\}$$

a. Use the roster method to specify S, $\text{Dom}(S)$, and $\text{Rng}(S)$.

b. Use the roster method to specify T, $\text{Dom}(T)$, and $\text{Rng}(T)$.

SOLUTION

a. Since $1 \in A$ and $1 < 2 \in B$, $1 < 3 \in B$, and $1 < 4 \in B$, the ordered pairs $(1,2)$, $(1,3)$, and $(1,4)$ are elements of S. Likewise, since $2 \in A$ but $2 \not< 2 \in B$, $(2,2) \notin S$. However, since $2 < 3 \in B$ and $2 < 4 \in B$, $(2,3)$ and $(2,4)$ are elements of S. Since $3 \in A$, $3 \not< 2 \in B$ and $3 \not< 3 \in B$, $(3,2) \notin S$ and $(3,3) \notin S$. But since $3 < 4 \in B$, $(3,4) \in S$. Hence,

$$S = \{(1,2),(1,3),(1,4),(2,3),(2,4),(3,4)\}$$

Because the first coordinates of the ordered pairs of S are 1, 2, 3, $\text{Dom}(S) = \{1,2,3\} = A$. Because the second coordinates of the ordered pairs of S are 2, 3, 4, $\text{Rng}(S) = \{2,3,4\} = B$.

b. Since $2 \in B$ and $2 < 3 \in A$ but $b < a$ is false for all other $b \in B$ and $a \in A$, $T = \{(2,3)\}$, $\text{Dom}(T) = \{2\} \subset B$, and $\text{Rng}(T) = \{3\} \subset A$.

A graph of the relation $S = \{(x,y) \in A \times B \mid x < y\}$ where $A = \{1,2,3\}$, and $B = \{2,3,4\}$ is displayed in Figure 4.1.a and a graph of the complement of S, $S' = \{(x,y) \in A \times B \mid \neg(x < y)\} = \{(x,y) \in A \times B \mid x \geq y\}$ is displayed in Figure 4.1.b. Observe that $A \times B = S \cup S'$.

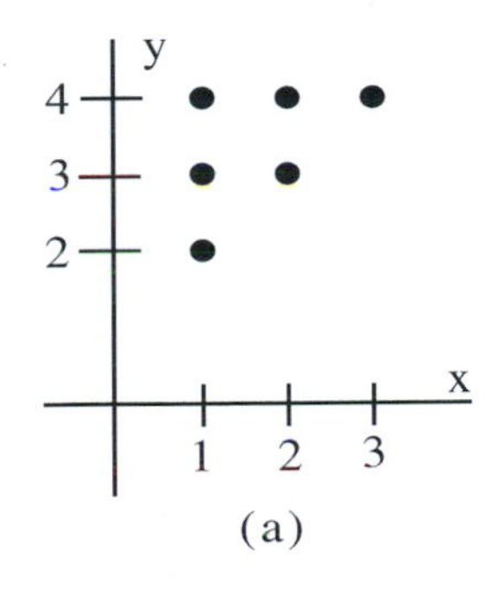

y
4
3
2
x
1 2 3
(b)

FIGURE 4.1: (a) Graph of S (b) Graph of S'

If we interchange the components of the ordered pairs in the relation S of example 1, we obtain the set $\{(2,1),(3,1),(4,1),(3,2),(4,2),(4,3)\}$. This relation is designated by S^{-1} and is called the **inverse relation of** S. Observe that $\text{Dom}(S^{-1}) = \{2,3,4\} = \text{Rng}(S)$ and $\text{Rng}(S^{-1}) = \{1,2,3\} = \text{Dom}(S)$. Now let $U = \text{Dom}(S) \cup \text{Rng}(S) = \{1,2,3,4\}$. Graphing the ordered pairs in S using open dots, $\circ$, and graphing the ordered pairs in S^{-1} using closed dots, $\bullet$, we obtain the graph shown in Figure 4.2. Notice that corresponding to each point $(a,b) \in S$ there corresponds exactly one point $(b,a) \in S^{-1}$. For example, $(1,2)$ in S corresponds to $(2,1)$ in S^{-1}, $(2,3)$ in S corresponds to $(3,2)$ in S^{-1}, and so forth. If the "diagonal line $y = x$" is considered to be a mirror, then corresponding to each point $(a, b) \in S$ is the "mirror image" $(b, a) \in S^{-1}$ and vice versa. Hence, the relation S^{-1} is the "mirror image" of the relation S with respect to the "line $y = x$" and, likewise, S is the "mirror image" of S^{-1}. Also observe from Figure 4.1.b and Figure 4.2 that $S^{-1} \neq S'$—that is, the inverse of a relation is not equal to its complement.

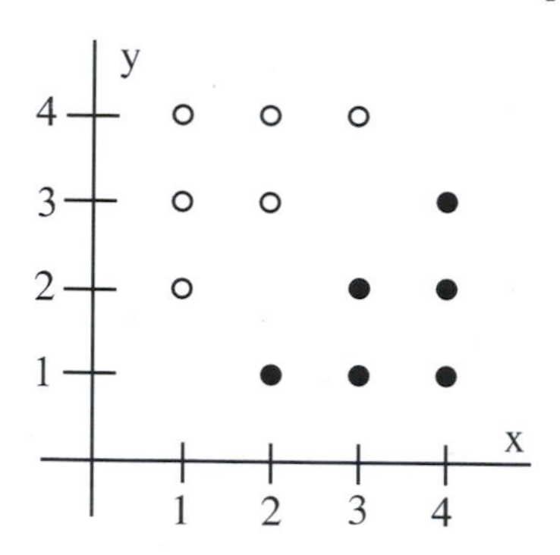

FIGURE 4.2: A Graph of the Relations S and S^{-1}

In set-builder notation, $S = \{(x, y) \in A \times B \mid x < y\}$ and as indicted by Figure 4.2 the inverse relation $S^{-1} = \{(x, y) \in B \times A \mid y < x\}$. The condition $y < x$ used to define S^{-1} is obtained from the condition $x < y$ used to define S by interchanging the variables x and y. Thus, if a relation is defined by some condition on the variables x and y, then the condition which defines the inverse relation is obtained from the original condition by interchanging the variables x and y. In general, we define an inverse relation in the following manner.

DEFINITION Inverse Relation

If R is a relation from A to B, then the **inverse relation from B to A** is the relation

$$R^{-1} = \{(x, y) \in B \times A \mid (y, x) \in R\}$$

An important family of relations is the family of identity relations—one relation for each nonempty set A. This family is defined as follows.

DEFINITION Identity Relation on a Set A

Let A be a nonempty set. The **identity relation on A** is the set

$$I_A = \{(x, x) \in A \times A \mid x \in A\}$$

From the definition of the identity relation, it is clear that $\text{Dom}(I_A) = \text{Rng}(I_A) = A$. If $C = \{1, 2, 3, 4\}$, then $I_C = \{(1, 1), (2, 2), (3, 3), (4, 4)\}$. A graph of I_C is shown if Figure 4.3. The points on I_C are points on the "line" about which a relation S on C is reflected in order to obtain the inverse relation S^{-1} on C. (Locate the points of the relation I_C in Figure 4.2 and notice that the point $(1, 3)$ is the reflection of $(3, 1)$ about the point $(2, 2) \in I_C$ and that the point $(2, 4)$ is the reflection of $(4, 2)$ about the point $(3, 3) \in I_C$.)

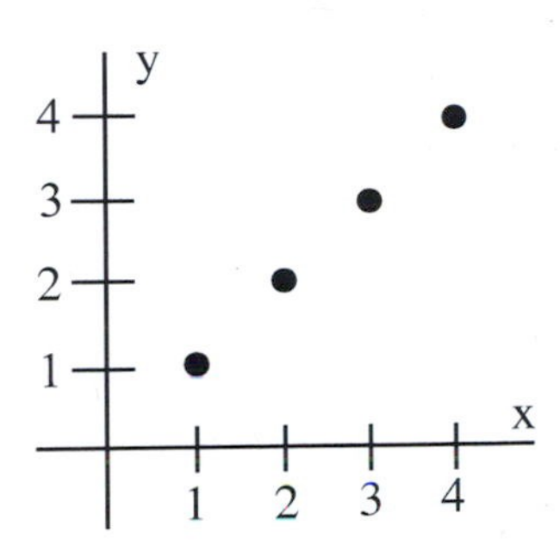

FIGURE 4.3: A Graph of the Relations I_C

DEFINITION Equality for Relations

Let R be a relation from A to B and let S be a relation from C to D. The relation R **equals** S, which is denoted by $R = S$, if and only if $A = C$, $B = D$, and $[(x, y) \in R \Leftrightarrow (x, y) \in S]$.

The next two examples should help clarify this definition.

The identity relation on the natural numbers is

$$I_{\mathbf{N}} = \{(x, x) \in \mathbf{N} \times \mathbf{N} \mid x = x\}$$

and the identity relation on the integers is

$$I_{\mathbf{Z}} = \{(x, x) \in \mathbf{Z} \times \mathbf{Z} \mid x = x\}.$$

The identity relation $I_{\mathbf{N}} \neq I_{\mathbf{Z}}$ because $\text{Dom}(I_{\mathbf{N}}) = \mathbf{N} \neq \mathbf{Z} = \text{Dom}(I_{\mathbf{Z}})$.

Let $A = \{1, 2\} = C$, $B = \{a, b\} = D$, $R = \{(1, a), (2, b)\}$, and $S = \{(1, b), (2, a)\}$. Then $\text{Dom}(R) = \{1, 2\} = \text{Dom}(S)$ and $\text{Rng}(R) = \{a, b\} = \text{Rng}(S)$; however, $R \neq S$ because $(1, a) \in R$ but $(1, a) \notin S$.

Since relations are sets, we can perform the usual unary set operation of complementation and the binary set operations of union, intersection, and set difference on relations. Moreover, because relations are sets of ordered pairs we can perform additional binary operations such as composition.

DEFINITION Composition of Two Relations

Let R be a relation from A to B and let S be a relation from B to C. The **composition of S and R** is the relation

$$S \circ R = \{(a, c) \in A \times C \mid (\exists b \in B)[((a, b) \in R) \wedge ((b, c) \in S)]\}$$

EXAMPLE 2 Determining a Composition

Let $A = \{a, b, c, d\}$, $B = \{1, 2, 3, 4, 5\}$, $C = \{w, x, y, z\}$, $R = \{(a, 1), (a, 2), (b, 2), (c, 3)\}$, and $S = \{(1, x), (2, y), (4, w)\}$. Determine

a. $S \circ R$ b. $\text{Dom}(S \circ R)$ c. $\text{Rng}(S \circ R)$

SOLUTION

Perhaps the best way to visualize the composition of S and R is through the **arrow diagram** shown in Figure 4.4. First, we draw regions to represent the sets A, B, and C and select points in the regions to represent the elements of the sets. Since the ordered pair $(a, 1)$ is in the relation R, we draw an arrow from the point a in set A to the point 1 in the set B. Since $(a, 2)$ is in the relation R, we draw an arrow from the point a to 2, and so forth. Once we have completed drawing all of the arrows from A to B which represent the relation R, we draw the arrows from points in the set B to points in the set C which represent the relation S. To determine the set of ordered pairs in the composition $S \circ R$, we select, in turn, elements in set A and find all paths, if any, which lead from that element to some element in C. The set of all ordered pairs found in this manner constitutes the elements of $S \circ R$. For instance, $a \in A$ and there is a path from a through $1 \in B$ to $x \in C$. Hence, $(a, x) \in S \circ R$. There is another path from $a \in A$ through $2 \in B$ to $y \in C$. Thus, $(a, y) \in S \circ R$. Using Figure 4.4 and proceeding in this fashion, we find

$$S \circ R = \{(a, x), (a, y), (b, y)\}$$

Hence, $\text{Dom}(S \circ R) = \{a, b\} \subset A$ and $\text{Rng}(S \circ C) = \{x, y\} \subset C$.

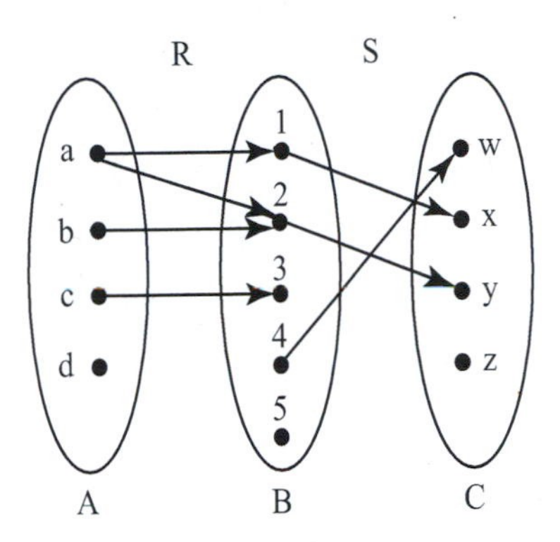

FIGURE 4.4: Arrow Diagram for $S \circ R$

Observe that for relations R and S as defined in example 2 the composition $R \circ S$ is not defined, since S is a relation from B to C and R is a relation from A to B. Hence, the binary operation of composition, $\circ$, is not commutative—that is, $S \circ R \neq R \circ S$. Even if S and R are both relations on the same set A and $S \circ R$ and $R \circ S$ are both defined, $S \circ R$ may not equal $R \circ S$. For example, let $A = \{1, b\}$, $R = \{(1, b)\}$, and $S = \{(b, 1)\}$. Then $S \circ R = \{(1, 1)\}$, $R \circ S = \{(b, b)\}$, and $S \circ R \neq R \circ S$.

Theorem 4.1 Let R be a relation from A to B, then

$$\text{a. } I_B \circ R = R \qquad \text{and} \qquad \text{b. } R \circ I_A = R.$$

Proof: a. Assume $(x, y) \in R$. Then $y \in B$ and $(y, y) \in I_B$. From the definition of composition, since $(x, y) \in R$ and $(y, y) \in I_B$, it follows that $(x, y) \in I_B \circ R$. Hence, $R \subseteq I_B \circ R$.

Now assume $(x, y) \in I_B \circ R$. By definition of $I_B \circ R$ there exists a $b \in B$ such that $(x, b) \in R$ and $(b, y) \in I_B$. Since $(b, y) \in I_B$, $b = y$ and since $(x, b) \in R$, $(x, y) \in R$. Thus, $I_B \circ R \subseteq R$. Consequently, $I_B \circ R = R$.

b. Proving that $R \circ I_A = R$ is an exercise.

Theorem 4.1 states that a relation R from A to B has a "left" identity—the identity relation I_B such that $I_B \circ R = R$—and a "right" identity—the identity relation I_A such that $R \circ I_A = R$. But unless $A = B$, the relation R does not have an identity I such that $I \circ R = R \circ I = R$. When $A = B$, the identity for the relation R on A is the relation I_A.

In elementary algebra, for all real numbers $x \neq 0$, there exists a real number called the multiplicative inverse of x and denoted by x^{-1} with the property that $x \cdot x^{-1} = x^{-1} \cdot x = 1$, the multiplicative identity for the set of real numbers. Example 3 provides an example of a relation R for which neither the composition $R \circ R^{-1}$ nor the composition $R^{-1} \circ R$ is an identity relation.

EXAMPLE 3 Computation of $R \circ R^{-1}$ and $R^{-1} \circ R$ for a Particular Relation

Let $R = \{(1,2), (1,3), (3,2)\}$. Calculate $R \circ R^{-1}$ and $R^{-1} \circ R$.

SOLUTION

The inverse relation is $R^{-1} = \{(2,1), (3,1), (2,3)\}$. Since $(2,1) \in R^{-1}$ and $(1,2), (1,3) \in R$, $(2,2), (2,3) \in R \circ R^{-1}$. Since $(3,1) \in R^{-1}$ and $(1,2), (1,3) \in R$, $(3,2), (3,3) \in R \circ R^{-1}$. And since $(2,3) \in R^{-1}$ and $(3,2) \in R$, $(2,2) \in R \circ R^{-1}$. Hence, $R \circ R^{-1} = \{(2,2), (2,3), (3,2), (3,3)\}$ which is not an identity relation because $(2,3) \in R \circ R^{-1}$.

In a like manner, we find $R^{-1} \circ R = \{(1,1), (1,3), (3,1), (3,3)\}$ which is not an identity relation because $(1,3) \in R^{-1} \circ R$.

In example 4 we illustrate how to specify the composition of two relations when they are both written in set-builder notation.

EXAMPLE 4 Computing a Composition Using Set-Builder Notation

Let

$$T = \{(x,y) \in \mathbf{R} \times \mathbf{R} \mid x^2 + y^2 = 4\}$$

and let

$$V = \{(x,y) \in \mathbf{R} \times \mathbf{R} \mid x^2 + (y-2)^2 = 9\}.$$

Find

a. $\mathrm{Dom}(T)$ b. $\mathrm{Rng}(T)$ c. $\mathrm{Dom}(V)$ d. $\mathrm{Rng}(V)$

e. $V \circ T$ f. $\mathrm{Dom}(V \circ T)$ g. $\mathrm{Rng}(V \circ T)$

SOLUTION

a.-d. The graph of the relation T is a circle with center at the origin and radius 2. Hence, $\mathrm{Dom}(T) = [-2,2] = \mathrm{Rng}(T)$. The graph of the relation V is a circle with center at $(0,2)$ and radius 3. Therefore, $\mathrm{Dom}(V) = [-3,3]$ and $\mathrm{Rng}(V) = [-1,5]$.

$$\begin{aligned} e.\ V \circ T &= \{(x,y) \in \text{Dom}(T) \times \text{Rng}(V) \mid (\exists z \in \text{Rng}(T))[((x,z) \in T) \wedge \\ &\qquad ((z,y) \in V)]\} \\ &= \{(x,y) \in [-2,2] \times [-1,5] \mid (\exists z \in [-2,2])[(x^2 + z^2 = 4) \wedge \\ &\qquad (z^2 + (y-2)^2 = 9)]\} \\ &= \{(x,y) \in [-2,2] \times [-1,5] \mid 4 - x^2 + (y-2)^2 = 9\} \\ &= \{(x,y) \in [-2,2] \times [-1,5] \mid (y-2)^2 - x^2 = 5\} \end{aligned}$$

f.-g. The graph of $(y-2)^2 - x^2 = 5$ is a hyperbola with center at $(0,2)$ and transverse axis the y-axis. See Figure 4.5. Since by definition of composition, $\text{Dom}(V \circ T) \subseteq \text{Dom}(T) = [-2,2]$ and since $(y-2)^2 - x^2 = 5$ is defined for all $x \in [-2,2]$, $\text{Dom}(V \circ T) = [-2,2]$. From Figure 4.5, we see that

$$\text{Rng}(V \circ T) = [-1, 2 - \sqrt{5}] \cup [2 + \sqrt{5}, 5] \subset [-1,5] = \text{Rng}(V).$$

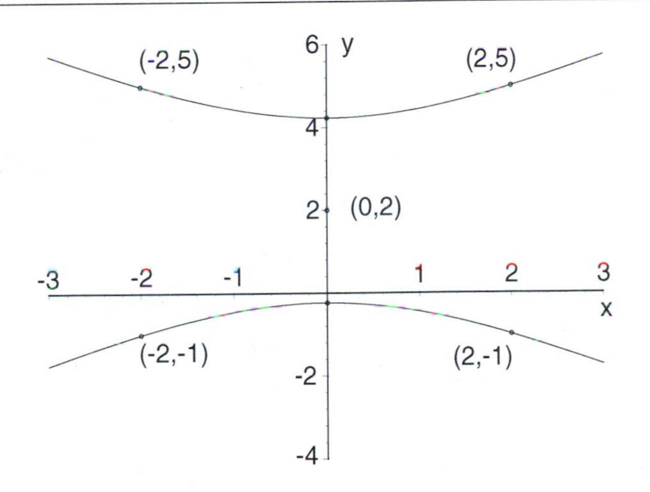

FIGURE 4.5: Graph of $(y-2)^2 - x^2 = 5$

Let $A = \{1,2\}$, $B = \{a,b,c\}$, $C = \{u,v,w\}$, $R = \{(1,a),(2,b),(2,c)\}$, and $S = \{(a,v),(b,w),(c,u)\}$. Then $S \circ R = \{(1,v),(2,w),(2,u)\}$ and $(S \circ R)^{-1} = \{(v,1),(w,2),(u,2)\}$. Notice that $R \circ S$ and, therefore, $(R \circ S)^{-1}$ do not exist. We easily compute

$$R^{-1} = \{(a,1),(b,2),(c,2)\} \quad \text{and} \quad S^{-1} = \{(v,a),(w,b),(u,c)\}$$

and observe that $S^{-1} \circ R^{-1}$ does not exist; however,

$$R^{-1} \circ S^{-1} = \{(v,1),(w,2),(u,2)\} = (S \circ R)^{-1}.$$

This leads us to conjecture and prove theorem 4.2.

Theorem 4.2 Let R be a relation from A to B and let S be a relation from B to C. Then the relation from C to A, $(S \circ R)^{-1} = R^{-1} \circ S^{-1}$.

Proof: Let $(c, a) \in C \times A$. Then

$(c, a) \in (S \circ R)^{-1} \Leftrightarrow (a, c) \in S \circ R$
Definition of inverse relation

$\Leftrightarrow (\exists b \in B)[((a, b) \in R) \wedge ((b, c) \in S)]$
Definition of composition

$\Leftrightarrow (\exists b \in B)[((b, a) \in R^{-1}) \wedge ((c, b) \in S^{-1})]$
Definition of inverse relation

$\Leftrightarrow (\exists b \in B)[((c, b) \in S^{-1}) \wedge ((b, a) \in R^{-1})]$
Commutative Law of Conjunction

$\Leftrightarrow R^{-1} \circ S^{-1}$
Definition of composition

Theorem 4.2 states that the inverse of the composition of two relations, $(S \circ R)^{-1}$, is the composition of the inverse of the second relation with the inverse of the first relation, $R^{-1} \circ S^{-1}$.

We proved earlier by example that the operation of composition on relations is not commutative. That is, we showed that, in general, for relations R and S, $R \circ S \neq S \circ R$. However, the operation of composition on relations is associative as we state and prove in the next theorem.

Theorem 4.3 Let R be a relation from A to B, let S be a relation from B to C, and let T be a relation from C to D. Then $T \circ (S \circ R) = (T \circ S) \circ R$.

Proof: Let $(a, d) \in A \times D$. Then

$(a, d) \in T \circ (S \circ R)$

$\Leftrightarrow (\exists c \in C)[((a, c) \in (S \circ R)) \wedge ((c, d) \in T)]$
Definition of composition

$\Leftrightarrow (\exists c \in C)[(\exists b \in B)[((a, b) \in R) \wedge ((b, c) \in S)] \wedge ((c, d) \in T)]$
Definition of composition

$\Leftrightarrow (\exists c \in C)(\exists b \in B)[((a, b) \in R) \wedge [((b, c) \in S) \wedge ((c, d) \in T)]]$
Associative Law for Conjunction

$\Leftrightarrow (\exists b \in B)(\exists c \in C)[((a, b) \in R) \wedge [((b, c) \in S) \wedge ((c, d) \in T)]]$
A property of quantifiers

$\Leftrightarrow (\exists b \in B)[((a,b) \in R) \wedge [(\exists c \in C)((b,c) \in S) \wedge ((c,d) \in T)]]$
A property of quantifiers

$\Leftrightarrow (\exists b \in B)[((a,b) \in R) \wedge ((b,d) \in (T \circ S))]$
Definition of composition

$\Leftrightarrow (a,d) \in (T \circ S) \circ R$
Definition of composition

EXERCISES 4.1

1. Let $A = \{1, 2, 3\}$ and $B = \{a\}$

   a. How many elements are there in $A \times B$? List them.

   b. How many elements are there in the power set $\mathcal{P}(A \times B)$? List them.

   c. How many relations are there from A to B?

2. Suppose A has m elements and B has n elements.

   a. How many elements are there in $A \times B$?

   b. How many elements are there in $\mathcal{P}(A \times B)$?

   c. How many relations are there from A to B?

   d. How many elements are there in $A \times A$?

   e. How many elements are there in $\mathcal{P}(A \times A)$?

   f. How many relations are there on A?

3. Let $A = \{2, 3, 4\}$, let $B = \{2, 6, 12, 17\}$, and let

$$R = \{(x, y) \in A \times B \mid x \text{ divides } y\}.$$

   Write R using the roster method.

4. Let $A = \{1, 2, 3, 4\}$ and let

$$R = \{(x, y) \in A \times A \mid y - x \text{ is an even natural number}\}.$$

Write R using the roster method.

5. Let S be the relation $S = \{(c, 1), (b, 3), (3, e), (2, b), (a, f), (b, 6)\}$. Find

a. $\text{Dom}(S)$ b. $\text{Rng}(S)$ c. S^{-1} d. $(S^{-1})^{-1}$

6. Find the domain and range of the following relations on $\mathbf{R}$.

a. $R_1 = \{(x, y) \in \mathbf{R} \times \mathbf{R} \mid y = -2x^2 + 3\}$

b. $R_2 = \{(x, y) \in \mathbf{R} \times \mathbf{R} \mid y = \sqrt{1 - x^2}\}$

c. $R_3 = \{(x, y) \in \mathbf{R} \times \mathbf{R} \mid (x = -3) \vee (|y| < 4)\}$

d. $R_4 = \{(x, y) \in \mathbf{R} \times \mathbf{R} \mid (x = -3) \wedge (|y| < 4)\}$

e. $R_5 = \{(x, y) \in \mathbf{R} \times \mathbf{R} \mid x^2 + y^2 < 9\}$

f. $R_6 = \{(x, y) \in \mathbf{R} \times \mathbf{R} \mid |x| + |y| \leq 9\}$

7. Graph the relations of exercise 6.

8. For the following relations determine the inverse relation.

a. $R = \{(a, 1), (2, b), (3, 4), (x, y)\}$

b. $S = \{(x, y) \in \mathbf{Z} \times \mathbf{Z} \mid x^2 + y^2 = 1\}$

c. $T = \{(x, y) \in \mathbf{R} \times \mathbf{R} \mid 3x^2 - 4y^2 = 9\}$

d. $V = \{(x, y) \in \mathbf{R} \times \mathbf{R} \mid y < 2x - 5\}$

e. $W = \{(x, y) \in \mathbf{R} \times \mathbf{R} \mid y(x + 3) = x\}$

9. Let $R = \{(1, 4), (2, 3), (5, 4), (3, 2)\}$, let $S = \{(5, 1), (2, 4), (3, 3)\}$, and let $T = \{(1, 2), (2, 1)\}$. Determine

a. $R \circ S$ b. $S \circ T$ c. $R \circ R$ d. $T \circ T$

e. $R \circ (S \circ T)$ f. $(R \circ S) \circ T$ g. $(R \circ S)^{-1}$ h. $R^{-1} \circ S^{-1}$

10. Use set-builder notation to write $S \circ R$ for the given relations R and S.

    a. $R = \{(x, y) \in \mathbf{R} \times \mathbf{R} \mid y = 2x - 1\}$
    and
    $S = \{(x, y) \in \mathbf{R} \times \mathbf{R} \mid 2x^2 + 3y^2 = 5\}$

    b. $R = \{(x, y) \in \mathbf{R} \times \mathbf{R} \mid y = \sqrt{x}\}$
    and
    $S = \{(x, y) \in \mathbf{R} \times \mathbf{R} \mid y = \sin x\}$

11. Let P be the set of all living people,

    let $B = \{(x, y) \in P \times P \mid y \text{ is the brother of } x\}$,

    let $F = \{(x, y) \in P \times P \mid y \text{ is the father of } x\}$,

    let $M = \{(x, y) \in P \times P \mid y \text{ is the mother of } x\}$,

    and let $S = \{(x, y) \in P \times P \mid y \text{ is the sister of } x\}$.

    Determine the following compositions.

    a. $F \circ F$ b. $M \circ F$ c. $F \circ M$ d. $M \circ B$

    e. $B \circ M$ f. $F \circ S$ g. $S \circ M$ h. $M \circ S$

12. Prove theorem 4.1.b. That is, prove if R is a relation from A to B, then $R \circ I_A = R$.

13. Let R and S be relations on a set A. Provide counterexamples to the following statements.

    a. $\text{Dom}(R) \subseteq \text{Dom}(S \circ R)$ *b.* $\text{Rng}(R) \subseteq \text{Rng}(S \circ R)$

4.2 The Order Relations $<$, $\leq$, $>$, $\geq$

In addition to the operations of addition, $+$, and multiplication, $\cdot$, the sets of natural numbers, integers, rational numbers, and real numbers have an order relation $<$ (read "less than") which satisfies certain axioms. These axioms can be expressed in terms of the undefined concept of **positiveness**. A set of numbers S is **ordered** if there exists a subset of **positive** numbers S_p

which satisfy the axioms:

O1. (Trichotomy Law) For all $x \in S$ exactly one of the following three statements is true:

$$x = 0, \quad x \in S_p, \quad -x \in S_p.$$

O2. For all $x, y \in S_p$, $\quad x + y \in S_p$.

O3. For all $x, y \in S_p$, $\quad x \cdot y \in S_p$.

DEFINITIONS Negative, $<$, $\leq$, $>$, $\geq$

The number x is **negative** if and only if $-x$ is positive.

The **order relation** $<$ (read "is less than") is defined by $x < y \Leftrightarrow y - x$ is positive.

The **order relation** $>$ (read "is greater than") is defined by $x > y \Leftrightarrow x - y$ is positive.

The **order relation** $\leq$ (read "is less than or equal to") is defined by $x \leq y \Leftrightarrow x < y$ or $x = y$.

The **order relation** $\geq$ (read "is greater than or equal to") is defined by $x \geq y \Leftrightarrow x > y$ or $x = y$.

Observe that $x < y$ and $y > x$ are equivalent statements as are $x \leq y$ and $y \geq x$. Likewise, the conjunction $x \leq y$ and $y \leq z$ can be abbreviated by the single statement $x \leq y \leq z$. Similarly, $x > y > z \Leftrightarrow x > y$ and $y > z$, and $x \geq y \geq z \Leftrightarrow x \geq y$ and $y \geq z$. The natural numbers, the integers, the rational numbers, and the real numbers with the usual operations of addition, $+$, and multiplication, $\cdot$, and the order relation, $<$, are all ordered sets. The integers are an example of an integral domain and the rational numbers and the real numbers are examples of ordered fields.

Using the axioms and definitions stated above, for $w, x, y, z \in \mathbf{R}$ the following theorems can be proved.

T1. $0 < x \Leftrightarrow x$ is positive.

T2. The transitive law holds for $<$: If $x < y$ and $y < z$, then $x < z$.

T3. The trichotomy law holds for $<$: If $x, y \in \mathbf{R}$ exactly one of the following statements is true:

$$x = y, \quad x < y, \quad y < x.$$

T4. $x < y \Rightarrow x + z < y + z$.

T5. $x < 0 \Leftrightarrow x$ is negative.

T6. If $x < 0$ and $y < 0$, then $x + y < 0$. (The sum of two negative numbers is a negative number.)

T7. If $x < 0$ and $y < 0$, then $x \cdot y > 0$. (The product of two negative numbers is a positive number.)

T8. If $x \neq 0$, then $0 < x^2$.

T9. $1 > 0$.

T10. If $0 < x < y$, then $0 < \frac{1}{y} < \frac{1}{x}$.

T11. If $x < y$ and $0 < z$, then $xz < yz$.

T12. If $x < y$ and $z < 0$, then $yz < xz$.

T13. If $x < y$, then $-y < -x$.

T14. If $x < y$ and $z < w$, then $x + z < y + w$.

T15. If $0 < x < y$ and $0 < z < w$, then $xz < yw$ and $\frac{x}{w} < \frac{y}{z}$.

T16. If $x \leq y$ and $y \leq z$, then $x \leq z$.

T17. If $x \leq y$ and $y \leq x$, then $x = y$.

T18. If $x \leq y$ and $0 < z$, then $xz \leq yz$.

T19. If $x \leq y$ and $z < 0$, then $yz \leq xz$.

DEFINITIONS Least Element of a Set and the Greatest Element of a Set

Let A be a nonempty set of real numbers.

The number $m \in A$ is the **least element of** A (**smallest element of** A or **minimum of** A) if and only if for every $x \in A$, $m \leq x$.

The number $M \in A$ is the **greatest element of** A (**largest element of** A or **maximum of** A) if and only if for every $x \in A$, $M \geq x$.

Some nonempty subsets of **R** have least elements; others do not. For example, the set of natural numbers **N** has a least element which is 1. The set of integers **Z** does not have a least element. The open interval $(-2, 3)$ does not have a least element. Although for all $x \in (-2, 3)$, $-2 \leq x$, -2 is not a least element of $(-2,3)$, because $-2 \notin (-2,3)$. On the other hand, -2 is the least element of the interval $[-2,3)$, because $-2 \leq x$ for all $x \in [-2,3)$ and $-2 \in [-2,3)$. The following theorem proves if a set of real numbers has a least element, it is unique.

Theorem 4.4 The least element of a set A of real numbers is unique.

Proof: Suppose x and y are distinct least elements of A. Since x is a least element of A and $y \in A$, $x \leq y$. Also since y is a least element of A and $x \in A$, $y \leq x$. By theorem T17, $x = y$, and the least element of A is unique.

The **Archimedean property** of Euclidean geometry states that any length no matter how large can be exceeded by repeatedly "marking off" a given length no matter how small. The following is the algebraic formulation of the Archimedean property for the set of natural numbers.

Theorem 4.5 The Archimedean Property for the Natural Numbers

For all $m, n \in \mathbf{N}$ there exists a $k \in \mathbf{N}$ such that $m < kn$.

Proof: By T9, $0 < 1$. Adding m and using T4, we find that

$$m = 0 + m < 1 + m = m + 1.$$

That is, (1) $m < m + 1$. Since 1 is the least element of $\mathbf{N}$ and since $0 < 1$, (2) $0 < 1 \leq m$ and (3) $1 \leq n$. From (1) and (2), we have (4) $0 < m < m+1$. Multiplying (3) by $m + 1 > 0$, yields (5) $1 \cdot (m+1) \leq n \cdot (m+1)$ by T18. It follows from (4) and (5) that $m < m + 1 \leq (m+1) \cdot n$. That is, the value $k = m + 1 \in \mathbf{N}$ makes the conclusion of the Archimedean property true.

DEFINITION Well-Ordered

A nonempty set A of real numbers is **well-ordered** if and only if every nonempty subset of A has a least element.

For instance, the nonempty set $A = \{-1, \pi, e\}$ is well-ordered, because each of the nonempty subsets $\{-1\}$, $\{\pi\}$, $\{e\}$, $\{-1, \pi\}$, $\{-1, e\}$, $\{\pi, e\}$, and A all have a least element. It is fairly evident from this example that any nonempty, finite subset of real numbers is well-ordered. The open interval $(-2,3)$ is not well-ordered, since it has no least element. Although, the interval $[-2,3)$ has a least element, it is not well-ordered since the subset $(-2,3)$ has no least element. The sets $\mathbf{Z}$, $\mathbf{Q}$, and $\mathbf{R}$ are not well-ordered, because none of these sets has a least element. Even though it appears obvious that the set of natural numbers $\mathbf{N}$ is well-ordered, a proof of this fact requires the use of the Axiom of Induction which we will not study until chapter 6. Thus, for now, we state the well-ordering principle without proof.

> **The Well-Ordered Principle:** The set of natural numbers is well-ordered by the relation $<$.

It follows easily from the well-ordering principle that the set of whole numbers $\mathbf{W} = \mathbf{N} \cup \{0\}$ is well-ordered by $<$, since any nonempty set of $\mathbf{W}$ which contains the element 0 has 0 as its least element and since any nonempty set of $\mathbf{W}$ which does not contain 0 is a nonempty subset of $\mathbf{N}$ and, therefore, has a least element by the well-ordering principle.

In elementary school, you learned how to use a process called "long division" to divide one natural number b, called the **dividend**, by a natural number a, called the **divisor**, to obtain two whole numbers q, called the **quotient**, and r, called the **remainder**. The whole numbers q and r are unique and satisfy the equation $b = aq + r$ where $0 \le r < a$. We now state and prove a theorem called the division algorithm for natural numbers.

Theorem 4.6 The Division Algorithm for the Natural Numbers

For all $a, b \in \mathbf{N}$ there exists unique $q, r \in \mathbf{W}$ such that $b = aq + r$ and $0 \le r < a$.

Proof: Let $a, b \in \mathbf{N}$. Consider the set $S = \{b - ax \mid x \in \mathbf{W} \text{ and } b - ax \ge 0\}$. Since $x = 0 \in \mathbf{W}$ and $b \in \mathbf{N}$, $b - a \cdot 0 = b \ge 0$. Thus, by definition of S, $b \in S$ and S is a nonempty subset of the whole numbers. As a consequence of the well-ordering principle, S has a smallest element r and by definition of S, $r \ge 0$. Since $r \in S$, there exists a $q \in \mathbf{W}$ such that $r = b - aq$. That is, there exist whole numbers q and r such that $b = aq + r$ and $r \ge 0$.

We now show that $r < a$ by contradiction. Assume to the contrary $a \le r$ and let $d = r - a \ge 0$. Since $a \in \mathbf{N}$, $a > 0$ and $d < r$. Furthermore,

$$d = r - a = (b - aq) - a = b - a(q + 1) \ge 0$$

Consequently, $d \in S$ and $d < r$ which contradicts r being the smallest element of S. Thus, we have $0 \le r < a$.

In order to show that the whole numbers q and r which satisfy $b = aq + r$ and $0 \le r < a$ are unique, assume there exist a second pair of whole numbers $q_2 \ne q$ and $r_2 \ne r$ such that $b = aq_2 + r_2$ and $0 \le r_2 < a$. Since $b = aq + r = aq_2 + r_2$, (1) $a(q - q_2) = r_2 - r$. By definition a divides $r_2 - r$, because a, $q - q_2$, and $r_2 - r$ are integers. Since $0 \le r < a$, $-a < -r \le 0$. Adding $0 \le r_2 < a$ to this last inequality, we obtain $-a < r_2 - r < a$. Assume $r_2 \ge r$, then (2) $0 \le r_2 - r < a$. <<Instead of assuming $r_2 \ge r$, we could assume $r \ge r_2$ and reach the same conclusions throughout the remainder of the proof by interchanging r and r_2.>> Since a divides the nonnegative integer $r_2 - r$ and since from (2) $0 \cdot a \le r_2 - r < 1 \cdot a$ either there is an integer between 0 and 1 or

$r_2 = r$. Because there is no integer between 0 and 1, (3) $r_2 = r$. Substituting (3) into (1), we find $a(q - q_2) = 0$ which implies $a = 0$ or $q - q_2 = 0$. Since $a \in \mathbf{N}$, $a \neq 0$ and, therefore, $q = q_2$.

The concept of the greatest lower bound of a set is similar to, but not the same as, the concept of the least element and the concept of a least upper bound of a set is similar to the concept of a greatest element. We define these new concepts below.

DEFINITIONS Lower Bound, Greatest Lower Bound, Upper Bound, and Least Upper Bound

Let A be a subset of the real numbers $\mathbf{R}$,

The number $m \in \mathbf{R}$ is a **lower bound for** $A \Leftrightarrow m \leq x$ for all $x \in A$.

The number $m \in \mathbf{R}$ is the **greatest lower bound for** A (**infimum of** A) $\Leftrightarrow$ (i) m is a lower bound for A and (ii) if ℓ is a lower bound for A, then $m \geq \ell$.

The number $u \in \mathbf{R}$ is an **upper bound for** $A \Leftrightarrow u \geq x$ for all $x \in A$.

The number $u \in \mathbf{R}$ is the **least upper bound for** A (**supremum of** A) $\Leftrightarrow$ (i) u is an upper bound for A and (ii) if t is an upper bound for A, then $u \leq t$.

The set A is **bounded** if it has a lower bound and an upper bound.

The set $A_1 = \{-3, -1, 3, 7\}$ has a lower bounds of -20, -8, -5 and -3. A_1 has a greatest lower bound of -3, and the least element of A_1 is -3. A_1 has upper bounds of 7, 11, 15, and 24. A_1 has least upper bound 7, and the greatest element of A_1 is 7. The set A_1 is bounded. The set $A_2 = (5, 11]$ has lower bounds of -2, 0, and 5. The greatest lower bound of A_2 is 5, but A_2 has no least element since $5 \notin A_2$. A_2 has upper bounds of 13, 17, and 24. The least upper bound of A_2 is 11, and the greatest element of A_2 is 11. A_2 is bounded. The set $A_3 = (-\infty, 4)$ is not bounded below and, therefore, has no least element. A_3 is bounded above, has least upper bound of 4, but has no greatest element.

DEFINITION Complete Ordered Field

A field F with addition $+$ and multiplication $\cdot$ which is ordered by the relation $<$ is a **complete ordered field** if and only if every nonempty subset of F which has an upper bound has a least upper bound in F.

Both the rational number system and the real number system are ordered fields. The real number system is a complete ordered field, but the rational number system is not a complete ordered field. This is one distinction between the field of rational numbers and the field of real numbers. We will not prove the ordered field of real numbers is complete; however, after we prove the Archimedean property of the real numbers we will prove the ordered field of rational numbers is not complete.

Theorem 4.7 The Archimedean Property for the Real Numbers

If $x, y \in \mathbf{R}$, $x > 0$, and $y > 0$, then there exists an $n \in \mathbf{N}$ such that $nx > y$.

Proof: If this theorem were false, the inequality $nx \le y$ would hold for all $n \in \mathbf{N}$. That is, the set $A = \{nx \mid n \in \mathbf{N},\ x \in \mathbf{R} \text{ and } x > 0\}$ would be bounded above by y. The set A is nonempty, since $x \in A$. Since the field of real numbers is complete, there exists a $u \in \mathbf{R}$ which is the least upper bound for A. Thus, $nx \le u$ for all $n \in \mathbf{N}$. Hence, $(n+1)x \le u$ for all $n \in \mathbf{N}$. Consequently, $u - x$ is an upper bound for A. But $u - x < u$ which contradicts u being the least upper bound for A.

Corollary 4.7.1 If $y \in \mathbf{R}$, there exists an $n \in \mathbf{N}$ such that $n > y$.

Proof: If $y \le 0$, let $n = 1$. If $y > 0$, let $x = 1$ in theorem 4.7.

Corollary 4.7.2 If $p \in \mathbf{R}$ and $p > 0$, there exists an $n \in \mathbf{N}$ such that $\frac{1}{n} < p$.

Proof: Let $x = 1$ and $y = \frac{1}{p} > 0$. Then by theorem 4.7 there exists an $n \in \mathbf{N}$ such that $n \cdot 1 > \frac{1}{p}$ or equivalently $\frac{1}{n} < p$.

Theorem 4.8 The ordered field of rational numbers, $\mathbf{Q}$, is not complete.

Proof Consider the set of rational numbers $A = \{x \in \mathbf{Q} \mid x < \sqrt{2}\}$. The set A is bounded above in $\mathbf{Q}$ by 2. Suppose $u \in \mathbf{Q}$ is the least upper bound of A.

Case 1. If $u < \sqrt{2}$, then $\sqrt{2} - u > 0$ and by corollary 4.7.2 there exists an $n \in \mathbf{N}$ such that $\frac{1}{n} < \sqrt{2} - u$. The number $u + \frac{1}{n}$ is rational, is less than $\sqrt{2}$, and is an element of A by definition. Since $u + \frac{1}{n} \in A$ and $u < u + \frac{1}{n}$, u is not an upper bound for A, which is a contradiction.

Case 2. If $u > \sqrt{2}$, then $u - \sqrt{2} > 0$ and by corollary 4.7.2 there exists an $m \in \mathbf{N}$ such that $\frac{1}{m} < u - \sqrt{2}$. The number $u - \frac{1}{m}$ is rational, is greater than $\sqrt{2}$, and is an upper bound for A. But $u - \frac{1}{m} < u$ which contradicts u being the least upper bound for A.

Since $u < \sqrt{2}$ is false and $u > \sqrt{2}$ is false, we conclude from the trichotomy law that $u = \sqrt{2} \notin \mathbf{Q}$. So A is a nonempty subset of the field of rational numbers which has a least upper bound which is not an element of $\mathbf{Q}$. Therefore, by definition, the ordered field of rational numbers is not complete.

The following theorem states that for all real numbers x there is a unique integer n such that $x \in [n, n+1)$.

Theorem 4.9 If $x \in \mathbf{R}$, there exists a unique $n \in \mathbf{Z}$ such that $n \leq x < n+1$.

Proof: **Case 1.** If x is an integer, let $n = x$ and the conclusion is true.

Case 2. If x is not an integer, by corollary 4.7.1 there exists a natural number k such that $k > x$. Also by corollary 4.7.1 there exists a natural number ℓ such that $\ell > -x$ which implies $-\ell < x$. That is, there are two integers k and $m = -\ell$ such that $m < x < k$. Since x is not an integer, x must lie between two consecutive integers in the finite set $S = \{m, m+1, m+2, \ldots k\}$. Suppose n_1 and n_2 are two distinct integers in S such that $n_1 < x < n_1 + 1$ and $n_2 < x < n_2 + 1$. Assume $n_1 < n_2$. Then $n_1 < n_2 < x < n_1 + 1$. That is, there is an integer (namely, n_2) between n_1 and $n_1 + 1$. Contradiction.

An important property of both the set of rational numbers and the set of irrational numbers is that they are both **dense** in the set of real numbers. That is, between any two distinct real numbers there is a rational number and an irrational number. As a matter of fact, between any two distinct real numbers there are an infinite number of rational numbers and an infinite number of irrational numbers. In the following theorem, we prove the set of rational numbers are dense in the set of real numbers.

Theorem 4.10 If $a, b \in \mathbf{R}$ and $a < b$, then there exists a rational number r such that $a < r < b$.

Proof: Since $a < b$, $b - a > 0$ and by corollary 4.7.2 there exists a natural number n such that $\frac{1}{n} < b - a$. Solving for b, we find (1) $a + \frac{1}{n} < b$. By theorem 4.9 there exists a unique integer m such that (2) $m \leq an + 1 < m + 1$. Multiplying the left-hand inequality of (2) by $\frac{1}{n} > 0$ yields (3) $\frac{m}{n} \leq a + \frac{1}{n}$.

Solving the right-hand inequality of (2) for a yields (4) $a < \frac{m}{n}$. Combining (1), (3), and (4), we have $a < \frac{m}{n} \leq a + \frac{1}{n} < b$. Observe that $r_1 = \frac{m}{n}$ is a rational number such that $a < r_1 < b$.

In the proof of theorem 4.10 the rational number r_1 satisfies $r_1 < b$. So by theorem 4.10 there exists a second rational number r_2 which satisfies $a < r_1 < r_2 < b$, and a third rational number r_3 which satisfies $a < r_1 < r_2 < r_3 < b$, etc. Thus, there are an infinite number of rational numbers between a and b.

We now show that between every two distinct rational numbers there is an irrational number.

Theorem 4.11 If $r, s \in \mathbf{Q}$ and $r < s$, then there exists an irrational number t such that $r < t < s$.

Proof: Claim: The number $t = r + (\sqrt{2} - 1)(s - r)$ is irrational and $r < t < s$. The number $\sqrt{2} - 1$ is irrational, since it is the sum of the irrational number, $\sqrt{2}$, and the rational number, -1. The number $s - r$ is rational, because s and r are both rational. Since $(\sqrt{2} - 1)(s - r)$ is the product of an irrational number and a rational number, it is irrational. Therefore, t, which is the sum of the rational number r and the irrational number $(\sqrt{2} - 1)(s - r)$ is irrational.

Since $\sqrt{2} - 1 > 0$ and $s - r > 0$ by **O3**, $(\sqrt{2} - 1)(s - r) > 0$ and $t - r = (\sqrt{2} - 1)(s - r) > 0$. Hence, $r < t$. Multiplying and rearranging, we find $t = r + (\sqrt{2} - 1)(s - r) = (2 - \sqrt{2})r + (\sqrt{2} - 1)s$. Since $r < s$ and $2 - \sqrt{2} > 0$, $(2 - \sqrt{2})r < (2 - \sqrt{2})s$. Consequently,

$$t = (2 - \sqrt{2})r + (\sqrt{2} - 1)s < (2 - \sqrt{2})s + (\sqrt{2} - 1)s = s$$

Therefore, $r < t < s$.

Finally, we show that between any two distinct real numbers there is an irrational number.

Theorem 4.12 If $a, b \in \mathbf{R}$ and $a < b$, then there exists an irrational number t such that $a < t < b$.

Proof: By theorem 4.10 there exists a rational number r such that $a < r < b$. Also by theorem 4.10 there exists a rational number s such that $a < r < s < b$. By theorem 4.11 there exists an irrational number t such that $a < r < t < s < b$.

EXERCISES 4.2

1. Prove theorems T1 through T19 of this section.

2. For real numbers x, y, z, ϵ prove the following theorems.

   a. If $x > 1$, then $x^2 > x$.

   b. If $0 < x < 1$, then $x^2 < 1$.

   c. If $xy > 0$, then either (i) $x > 0$ and $y > 0$ or (ii) $x < 0$ and $y < 0$

   d. If $x > 0$ and $xy > xz$, then $y > z$.

   e. If $x < y + \epsilon$ for all $\epsilon > 0$, then $x \leq y$.

3. For the following subsets of real numbers find, if they exist, the least element, the greatest element, the greatest lower bound, and the least upper bound.

   a. $\{e, \sqrt{2}\}$

   b. $(-3, 5]$

   c. $[-7, \infty)$

   d. $(-4, 4) \cup \{5\}$

   e. $[-4, 0) \cup (0, 3)$

   f. $\{x \in \mathbf{Q} \mid x^2 < 50\}$

   g. $\left\{\frac{2n+1}{n} \mid n \in \mathbf{N}\right\}$

   h. $\left\{\frac{2n+1}{n} \mid n \in \mathbf{Z} - \{0\}\right\}$

   i. $\left\{(-1)^n \left(\frac{n+1}{n}\right) \mid n \in \mathbf{N}\right\}$

   j. $\left\{(-1)^n \left(\frac{n+1}{n}\right) \mid n \in \mathbf{Z} - \{0\}\right\}$

   k. $\bigcup_{n=1}^{\infty} [2 + \frac{1}{n}, 6 - \frac{1}{n}]$

   l. $\bigcap_{n=1}^{\infty} (2 - \frac{1}{n}, 6 + \frac{1}{n})$

4. a. Prove the following Division Algorithm for the set of integers. For all $a, b \in \mathbf{Z}$ with $a \neq 0$, there exist unique integers q and r such that $b = aq + r$ where $0 \leq r < |a|$.

   b. For the following pairs of integers, find the quotient, q, and the remainder, r, when the first integer b is divided by the second integer a.

   (i) 6132, 53 (ii) 53, 6132 (iii) 988, 76

   (iv) -175, 16 (v) -342, -91

5. Let S be a nonempty set of real numbers that is bounded below and let m be the greatest lower bound. Prove for every $\epsilon > 0$ there exists an $x \in S$ such that $x < m + \epsilon$.

4.3 Reflexive, Symmetric, Transitive, and Equivalence Relations

Often relations are classified based upon one or more properties they satisfy. Usually, relations with the same classifications have additional like properties as well. Three important properties used to classify relations are the properties of being reflexive, symmetric, or transitive.

DEFINITIONS Reflexive Relation, Symmetric Relation, and Transitive Relation

Let R be a relation on a set A.

R **is a reflexive relation on** A if and only if $(\forall x \in A)[(x, x) \in R]$.

R **is a symmetric relation on** A if and only if

$$(\forall x, y \in A)[((x, y) \in R) \Rightarrow ((y, x) \in R)].$$

R **is a transitive relation on** A if and only if

$$(\forall x, y, z \in A)[(((x, y) \in R) \wedge ((y, z) \in R)) \Rightarrow ((x, z) \in R)].$$

DEFINITION Equivalence Relation

A relation R **is an equivalence relation on** A if and only if R is reflexive, symmetric, and transitive on A.

For a nonempty set A, the reflexive property is the only property which requires some ordered pair to be in the relation R. The symmetric and transitive properties merely specify conditions which the ordered pairs in the relation must satisfy to be symmetric or transitive. Since the identity relation on A is $I_A = \{(x, x) \mid x \in A\}$, a relation R is reflexive on A if and only if $I_A \subseteq R$.

Negating the above definitions, we find

A relation R on a set A is **not reflexive** $\Leftrightarrow (\exists x \in A)((x, x) \notin R)$.

A relation R on a set A is **not symmetric** $\Leftrightarrow$

$$(\exists x, y \in A)[((x, y) \in R) \wedge ((y, x) \notin R)].$$

A relation R on a set A is **not transitive** $\Leftrightarrow$

$$(\exists x, y, z \in A)[((x, y) \in R) \wedge ((y, z) \in R) \wedge ((x, z) \notin R)].$$

Thus, to show a relation R is not reflexive, we must find a single $x \in A$ such that $(x, x) \notin R$. To show that R is not symmetric, we must find an x and y in A such that $(x, y) \in R$ and $(y, x) \notin R$. To show R is not transitive, we must find some x, y, and z (not necessarily distinct) in A such that $(x, y) \in R$, $(y, z) \in R$, and $(x, z) \notin R$.

Example 1. Consider the relation $S = \{(1, 1), (2, 2), (3, 3), (1, 2), (2, 3)\}$ on the set $A = \{1, 2, 3\}$.

Reflexive: S is reflexive on A, since $I_A = \{(1, 1), (2, 2), (3, 3)\} \subset S$.

Symmetric: S is not symmetric on A, since $1, 2 \in A$, $(1, 2) \in S$, but $(2, 1) \notin S$.

Transitive: S is not transitive on A, since $1, 2, 3 \in A$, $(1, 2) \in S$ and $(2, 3) \in S$, but $(1, 3) \notin S$.

The relation S is reflexive, not symmetric, and not transitive.

Example 2. The set $T = \{(1, 2), (2, 1)\}$ is a relation on the set $B = \{1, 2\}$.

Reflexive: T is not reflexive on B, since $1 \in B$ but $(1, 1) \notin T$.

Symmetric: T is symmetric on B, because $(1, 2) \in T$ and $(2, 1) \in T$ and because there are no other ordered pairs in T.

Transitive: T is not transitive on B, since $1, 2 \in B$, $(1, 2) \in T$ and $(2, 1) \in T$, but $(1, 1) \notin T$.

The relation T is symmetric, not reflexive, and not transitive.

Example 3. Let $A = \{1, 2, 3\}$ and $V = \{(1, 2), (1, 3), (2, 3)\}$.

Reflexive: V is not reflexive on A, since $1 \in A$ but $(1, 1) \notin V$.

Symmetric: V is not symmetric on A, since $1, 2 \in A$ and $(1, 2) \in V$, but $(2, 1) \notin V$.

Transitive: V is transitive on A, since $(1, 2) \in V$, $(2, 3) \in V$ and $(1, 3) \in V$, and there are no two other ordered pairs in V of the form (a, b), (b, c).

The relation V is transitive, not reflexive, and not symmetric.

Example 4. Let W be the relation $W = \{(1, 1), (2, 2)\}$ on the set $B = \{1, 2\}$.

Reflexive: W is reflexive on B, since $I_B = W$.

Symmetric: W is symmetric since $(1,1) \in W$ implies $(1,1) \in W$ and $(2,2) \in W$ implies $(2,2) \in W$.

Transitive: The hypothesis of the implication $[((x,y) \in W) \wedge ((y,z) \in W)] \Rightarrow ((x,z) \in W)$ is false for the relation W. Hence, the implication is true, and W is transitive vacuously.

Since the relation W is reflexive, symmetric, and transitive on B, W is an equivalence relation on B.

The empty relation $\emptyset$, which has no ordered pairs, is another example of an equivalence relation. That is, the empty relation is reflexive, symmetric, and transitive.

Of course, relations do not need to be specified as a set of ordered pairs. When a relation is specified in another manner, logical arguments must be used in determining whether the relation is reflexive, symmetric, or transitive.

Example 5. Let L be the set of all lines in the Euclidean plane and let $\not\perp$ be the relation "is not perpendicular to."

Reflexive: Let $\ell \in L$. The statement "$\ell \not\perp \ell$." is true, so the relation $\not\perp$ is reflexive.

Symmetric: Let $\ell, m \in L$. The statement "If $\ell \not\perp m$, then $m \not\perp \ell$." is true, so $\not\perp$ is symmetric.

Transitive: Let $\ell, m, n \in L$. The statement "If $\ell \not\perp m$ and $m \not\perp n$, then $\ell \not\perp n$." is false, since ℓ could be perpendicular to n and the line m could bisect two of the right angles formed by the lines ℓ and m. Thus, $\not\perp$ is not transitive.

Consequently, the relation $\not\perp$ is reflexive and symmetric but not transitive.

Example 6. Consider the relation "is a multiple of" on the set of natural numbers.

Reflexive: For all $n \in \mathbf{N}$, $n = 1 \cdot n$, so the statement "n is a multiple of n." is true. Thus, "is a multiple of" is a reflexive relation.

Symmetric: Let $m, n \in \mathbf{N}$. The statement "If m is a multiple of n, then n is a multiple of m." is false. A counterexample is $m = 6$ and $n = 2$, because $m = 3 \cdot n$, but there is no natural number k such that $n = k \cdot m$—that is, there is no natural number k such that $2 = k \cdot 6$. Hence, the relation "is a multiple of" is not symmetric.

Transitive: Let $m, n, p \in \mathbf{N}$. "If m is a multiple of n and n is a multiple of p, then m is a multiple of p." is a true statement. This is a theorem which we proved earlier.

The relation "is a multiple of" is reflexive and transitive but not symmetric.

Example 7. Let M be the set of all living male people and consider the relation "is a brother of." In this example, we will call x a brother of y if and only if x and y have the same pair of parents.

Reflexive: For $a \in M$, "a is a brother of a." is a false statement. Thus, "is a brother of" is not reflexive.

Symmetric: Let $a, b \in M$. If a is a brother of b, then since $b \in M$, b is a brother of a. Hence, the relation "is a brother of" is a symmetric relation on the set M of all living male people. (If the set M were changed to the set of all living people P (male and female), then the relation "is a brother of" is not symmetric on the set P, since a could be a brother to b and b could be a sister to a.)

Transitive: Let $a, b, c \in M$. The statement "If a is a brother of b and b is a brother to c, then a is a brother to c." is true. Thus, the relation "is a brother to" is transitive on the set M.

The relation "is a brother of" on the set of all living male people M is not reflexive but is symmetric and transitive.

Example 8. Let P be the set of all living people and consider the relation "is the mother of."

Reflexive: For $a \in P$, "a is the mother of a." is a false statement. Hence, the relation "is the mother of" is not reflexive.

Symmetric: Let $a, b \in P$. The statement "If a is the mother of b, then b is the mother of a." is false. Thus, the relation "is the mother of" is not symmetric.

Transitive: Let $a, b, c \in P$. The statement "If a is the mother of b and b is the mother of c, then a is the mother of c." is false, since a is the maternal grandmother of c. Consequently, the relation "is the mother of" is not transitive.

The relation "is the mother of" is not reflexive, is not symmetric, and is not transitive on the set P.

We have defined three properties for relations—the reflexive property, the symmetric property, and the transitive property. Given a relation, it is reflexive or not, it is symmetric or not, and it is transitive or not. Consequently, a relation may have any one of eight different combinations of these properties. So far, we have presented eight relations—one which has each of the possible combination of properties. Displayed in Table 4.1 is each type of relation and an example which corresponds to that type. In the table, the symbol T

indicates the relation has the specified property, while the symbol F indicates it does not.

Reflexive	Symmetric	Transitive	Example
T	T	T	4
T	T	F	5
T	F	T	6
T	F	F	1
F	T	T	7
F	T	F	2
F	F	T	3
F	F	F	8

TABLE 4.1: Examples of Eight Combinations of Properties of Relations

EXAMPLE 9 Proving a Relation Is an Equivalence Relation

Let R be the relation defined on the set of integers, $\mathbf{Z}$, by $(a, b) \in R \Leftrightarrow |a| = |b|$. Prove R is an equivalence relation on $\mathbf{Z}$.

SOLUTION

Let $a \in \mathbf{Z}$. Since $|a| = |a|$, $(a, a) \in R$ and R is reflexive on $\mathbf{Z}$.

Suppose $(a, b) \in R$. Then by definition of R, $|a| = |b|$. Since equality is a reflexive relation, $|b| = |a|$. Hence, $(b, a) \in R$ and R is symmetric on $\mathbf{Z}$.

Suppose $(a, b) \in R$ and $(b, c) \in R$. Then by definition $|a| = |b|$ and $|b| = |c|$. Since equality is a transitive relation, $|a| = |c|$, $(a, c) \in R$, and R is transitive on $\mathbf{Z}$.

Since R is reflexive, symmetric, and transitive, R is an equivalence relation on $\mathbf{Z}$.

In previous examples, the elements of the set A were usually numbers, lines, or people. That is, the elements of A were singular entities. In the following example, the elements of A are ordered pairs. In general, the elements of A may be almost anything.

EXAMPLE 10 Proving a Relation Is an Equivalence Relation

Let $A = \mathbf{Z} \times \mathbf{Z}$, the Cartesian product of the set of integers. Define the relation P on $\mathbf{Z} \times \mathbf{Z}$ by $((a, b), (c, d)) \in P \Leftrightarrow ab = cd$. Prove P is an equivalence relation on A.

SOLUTION

For all $a, b \in \mathbf{Z}$, $ab = ab$. Consequently, for $a, b \in \mathbf{Z}$, $((a, b), (a, b)) \in P$ and P is reflexive.

Suppose $((a, b), (c, d)) \in P$. Then by definition of P, $ab = cd$. Since equality is a reflexive relation, $cd = ab$. Hence, $((c, d), (a, b)) \in P$ and P is symmetric.

Suppose $((a, b), (c, d)) \in P$ and $((c, d), (e, f)) \in P$. Then by definition of P, $ab = cd$ and $cd = ef$. Since equality is a transitive relation, $ab = ef$, and $((a, b), (e, f)) \in P$. Consequently, P is transitive.

Since P is reflexive, symmetric, and transitive, P is an equivalence relation on $\mathbf{Z} \times \mathbf{Z}$.

We now prove that the inverse relation of an equivalence relation is also an equivalence relation.

Theorem 4.13 If E is an equivalence relation on the set A, then E^{-1} is an equivalence relation on A.

Proof: Let $a \in A$. Since E is an equivalence relation on A, $(a, a) \in E$. By definition of the inverse relation of E, $(a, a) \in E^{-1}$ also. Hence, E^{-1} is reflexive on A.

Suppose $(a, b) \in E^{-1}$. Then $(b, a) \in E$. Since E is an equivalence relation on A, E is symmetric and $(a, b) \in E$. Therefore, $(b, a) \in E^{-1}$ and E^{-1} is symmetric on A.

Finally, suppose $(a, b) \in E^{-1}$ and $(b, c) \in E^{-1}$. From the definition of inverse relation it follows that $(b, a) \in E$ and $(c, b) \in E$. Since E is an equivalence relation, E is transitive and $(c, a) \in E$. Consequently, $(a, c) \in E^{-1}$ and E^{-1} is transitive on A.

Since E^{-1} is reflexive, symmetric, and transitive, E^{-1} is an equivalence relation on A.

EXERCISES 4.3

1. Let $A = \{a, b\}$. For each relation defined below, determine whether the relation is reflexive, symmetric, or transitive on A.

   a. $R_1 = \{(b, a)\}$
   b. $R_2 = \{(b, a), (a, b)\}$
   c. $R_3 = \{(b, a), (a, b), (a, a)\}$
   d. $R_4 = \{(b, a), (a, b), (a, a), (b, b)\}$
   e. $R_5 = \{(b, b), (a, a)\}$

2. Which relations in exercise 1 are equivalence relations?

3. Let $A = \{1, 2, 3, 4, 5\}$. Consider the following relations on A.

   $T_1 = \{(x, y) \mid x < y\}$
   $T_2 = \{(x, y) \mid x \leq y\}$
   $T_3 = \{(x, y) \mid x = y\}$
   $T_4 = \{(x, y) \mid x \text{ is a factor of } y\}$
   $T_5 = \{(x, y) \mid x \text{ is a prime factor of } y\}$

   For each relation $T_i, \ i = 1, 2, \ldots, 5$

   a. Determine $\text{Dom}(T_i)$ and $\text{Rng}(T_i)$.
   b. Graph T_i on $A \times A$.
   c. Write T_i as a set of ordered pairs.
   d. Determine if T_i is reflexive, symmetric, or transitive.
   e. Determine which T_i are equivalence relations on the set A.

4. Let A be the set of all living people. Consider the following relations on A.

   $U_1 = \{(x, y) \mid x \text{ lives on the same street as } y\}$
   $U_2 = \{(x, y) \mid x \text{ lives next door to } y\}$
   $U_3 = \{(x, y) \mid x \text{ is a descendant of } y\}$

   a. Determine whether U_1, U_2, and U_3 are reflexive, symmetric, or transitive on A.
   b. Which relations are equivalence relations?

5. Let $A = \mathbf{R} - \{0\}$ and $P = \{(x, y) \mid xy > 0\}$. Prove P is an equivalence relation on A.

6. Let $A = \mathbf{R}$ and $Q = \{(x, y) \mid xy \geq 0\}$. Is Q is an equivalence relation on $\mathbf{R}$? Justify your answer.

7. Prove $R = \{(x, y) \mid x + y \text{ is an even integer}\}$ is an equivalence relation on $\mathbf{Z}$.

8. Is $S = \{(x, y) \mid x + y \text{ is an odd integer}\}$ an equivalence relation on $\mathbf{Z}$? Justify your answer.

9. Prove $T = \{(x, y) \mid x - y = 3k \text{ for some } k \in \mathbf{Z}\}$ is an equivalence relation on $\mathbf{Z}$.

10. Is $U = \{(x, y) \mid x + y = 3k \text{ for some } k \in \mathbf{Z}\}$ an equivalence relation on $\mathbf{Z}$? Justify your answer.

11. Prove $V = \{(x, y) \mid x^2 = y^2\}$ is an equivalence relation on $\mathbf{N}$.

12. Prove $W = \{(x, y) \mid x^2 + y^2 = r^2 \text{ for some } r \in \mathbf{R}\}$ is an equivalence relation on $\mathbf{R}$.

13. Let $A = \mathbf{Z} \times (\mathbf{Z} - \{0\})$ and $X = \{((a, b), (c, d)) \mid ad = bc\}$. Prove X is an equivalence relation on A.

14. Let $A = \mathbf{R} \times \mathbf{R}$ and $Y = \{((a, b), (c, d)) \mid a - b = c - d\}$. Prove Y is an equivalence relation on $\mathbf{R} \times \mathbf{R}$.

15. Prove if E is an equivalence relation on the set A, then $E^{-1} = E$. (Recall in theorem 4.4 we proved if E is an equivalence relation on A, then E^{-1} is an equivalence relation on A.)

16. Prove or disprove the statement: If R and S are equivalence relations on a set A, then $R \cup S$ is an equivalence relation on A.

17. Let R and S be relations on A. Prove the following theorems.

    a. If R and S are reflexive on A, then $R \cap S$ is reflexive on A.

    b. If R and S are symmetric on A, then $R \cap S$ is symmetric on A.

    c. If R and S are transitive on A, then $R \cap S$ is transitive on A.

    d. If R and S are equivalence relations on A, then $R \cap S$ is an equivalence relation on A.

18. Prove or disprove the following statements.

    a. If R is an equivalence relation on A, then $R \circ R$ is an equivalence relation on A.

    b. If R and S are equivalence relations on A, then $R \circ S$ is an equivalence relation on A.

4.4 Equivalence Relations, Equivalence Classes, and Partitions

It is easy to verify that the relation $R = \{(1,1),(1,2),(2,1),(2,2),(3,3),(4,4),(4,5),(5,4),(4,6),(6,4),(5,5),(5,6),(6,5),(6,6)\}$ is an equivalence relation on the set $A = \{1,2,3,4,5,6\}$. Observe that the element $1 \in A$ is R-related to the elements 1 and 2, since $(1,1) \in R$, $(1,2) \in R$, and $(2,1) \in R$. Likewise, the element 2 is R-related to 1 and 2. The element 3 is R-related to 3, since $(3,3) \in R$. Because $(4,4)$, $(4,5)$, $(5,4)$, $(4,6)$, $(6,4) \in R$, the element 4 is R-related to 4, 5, and 6. We also find 5 is R-related to 4, 5, and 6 and 6 is R-related to 4, 5, and 6. Thus, associated with each element $x \in A$ there is a set S_x which consists of all elements of A that are R-related to x. The following table summarizes our findings for the given relation R on the given set A.

x	S_x
1	$\{1,2\}$
2	$\{1,2\}$
3	$\{3\}$
4	$\{4,5,6\}$
5	$\{4,5,6\}$
6	$\{4,5,6\}$

Observe that $S_1 = S_2$, $S_4 = S_5 = S_6$ and that $A = S_1 \cup S_3 \cup S_4$. Assuming $A \neq \emptyset$, for any $x \in A$ the set S_x of elements which are R-related to x is nonempty, because R is reflexive on A and, therefore, $(x,x) \in R$. That is, if $A \neq \emptyset$ and $x \in A$, then $x \in S_x$. Also notice that the equivalence relation R subdivides the set A into subsets of R-equivalent elements. We formalize our findings with the following two definitions.

DEFINITIONS Equivalence Class and A/R

Let R be an equivalence relation on the nonempty set A.

For $x \in A$, the **equivalence class of x determined by the relation R** is the set

$$[x]_R = \{y \in A \mid (x,y) \in R\}$$

The set of all equivalence classes is called **A modulo R** (or **A mod R**) and is denoted by

$$A/R = \{[x]_R \mid x \in A\}$$

The mathematical expression $[x]_R = \{y \in A \mid (x, y) \in R\}$ which defines the equivalence class of x modulo R is read "the equivalence class of x modulo R is the set of all y in A such that (x, y) is in R." Any element $y \in A$ which appears with x in any ordered pair of R is in the set $[x]_R$. When it is obvious what relation R we are discussing, we will omit the subscript R from the equivalence class notation—that is, we will abbreviate $[x]_R$ as $[x]$. In the example above, the equivalence classes are $[1] = [2] = \{1, 2\}$, $[3] = \{3\}$, and $[4] = [5] = [6] = \{4, 5, 6\}$; and the set of all equivalence classes is $A/R = \{\{1, 2\}, \{3\}, \{4, 5, 6\}\}$.

DEFINITION Partition

Let A be a nonempty set and let $\mathcal{A}$ be a collection of subsets of A. The collection of sets $\mathcal{A}$ is a **partition of** A if and only if

(1) $X \in \mathcal{A} \Rightarrow X \neq \emptyset$

(2) $X \in \mathcal{A}$ and $Y \in \mathcal{A} \Rightarrow$ either $X = Y$ or $X \cap Y = \emptyset$

(3) $\bigcup_{X \in \mathcal{A}} X = A$

On a specified date at a particular university, the set of undergraduate students is partitioned (1) by class standing (freshman, sophomore, junior, senior), (2) by major, (3) by age, (4) by sex, (5) by residency, etc. The United States is partitioned geographically in several different ways. It is partitioned into states, into time zones, into zip code areas by the postal service, and into area codes by the telephone companies.

In the example we have been studying, the set $A = \{1, 2, 3, 4, 5, 6\}$ is partitioned by the relation R into the collection of sets

$$\mathcal{A} = A/R = \{\{1, 2\}, \{3\}, \{4, 5, 6\}\}.$$

In this section, we will study properties of equivalence relations on a set A, properties of partitions of a set A, and the connection between equivalence relations and partitions. First, we prove the following theorem regarding equivalence classes.

Theorem 4.14 Let R be an equivalence relation on a nonempty set A and let $a, b \in A$. If $b \in [a]_R$, then $[b]_R = [a]_R$.

Proof: Assume $b \in [a]_R$. By the definition of an equivalence class $(a, b) \in R$. To show that $[b]_R = [a]_R$, we show $[b]_R \subseteq [a]_R$ and $[a]_R \subseteq [b]_R$. Let $x \in [b]_R$. By definition $(b, x) \in R$. Since $(a, b) \in R$ and R is transitive on A, $(a, x) \in R$ and, therefore, $x \in [a]_R$. Thus, $[b]_R \subseteq [a]_R$. Now let $y \in [a]_R$. By definition

$(a, y) \in R$. Since $(a, b) \in R$ and R is symmetric on A, $(b, a) \in R$. Because $(b, a) \in R$, $(a, y) \in R$, and R is transitive, $(b, y) \in R$. Consequently, $y \in [b]_R$. Hence, $[a]_R \subseteq [b]_R$. Since $[a]_R \subseteq [b]_R$ and $[b]_R \subseteq [a]_R$, $[b]_R = [a]_R$.

We state the following theorems for equivalence classes and leave them for you to prove as exercises.

Theorem 4.15 Let R be an equivalence relation on a nonempty set A and let $a, b \in A$. Then $[a]_R = [b]_R \Leftrightarrow (a, b) \in R$.

Theorem 4.16 Let R be an equivalence relation on a nonempty set A and let $a, b \in A$. If $[a]_R \cap [b]_R \neq \emptyset$, then $[a]_R = [b]_R$.

Theorem 4.17 Let R be an equivalence relation on a nonempty set A and let $a, b \in A$. If $[a]_R \cap [b]_R = \emptyset \Leftrightarrow (a, b) \notin R$.

Theorem 4.18 Let R be an equivalence relation on a nonempty set A and let $a, b, c, d \in A$. If $c \in [a]_R$, if $d \in [b]_R$, and if $[a]_R \neq [b]_R$, then $(c, d) \notin R$.

The following theorem states that an equivalence relation R on a nonempty set A produces a partition of A.

Theorem 4.19 Let R be an equivalence relation on a nonempty set A. Then A/R is a partition of A.

Proof: To show that the collection of all equivalence classes

$$A/R = \{[x]_R \mid x \in A\}$$

is a partition of A, we must show three things:

(1) $[x]_R \in A/R \Rightarrow [x]_R \neq \emptyset$
(2) $[x]_R \in A/R$ and $[y]_R \in A/R \Rightarrow$ either $[x]_R = [y]_R$ or $[x]_R \cap [y]_R = \emptyset$
(3) $\bigcup_{x \in A} [x]_R = A$

(1) For each $x \in A$, $x \in [x]_R$ by definition. Hence, each equivalence class $[x]_R$ in A/R is nonempty.

(2) For $[x]_R \in A/R$ and $[y]_R \in A/R$ either (i) $[x]_R \cap [y]_R \neq \emptyset$ or (ii) $[x]_R \cap [y]_R = \emptyset$. In case (i) $[x]_R \cap [y]_R \neq \emptyset \Leftrightarrow (x, y) \in R$ by the contrapositive of theorem 4.17. By theorem 4.15, $(x, y) \in R \Leftrightarrow [x]_R = [y]_R$. Hence, $[x]_R \cap [y]_R \neq \emptyset \Leftrightarrow [x]_R = [y]_R$.

(3) Consider $\bigcup_{x \in A} [x]_R$. Since $[x]_R$ is a subset of A, $\bigcup_{x \in A} [x]_R \subseteq A$. Let y be any element of A. Since $y \in [y]_R$, $y \in \bigcup_{x \in A} [x]_R$. Thus, $A \subseteq \bigcup_{x \in A} [x]_R$ and, therefore, $A = \bigcup_{x \in A} [x]_R$.

Consequently, the collection of equivalence classes A/R (A modulo R) is a partition of the set A.

EXAMPLE 1 Congruence Modulo d, $\equiv_d$

Let d be a positive integer and let $\equiv_d$ be the relation defined on the set of integers, **Z**, by

$$m \equiv_d n \Leftrightarrow \text{ there exists a } k \in \mathbf{Z} \text{ such that } m - n = kd.$$

The expression $m \equiv_d n$ is read "m is congruent to n modulo d" or simply "m is congruent to n mod d." Sometimes this relation is written in the form $m \equiv n (\text{mod } d)$ which is also read as stated above. An equivalent way to define this relation is

$$m \equiv_d n \Leftrightarrow m - n \text{ is divisible by } d.$$

a. Prove congruence modulo d is an equivalence relation on **Z**.

b. Write the distinct equivalence classes of the equivalence relation $m \equiv_4 n$.

c. Write the partition $\mathbf{Z}/\equiv_4$.

SOLUTION

a. Let d be a positive integer. Let $m \in \mathbf{Z}$. Since $m - m = 0 = 0 \cdot d$, the relation $\equiv_d$ is reflexive.

Let $m, n \in \mathbf{Z}$ and suppose $m \equiv_d n$. Then there exists a $k \in \mathbf{Z}$ such that $m - n = k \cdot d$. Hence, $n - m = (-k) \cdot d$. Since $-k \in \mathbf{Z}$, $n \equiv_d m$ and the relation $\equiv_d$ is symmetric.

Let $m, n, p \in \mathbf{Z}$ and suppose $m \equiv_d n$ and $n \equiv_d p$. Then there exist integers k_1 and k_2 such that $m - n = k_1 \cdot d$ and $n - p = k_2 \cdot d$. Adding these two equations, we find $m - p = k_1 \cdot d + k_2 \cdot d = (k_1 + k_2) \cdot d$. Since **Z** is closed under addition, $m \equiv_d p$. That is, $\equiv_d$ is transitive.

Since the relation $\equiv_d$ is reflexive, symmetric, and transitive on **Z**, $\equiv_d$ is an equivalence relation on **Z**.

b. For $d = 4$, the distinct equivalence classes for the equivalence relation $\equiv_4$ are
$[0] = \{\ldots, -12, -8, -4, 0, 4, 8, 12, \ldots\}$
$[1] = \{\ldots, -11, -7, -3, 1, 5, 9, 13, \ldots\}$
$[2] = \{\ldots, -10, -6, -2, 2, 6, 10, 14, \ldots\}$
$[3] = \{\ldots, -9, -5, -1, 3, 7, 11, 15, \ldots\}$

c. The partition of **Z** corresponding to the equivalence relation $\equiv_4$ is $\mathbf{Z}/\equiv_4 = \{[0], [1], [2], [3]\}$.

The collection $\mathcal{A} = \{\{a, c\}, \{b, d\}, \{e\}\}$ is a partition of the set $A = \{a, b, c, d, e\}$ because (1) each set in $\mathcal{A}$ is nonempty, (2) if X, $Y \in \mathcal{A}$, then either $X = Y$ or $X \cap Y = \emptyset$, and (3) $\bigcup_{X \in \mathcal{A}} X = A$. How do we obtain the relation R on the set A which corresponds to the partition $\mathcal{A}$? Consider the set $\{a, c\} \in \mathcal{A}$. For $\{a, c\}$ to be an equivalence class of R, a must be R-related to a, a must be R-related to c, c must be R-related to a, and c must be R-related to c. That is, it must be the case that $(a, a) \in R$, $(a, c) \in R$, $(c, a) \in R$, and $(c, c) \in R$. If we let $A_1 = \{a, c\}$, we see we must have $A_1 \times A_1 \subseteq R$. Likewise, we must have $\{b, d\} \times \{b, d\} \subseteq R$, and $\{e\} \times \{e\} \subseteq R$. Hence,

$$R = \{(a, a), (a, c), (c, a), (c, c), (b, b), (b, d), (d, b), (d, d), (e, e)\}$$

Consequently, for a partition $\mathcal{A}$ which is a finite collection of finite sets, we have a relative simple technique for constructing an equivalence relation R corresponding to the partition $\mathcal{A}$. For every set $A \in \mathcal{A}$ we form the Cartesian product $A \times A$. The union of these sets of ordered pairs is the required equivalence relation R.

Let $A = \mathbf{R} \times \mathbf{R}$, the Cartesian plane, and for $r \in [0, \infty)$ define $C_r = \{(x, y) \mid x^2 + y^2 = r^2\}$. The collection $\mathcal{C} = \{C_r \mid r \in [0, \infty)\}$ is a partition of the plane. For $r = 0$, $C_0 = \{(0, 0)\}$. That is, C_0 contains only the origin of the plane. For $r > 0$, C_r is the circle with center at the origin and radius r. The collection $\mathcal{C}$ is a partition of the plane, because

(1) no C_r is empty (for any $r \in [0, \infty)$, $(0, r) \in C_r$),

(2) $C_r, C_s \in \mathcal{C} \Rightarrow$ either $C_r = C_s$ (which occurs when $r = s$) or $C_r \cap C_s = \emptyset$ (which occurs when $r \neq s$),

(3) $\bigcup_{r \in [0, \infty)} C_r = \mathbf{R} \times \mathbf{R}$ (for any $(a, b) \in \mathbf{R} \times \mathbf{R}$, $(a, b) \in C_r$ where $r = \sqrt{a^2 + b^2}$).

In this instance, the partition $\mathcal{C}$ contains an infinite number of sets and each set C_r, except for C_0, contains an infinite number of elements. The equivalence relation R on $\mathbf{R} \times \mathbf{R}$ corresponding to the partition $\mathcal{C}$ is defined as follows: Let (x, y) and (u, v) be points in the plane. Define

$$((x, y), (u, v)) \in R \Leftrightarrow x^2 + y^2 = u^2 + v^2.$$

That is, (x, y) is R-equivalent to (u, v) if and only if (x, y) and (u, v) both lie on the same circle whose center is at the origin.

By theorem 4.19 an equivalence relation R on a set A yields a partition of the set A. The next theorem, which is the converse of theorem 4.19, states that a partition of a set yields an equivalence relation on the set.

Theorem 4.20 Let $\mathcal{P}$ be a partition of a nonempty set A. For $x, y \in A$ define the set R as follows:

$$(x, y) \in R \Leftrightarrow \text{ there exists an } S \in \mathcal{P} \text{ such that } x \in S \text{ and } y \in S.$$

Then R is an equivalence relation on A.

Proof: Let $a \in A$. Since $A = \bigcup_{S \in \mathcal{P}} S$, there exists an $S \in \mathcal{P}$ such that $a \in S$. Consequently, $(a, a) \in R$ and R is reflexive.

Suppose $(a, b) \in R$. By definition $a, b \in S$ for some $S \in \mathcal{P}$. Obviously, $b, a \in S$ and, therefore, $(b, a) \in R$. Thus, R is symmetric.

Assume $(a, b) \in R$ and $(b, c) \in R$. By definition there exists an $S \in \mathcal{P}$ such that $a, b \in S$ and there exists a $T \in \mathcal{P}$ such that $b, c \in T$. Since $b \in S \cap T$, $S \cap T \neq \emptyset$ which by (2) in the definition of a partition implies $S = T$. Thus, $a, b, c \in S$ and, therefore, $(a, c) \in R$. That is, R is transitive.

From theorems 4.19 and 4.20 it follows that a partition of a nonempty set A is equivalent to an equivalence relation on the set A. Consequently, in any given situation we may select and use the concept which most readily suits our needs.

EXERCISES 4.4

1. The relation

   $$R = \{(a, a), (a, c), (c, a), (c, c), (b, b), (b, e), (e, b), (e, e), (d, d), (f, f)\}$$

   is an equivalence relation on the set $A = \{a, b, c, d, e, f\}$.

   a. List the distinct equivalence classes.

   b. Write the partition of A which corresponds to the equivalence relation R.

2. Let W be the set of English words for the days of the week. That is, let

   $$W = \{\text{Sunday, Monday, Tuesday, Wednesday, Thursday, Friday, Saturday}\}.$$

   For each of the equivalence relations defined below
   (1) list the distinct equivalence classes and
   (2) write the partition which corresponds to the equivalence relation.

   a. $x, y \in W$ are S-equivalent $\Leftrightarrow$ x and y have the same first letter.

   b. $x, y \in W$ are T-equivalent $\Leftrightarrow$ x and y have the same last letter.

c. $x, y \in W$ are U-equivalent $\Leftrightarrow$ x and y have the same number of vowels.

d. $x, y \in W$ are V-equivalent $\Leftrightarrow$ x and y have the same second vowel.

3. For $m, n \in \mathbf{Z}$ let R be the equivalence relation defined on $\mathbf{Z}$ by

$$(m, n) \in R \Leftrightarrow m + n \text{ is even.}$$

   a. What is the set $[0]_R$?

   b. What is the set $[1]_R$?

   c. Write the partition of $\mathbf{Z}$ which corresponds to the equivalence relation R.

4. Let S be the equivalence relation on $\mathbf{R} \times \mathbf{R}$ defined by (x, y) is S-related to (u, v) if and only if $x = u$.

   a. Write the equivalence class $[(2,3)]_S$ using set notation. Geometrically, what is $[(2,3)]_S$?

   b. Write the partition of $\mathbf{R} \times \mathbf{R}$ corresponding to the relation S.

5. Let T be the equivalence relation on $\mathbf{R} \times \mathbf{R}$ defined by (x, y) is T-related to (u, v) if and only if $y - x = v - u$.

   a. Write the equivalence class $[(2,3)]_T$ using set notation. Geometrically, what is $[(2,3)]_T$?

   b. Write the partition of $\mathbf{R} \times \mathbf{R}$ corresponding to the relation T.

6. Let U be the equivalence relation on $(\mathbf{R} - \{0\}) \times \mathbf{R}$ defined by (x, y) is U-related to (u, v) if and only if $xv = yu$.

   a. Write the equivalence class $[(2,3)]_U$ using set notation. Geometrically, what is $[(2,3)]_U$?

   b. Write the partition of $(\mathbf{R} - \{0\}) \times \mathbf{R}$ corresponding to the relation U.

7. For each of the following equivalence relations on $\mathbf{Z}$
   (1) write the distinct equivalence classes and
   (2) write the corresponding partition of $\mathbf{Z}$.

   a. $\equiv_1$ *b.* $\equiv_2$ *c.* $\equiv_3$ *d.* $\equiv_5$

8. Determine which of the following statements is true and which is false.

   a. $14 \equiv 7 \pmod 3$ b. $-5 \equiv 4 \pmod 3$
   c. $15 \equiv -7 \pmod{11}$ d. $-10 \equiv -2 \pmod 5$

9. Let R be the equivalence relation on the set $\mathbf{Z}$ defined by

$$(m, n) \in R \Leftrightarrow m^2 \equiv n^2 \pmod{4}.$$

   a. What is $[0]_R$?

   b. What is $[1]_R$?

   c. Write the partition of $\mathbf{Z}$ corresponding to R.

10. For each partition of the set $A = \{a, b, c, d, e\}$ defined below, write the corresponding equivalence relation on A.

   a. $\mathcal{P}_1 = \{\{a\}, \{b\}, \{c\}, \{d\}, \{e\}\}$

   b. $\mathcal{P}_2 = \{\{a, d, e\}, \{b, c\}\}$

   c. $\mathcal{P}_3 = \{\{a, e\}, \{b, c\}, \{d\}\}$

   d. $\mathcal{P}_4 = \{\{a, b, c, e\}, \{d\}\}$

11. Prove $\mathcal{P} = \{[n, n+1) \mid n \in \mathbf{Z}\}$ is a partition of $\mathbf{R}$.

12. For each $r \in \mathbf{R}$ define $P_r = \{(x, y) \in (\mathbf{R} - \{0\}) \times \mathbf{R} \mid y = rx^2\}$

   a. Sketch P_{-1}, P_0, and P_1.

   b. Geometrically describe the equivalence classes corresponding to this partition of the plane.

   c. Prove $\mathcal{P} = \{P_r \mid r \in \mathbf{R}\}$ is a partition of $(\mathbf{R} - \{0\}) \times \mathbf{R}$.

   d. Define the equivalence relation S corresponding to the partition $\mathcal{P}$.

13. For each $n \in \mathbf{N}$ define $S_n = \{(x, y) \in \mathbf{R} \times \mathbf{R} \mid n - 1 \leq \sqrt{x^2 + y^2} < n\}$

   a. Prove $\mathcal{P} = \{S_n \mid n \in \mathbf{N}\}$ is a partition of $\mathbf{R} \times \mathbf{R}$.

   b. What is the equivalence relation T corresponding to $\mathcal{P}$?

14. Let R and S be equivalence relations on A. Prove $[x]_{R \cap S} = [x]_R \cap [x]_S$.

15. Prove theorem 4.15.

16. Prove theorem 4.16.

17. Prove theorem 4.17.

18. Prove theorem 4.18.

4.5 Chapter Review

Definitions

Let A, B, C, and D be sets.

A **relation from** A **to** B is any subset of $A \times B$. In particular, when $B = A$, a relation R from A to A is called a **relation on** A.

The **domain** of a relation R from A to B is the set

$$\text{Dom}(R) = \{x \in A \mid (\exists y \in B)((x, y) \in R)\}$$

The **range** of a relation R from A to B is the set

$$\text{Rng}(R) = \{y \in B \mid (\exists x \in A)((x, y) \in R)\}$$

If R is a relation from A to B, then the **inverse relation from** B **to** A is the relation

$$R^{-1} = \{(x, y) \in B \times A \mid (y, x) \in R\}$$

The **identity relation on** A is the set

$$I_A = \{(x, x) \in A \times A \mid x \in A\}$$

Let R be a relation from A to B and let S be a relation from C to D. The relation R **equals** S, which is denoted by $R = S$, if and only if $A = C$, $B = D$, and $[(x, y) \in R \Leftrightarrow (x, y) \in S]$.

Let R be a relation from A to B and let S be a relation from B to C. The **composition of** S **and** R is the relation

$$S \circ R = \{(a, c) \in A \times C \mid (\exists b \in B)[((a, b) \in R) \wedge ((b, c) \in S)]\}$$

A set of numbers S is **ordered** if there exists a subset of **positive** numbers S_p which satisfy the axioms:

O1. (Trichotomy Law) For all $x \in S$ exactly one of the following three statements is true:

$$x = 0, \quad x \in S_p, \quad -x \in S_p.$$

O2. For all $x, y \in S_p$, $x + y \in S_p$.

O3. For all $x, y \in S_p$, $x \cdot y \in S_p$.

The number x is **negative** if and only if $-x$ is positive.

The **order relation** $<$ (read "is less than") is defined by $x < y \Leftrightarrow y - x$ is positive.

The **order relation** $>$ (read "is greater than") is defined by $x > y \Leftrightarrow x - y$ is positive.

The **order relation** $\leq$ (read "is less than or equal to") is defined by $x \leq y \Leftrightarrow x < y$ or $x = y$.

The **order relation** $\geq$ (read "is greater than or equal to") is defined by $x \geq y \Leftrightarrow x > y$ or $x = y$.

Let A be a nonempty set of real numbers.

The number $m \in A$ is the **least element of** A (**smallest element of** A or **minimum of** A) if and only if for every $x \in A$, $m \leq x$.

The number $M \in A$ is the **greatest element of** A (**largest element of** A or **maximum of** A) if and only if for every $x \in A$, $M \geq x$.

Let A be a subset of the real numbers $\mathbf{R}$,

The number $m \in \mathbf{R}$ is a **lower bound for** $A \Leftrightarrow m \leq x$ for all $x \in A$.

The number $m \in \mathbf{R}$ is the **greatest lower bound for** A (**infimum of** A) $\Leftrightarrow$ (i) m is a lower bound for A and (ii) if ℓ is a lower bound for A, then $m \geq \ell$.

A nonempty set A of real numbers is **well-ordered** if and only if every nonempty subset of A has a least element.

The number $u \in \mathbf{R}$ is an **upper bound for** $A \Leftrightarrow u \geq x$ for all $x \in A$.

The number $u \in \mathbf{R}$ is the **least upper bound for** A (**supremum of** A) $\Leftrightarrow$ (i) u is an upper bound for A and (ii) if t is an upper bound for A, then $u \leq t$.

The set A is **bounded** if it has a lower bound and an upper bound.

A field F with addition $+$ and multiplication $\cdot$ which is ordered by the relation $<$ is a **complete ordered field** if and only if every nonempty subset of F which has an upper bound has a least upper bound in F.

Let R be a relation on a set A.

R **is a reflexive relation on** A if and only if $(\forall x \in A)[(x, x) \in R]$.

R **is a symmetric relation on** A if and only if

$$(\forall x, y \in A)[((x, y) \in R) \Rightarrow ((y, x) \in R)].$$

R **is a transitive relation on** A if and only if

$$(\forall x, y, z \in A)[(((x,y) \in R) \wedge ((y,z) \in R)) \Rightarrow ((x,z) \in R)].$$

A relation R **is an equivalence relation on** A if and only if R is reflexive, symmetric, and transitive on A.

A relation R on a set A is **not reflexive** $\Leftrightarrow (\exists x \in A)((x,x) \notin R)$.

A relation R on a set A is **not symmetric** $\Leftrightarrow$

$$(\exists x, y \in A)[((x,y) \in R) \wedge ((y,x) \notin R)].$$

A relation R on a set A is **not transitive** $\Leftrightarrow$

$$(\exists x, y, z \in A)[((x,y) \in R) \wedge ((y,z) \in R) \wedge ((x,z) \notin R)].$$

Let R be an equivalence relation on the nonempty set A. For $x \in A$, the **equivalence class of** x **determined by the relation** R is the set

$$[x]_R = \{y \in A \mid (x,y) \in R\}$$

The set of all equivalence classes is called **A modulo R** (or **A mod R**) and is denoted by

$$A/R = \{[x]_R \mid x \in A\}$$

Let A be a nonempty set and let $\mathcal{A}$ be a collection of subsets of A. The collection of sets $\mathcal{A}$ is a **partition of** A if and only if

(1) $X \in \mathcal{A} \Rightarrow X \neq \emptyset$

(2) $X \in \mathcal{A}$ and $Y \in \mathcal{A} \Rightarrow$ either $X = Y$ or $X \cap Y = \emptyset$

(3) $\bigcup_{X \in \mathcal{A}} X = A$

Named Theorems and Corollaries Mentioned in This Chapter

Theorem 4.5 The Archimedean property for the Natural Numbers

For all $m, n \in \mathbf{N}$ there exists a $k \in \mathbf{N}$ such that $m < kn$.

The Well-Ordered Principle: The set of natural numbers is well-ordered by the relation $<$.

Theorem 4.6 The Division Algorithm for the Natural Numbers

For all $a, b \in \mathbf{N}$ there exists unique $q, r \in \mathbf{W}$ such that $b = aq + r$ and $0 \leq r < a$.

Theorem 4.7 The Archimedean Property for the Real Numbers

If $x, y \in \mathbf{R}$, $x > 0$, and $y > 0$, then there exists an $n \in \mathbf{N}$ such that $nx > y$.

Corollary 4.7.1 If $y \in \mathbf{R}$, there exists an $n \in \mathbf{N}$ such that $n > y$.

Corollary 4.7.2 If $p \in \mathbf{R}$ and $p > 0$, there exists an $n \in \mathbf{N}$ such that $\frac{1}{n} < p$.

Review Exercises

1. Let R be the relation $R = \{(a, 2), (b, 1), (1, c), (x, d), (3, y)\}$. Determine

   a. $\text{Dom}(R)$ b. $\text{Rng}(R)$ c. R^{-1}

2. Find the domain and range of the relation

   $S = \{(x, y) \mid (x - 2)^2 + (y + 3)^2 = 16\}$.

3. Write the inverse of the relation $T = \{(x, y) \mid 3x + 2y = 4\}$.

4. Let $U = \{(2, 3), (4, 1), (5, 5)\}$ and $V = \{(3, 1), (5, 4), (2, 3)\}$. Calculate

   a. $U \circ V$ b. $V \circ U$ c. $U^{-1} \circ V^{-1}$ d. $(V \circ U)^{-1}$

5. Let $W = \{(x, y) \in \mathbf{R} \times \mathbf{R} \mid y = \cos x\}$ and let $X = \{(x, y) \in \mathbf{R} \times \mathbf{R} \mid y = \ln x\}$

   a. Write $W \circ X$ and $X \circ W$ in set-builder notation.

   b. Determine the domain and range of $W \circ X$ and $X \circ W$.

6. Find, if they exist, the least element, the greatest element, the greatest lower bound, and the least upper bound of the following subsets of real numbers.

   a. $\{-1, \pi, e\}$ b. $[-4, 6)$ c. $\{x \in \mathbf{Q} \mid x^2 > 2\}$ d. $\{\frac{3n - 1}{n + 1} \mid n \in \mathbf{N}\}$

7. Let $S = \{1, 2, 3\}$. For each of the following relations determine if the relation is reflexive, symmetric, or transitive on S and if the relation is an equivalence relation.

   a. $R_1 = \{(1, 1), (2, 2), (1, 2)\}$

   b. $R_2 = \{(1, 1), (1, 2), (2, 2), (2, 1), (3, 3)\}$

   c. $R_3 = \{(1, 1), (2, 2), (3, 3), (1, 2), (2, 3)\}$

   d. $R_4 = \{(x, y) \in \mathbf{R} \times \mathbf{R} \mid y = 1/x\}$

8. Given that $R = \{(a, a), (b, b), (c, c), (d, d), (a, d), (b, c), (c, b), (d, a)\}$ is an equivalence relation on the set $S = \{a, b, c, d\}$, find

   a. $[a]$ b. $[c]$ c. Write the partition of S that corresponds to R.

9. Write the equivalence relation R on the set $S = \{1, 2, 3, 4, 5, 6\}$ that corresponds to the partition $\mathcal{P} = \{\{1, 3\}, \{2, 4, 6\}, \{5\}\}$.

10. Write the partition $\mathbf{Z}/\equiv_6$.

11. Is $-15 \equiv 3 \pmod 4$?

12. Let R be a relation on the set S. Prove R is symmetric if and only if $R = R^{-1}$.

13. Prove that if R is a symmetric and transitive relation on the set A and if $\text{Dom}(R) = A$, then R is reflexive on A.

14. Give an example of a relation R from A to B and a relation S from B to C such that $\text{Dom}(S \circ R) \neq \text{Dom}(R)$.

15. Prove that $R = \{(x, y) \mid y = x + 5k \text{ for some } k \in \mathbf{Z}\}$ is an equivalence relation on $\mathbf{Z}$.

16. Prove that $T = \{(x, y) \in \mathbf{R} \times \mathbf{R} \mid xy \geq 0\}$ is not an equivalence relation on $\mathbf{R}$.

Chapter 5

Functions

Historically, the concept of a function evolved over time. The word "function" (or more precisely, its Latin equivalent) was used in mathematics for the first time in 1694 by Gottfried Leibniz. Initially, a function denoted any quantity associated with a curve such as the coordinates of a point on the curve, the slope of the curve, etc. By 1718 at the latest, Johann Bernoulli (1667-1748) viewed a function as an expression which involves a variable and some constants. Later, Euler thought of a function as an equation or formula consisting of variables and constants. In 1734, Euler introduced the now familiar notation $f(x)$ to denote a function in one variable. The present set theoretic definition of a function is due mainly to Gustav Peter Lejeune Dirichlet (1805-1859).

5.1 Functions

One of the more basic concepts, which appears in virtually every branch of mathematics is the concept of a function. Functions are a very special kind of relation—that is, functions are very special kinds of sets of ordered pairs. Let A and B be nonempty sets. Recall that a relation R from A to B is any subset of the Cartesian product $A \times B$. Given an element $a \in A$ there may be no ordered pair in R which has a as first coordinate, or there may be exactly one ordered pair in R which has a as first coordinate, or there may be any number (up to the number of elements in B) of ordered pairs in R which have a as first coordinate. A function f from A to B is a relation from A to B such that every $a \in A$ has a unique $b \in B$ such that the ordered pair $(a, b) \in f$. That is, for each element $a \in A$ there is a uniquely determined ordered pair $(a, b) \in f$. We state the following definitions of a function, domain of a function, codomain of a function, and range of a function.

DEFINITION Function, Domain, Codomain, and Range

Let A and B be sets. A **function** f **from** A **to** B, denoted by $f : A \to B$, is a relation from A to B such that

(1) $\text{Dom}(f) = A$

(2) If $(a, b) \in f$ and $(a, c) \in f$, then $b = c$.

The **domain** of a function f from A to B is the set

$$\text{Dom}(f) = \{x \in A \mid (\exists y \in B)((x, y) \in f)\}$$

The **codomain** of the function f from A to B is the set B.

The **range** of a function f from A to B is the set

$$\text{Rng}(f) = \{y \in B \mid (\exists x \in A)((x, y) \in f)\}$$

From the definitions above, for a set of ordered pairs f to be a function from A to B it is necessary that $\text{Dom}(f) = A$ and $\text{Rng}(f) \subseteq B$. Furthermore, any set B which contains the $\text{Rng}(f)$ may serve as the codomain of f. **To prove a set of ordered pairs f is not a function from A to B, we must show either (i) $\text{Dom}(f) \neq A$ or (ii) there are two ordered pairs (a, b) and (a, c) in f for which $b \neq c$.**

Let us consider the following four relations from the set $A = \{a, b, c\}$ to $B = \{1, 2, 3\}$.

$R_1 = \{(a, 1), (b, 1)\}$

$R_2 = \{(a, 1), (b, 1), (c, 2)\}$

$R_3 = \{(a, 1), (a, 2), (b, 2), (c, 3)\}$

$R_4 = \{(a, 1), (a, 1), (b, 2), (c, 3)\}$

R_1 is a relation from A to B, but R_1 is not a function from A to B, because $\text{Dom}(R_1) = \{a, b\} \neq A$.

R_2 is a relation from A to B and $\text{Dom}(R_2) = A$. Only one ordered pair in R_2 has first coordinate a—the ordered pair $(a, 1)$. Likewise, there is only one ordered pair in R_2 with first coordinate b and there is only one ordered pair with first coordinate c. Thus, by properties (1) and (2) in the definition of a function, R_2 is a function from A to B. The range of the function R_2 is the set $\{1, 2\}$, which is a subset of the codomain B.

R_3 is a relation from A to B, but R_3 is not a function. Although $\mathrm{Dom}(R_3) = A$, there exist two distinct ordered pairs in R_3 with the same first coordinate and different second coordinates—namely, the pairs $(a, 1)$ and $(a, 2)$. So R_3 is not a function.

R_4 is a relation from A to B and $\mathrm{Dom}(R_4) = A$. There are exactly two ordered pairs in R_4 with the same first coordinate—namely, the coordinate a. However, they are not distinct ordered pairs. They are identical pairs. Thus, R_4 is a function. In this case, the codomain of R_4 is $B = \mathrm{Rng}(R_4)$. That is, for this function the codomain and range are the same set.

Arrow diagrams for the relations R_1, R_2, R_3, and R_4 are displayed in Figures 5.1, 5.2, 5.3, and 5.4 respectively.

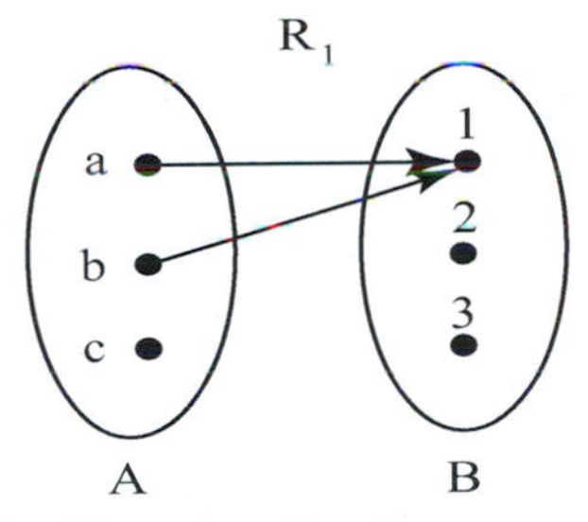

FIGURE 5.1: An Arrow Diagram for R_1

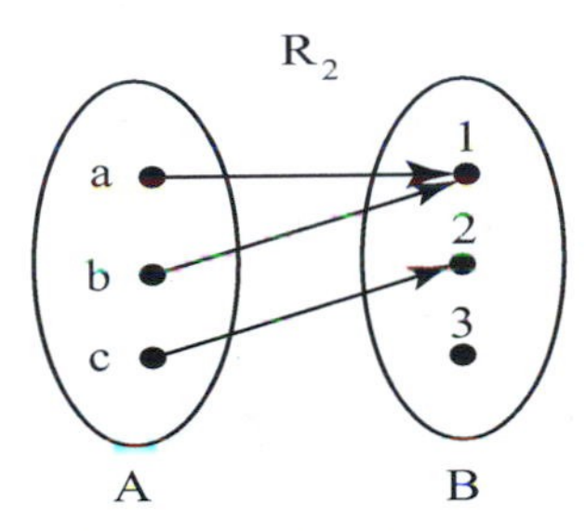

FIGURE 5.2: An Arrow Diagram for R_2

Notice in Figure 5.1 there is no arrow which has its tail at $c \in A$, so the relation R_1 is not a function from A to B, because $\mathrm{Dom}(R_1) \neq A$. In Figure 5.2, for each element of A there is an arrow which has its tail at that element, so $\mathrm{Dom}(R_2) = A$. Since there is only one arrow with its tail at each element in A, there are no two ordered pairs in R_2 with the same first coordinate and different second coordinate. That is, R_2 is a function from A to B. We also see from Figure 5.2 that $\mathrm{Rng}(R_2) = \{1, 2\} \subset B$.

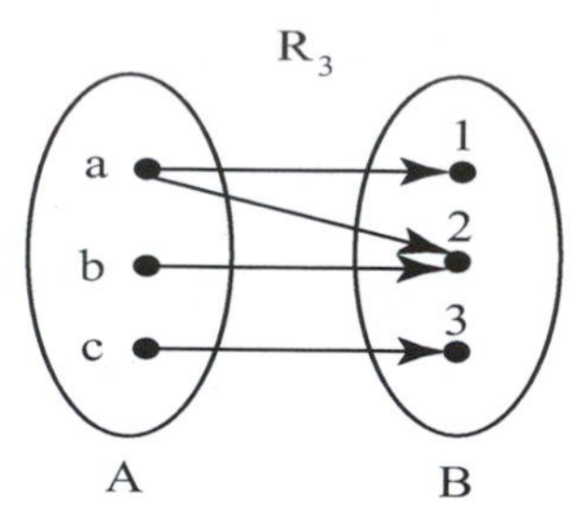

FIGURE 5.3: An Arrow Diagram for R_3

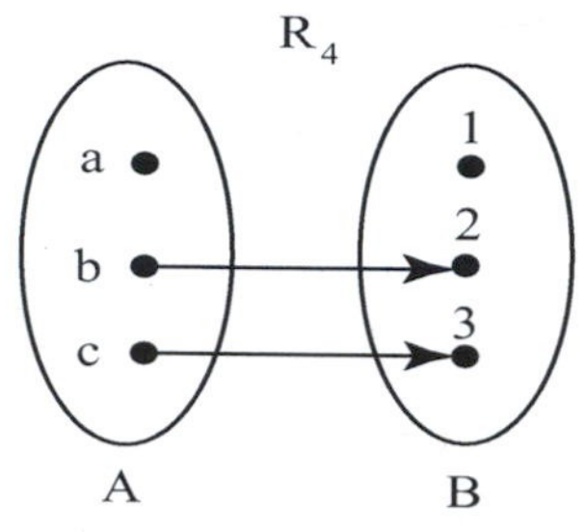

FIGURE 5.4: An Arrow Diagram for R_4

In Figure 5.3, we observe that each element of A has at least one arrow with its tail at that element. Thus, $\text{Dom}(R_3) = A$. However, the element a has two arrows with their tails at a. One of those arrows has its head at 1 and the other has its head at 2. Since $1 \neq 2$, the relation R_3 is not a function. Looking at Figure 5.4, we see $\text{Dom}(R_4) = A$, since for each element in A there is at least one arrow with its tail at the element. In this case, there are two arrows with their tails at a, but the heads of both of those arrows are at 1. So R_4 is a function from A to B.

Let f be a function which has ordered pairs (x, y). Since f is a function, it assigns to each x in its domain a unique y in its range. Therefore, we denote y by $f(x)$, read "f of x," and we write $y = f(x)$. Hence, **a function is a relation (a set of ordered pairs) which assigns to each element x in its domain the unique value $f(x)$ in its range.** For this reason, we can write the function $f : A \to B$ symbolically as follows:

$$f = \{(x, f(x)) \mid x \in A\}$$

It is important to understand the difference between the function f and its value at x, $f(x)$. The symbol f represents a set of ordered pairs, while $f(x)$ represents the value in the range of f assigned to the element x in the domain of f. The following definitions should clarify this even more.

DEFINITION Image and Pre-image

Let f be a function from A to B, $f : A \to B$, and let (x, y) denote the elements of f.

The **image of** x **under** f is $f(x)$. The image of x under f is also called the **value of** f **at** x.

The **pre-image of** y **under** f is x.

Notice that the image of x under f is a single element in the range of f, while the pre-image of y under f is one or more elements in the domain of f. Consider, for example, the function $f = \{(1, c), (2, a), (3, a)\}$. The image of 1 under f is $f(1) = c$, the image of 2 under f is $f(2) = a$, and the image of 3 under f is $f(3) = a$. The pre-image of c under f is 1, while the pre-images of a under f are 2 and 3.

Let A be any subset of the universe U. The **characteristic function of the set** A, denoted by χ_A, is the function $\chi_A : U \to [0, 1]$ defined by

$$\chi_A(x) = \begin{cases} 1, & \text{if } x \in A \\ 0, & \text{if } x \notin A \end{cases}$$

The image of any element in the set A is 1 and the image of any element in the complement of A, the set A', is 0. The pre-image of 1 is the set A and the pre-image of 0 is the set A'. The symbol χ is the lower case Greek letter "chi."

DEFINITION Equality for Functions

Two functions f and g are **equal**, written $f = g$, if and only if $\mathrm{Dom}(f) = \mathrm{Dom}(g)$ and for all $x \in \mathrm{Dom}(f)$, $f(x) = g(x)$.

Recall from chapter 4 that the identity relation on any nonempty set A is

$$I_A = \{(x, x) \in A \times A \mid x \in A\}$$

Since the unique element associated with each element $x \in A$ is x itself—that is, since $I_A(x) = x$, the identity relation is a function from A to A. The identity function on the natural numbers is $I_{\mathbf{N}} = \{(x, x) \in \mathbf{N} \times \mathbf{N} \mid x = x\}$

and the identity function on the integers is $I_{\mathbf{Z}} = \{(x, x) \in \mathbf{Z} \times \mathbf{Z} \mid x = x\}$. The identity function $I_{\mathbf{N}} \neq I_{\mathbf{Z}}$, because $\text{Dom}(I_N) = \mathbf{N} \neq \mathbf{Z} = \text{Dom}(I_{\mathbf{Z}})$.

Let f be a function with ordered pairs (x, y). The variable x is called the **independent variable** and x may assume any value in the domain of f. The variable x is also called the **argument** of f. The variable y is called the **dependent variable**. The dependent variable y represents some value in the range of f and "depends" on the value chosen for the independent variable. A **real-valued function** is a function whose codomain is a subset of the real numbers, $\mathbf{R}$. When the domain of a real-valued function is also a subset of the real numbers the function is said to be "a real-valued function of a real variable." In calculus, many functions are real-valued functions of a real variable.

Let $A = \{-1, 0, 1\}$ and $B = \{0, 1\}$. Define the real valued function $f : A \to B$ by $f(x) = |x|$ and define the real valued function $g : A \to B$ by $g(x) = x^2$. The domain of both functions f and g is the set A and for any $x \in A$, $f(x) = g(x)$, so $f = g$ on the domain A. Next, let C be the interval $[-1, 1]$ and D be the interval $[0, 1]$. Define $h : C \to D$ to be the function $h(x) = |x|$ and define $k : C \to D$ to be the function $k(x) = x^2$. The domain of both functions h and k is the set C and for $\frac{1}{2} \in A$, $h(\frac{1}{2}) = \frac{1}{2} \neq \frac{1}{4} = k(\frac{1}{2})$, so the function $h \neq k$ on the domain C. Consequently, whether one function equals another function, or not, depends on the domain of the functions.

When a function is a real-valued function of a real variable, we can graph it on a Cartesian (rectangular) coordinate system. Let $f : \mathbf{R} \to \mathbf{R}$ be defined by $f(x) = x^2$. That is, let

$$f = \{(x, y) \in \mathbf{R} \times \mathbf{R} \mid y = x^2\} = \{(x, x^2) \mid x \in \mathbf{R}\}$$

A graph of the function f is displayed in Figure 5.5. Observe that $\text{Dom}(f) = \mathbf{R}$, $\text{Rng}(f) = [0, \infty]$, the image of 2 is 4, and the pre-image of 4 is -2 and 2.

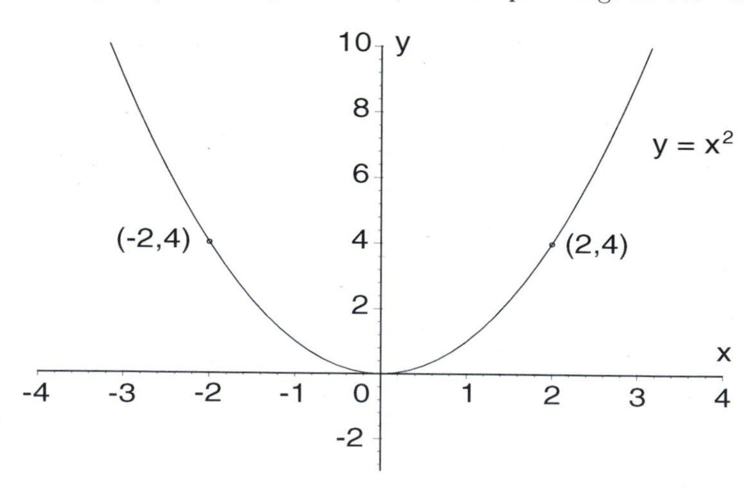

FIGURE 5.5: A Graph of the Function $f = \{(x, y) \in \mathbf{R} \times \mathbf{R} \mid y = x^2\}$

Often a real valued function of a real variable is defined by a formula and no information regarding the domain, range, or codomain is specified. In this case, the usual convention is to assume that the domain of the function is the set of all real numbers x for which the defining expression $f(x)$ is a real value. For example, the function defined by

$$f(x) = \frac{x+2}{x-3}$$

is defined for all real numbers $x \neq 3$, so the domain of this function is $\mathbf{R} - \{3\}$ or $(-\infty, 3) \cup (3, \infty)$. To determine the range, we solve the expression

$$y = \frac{x+2}{x-3} \quad \text{for } x \text{ and find} \quad x = \frac{3y+2}{y-1}$$

Since the last equation is undefined for $y = 1$, there is no pre-image of 1 and the range of the function is $\mathbf{R} - \{1\} = (-\infty, 1) \cup (1, \infty)$. Likewise, the function defined by $g(x) = \sqrt{3x-6}$ has domain $[2, \infty)$ and range $[0, \infty)$. The function defined by $h(x) = \ln(9 - 3x)$ has domain $(-\infty, 3)$ and range $\mathbf{R}$. A graph of the relation $g\{(x, y) \in \mathbf{R} \times \mathbf{R} \mid y^2 = x\}$ is shown in Figure 5.6. Observe that Dom $(g) = [0, \infty)$. Also notice that the vertical line $x = 4$ intersects the graph of g at the points $(4, -2)$ and $(4, 2)$, since $(-2)^2 = 4$ and $(2)^2 = 4$. Consequently, g is a relation but not a function.

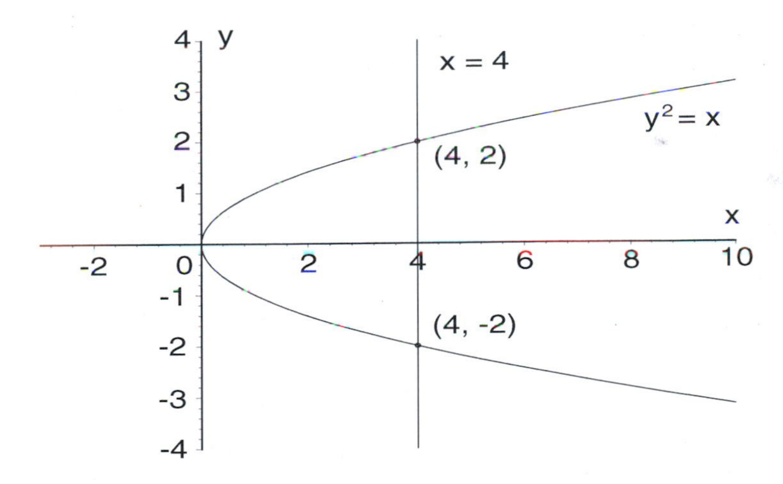

FIGURE 5.6: A Graph of the Relation $g = \{(x, y) \in \mathbf{R} \times \mathbf{R} \mid y^2 = x\}$

The following vertical line test is often useful in helping determine if a relation on $\mathbf{R}$ is a function or not. However, the vertical line test does not constitute a proof that a relation is or is not a function.

VERTICAL LINE TEST FOR A FUNCTION

Let f be a relation on $\mathbf{R}$.

If any vertical line intersects the graph of f in two or more points, then the relation f is not a function.

If no vertical line intersects the graph of f in more than one point, then the relation f is a function.

The next theorem states that the composition of two functions is a function.

Theorem 5.1 Let f be a function from A to B and let g be a function from B to C, then $g \circ f$ is a function from A to C.

Proof: Since f and g are functions, they are relations. By definition of composition for relations, $g \circ f$ is a relation from A to C. From the definition of a function, it follows that $\text{Dom}(f) = A$, $\text{Rng}(f) \subseteq B$, $\text{Dom}(g) = B$, and $\text{Rng}(g) \subseteq C$. Assume $x \in A$. Since $\text{Rng}(f) \subseteq B$, there exists a $y \in B$ such that $(x, y) \in f$. Since $\text{Dom}(g) = B$ and since $\text{Rng}(g) \subseteq C$, there exists a $z \in C$ such that $(y, z) \in g$. Thus, by the definition of composition for relations, the ordered pair $(x, z) \in g \circ f$. Therefore, $\text{Dom}(g \circ f) = A = \text{Dom}(f)$. Now suppose $(x, y), (x, z) \in g \circ f$. Hence, there exists a $u \in B$ such that $(x, u) \in f$ and $(u, y) \in g$ and there exists a $v \in B$ such that $(x, v) \in f$ and $(v, z) \in g$. Since f is a function, $(x, u) \in f$, and $(x, v) \in f$, it follows that $u = v$. Since g is a function, $(u, y) \in g$, and $(v, z) = (u, z) \in g$, it follows that $y = z$. Therefore, $g \circ f$ is a function from A to C, because (1) $\text{Dom}(g \circ f) = A$ and (2) $(x, y), (x, z) \in g \circ f$ implies $y = z$.

Let $f : A \to B$ and $g : B \to C$ be functions. Then $g \circ f$ is the function

$$g \circ f = \{(a, c) \in A \times C \mid (\exists b \in B)[(a, b) \in f) \wedge ((b, c) \in g)]\}$$

Let $(a, c) \in g \circ f$. Since f and g are functions, we may rewrite $(a, b) \in f$ as $b = f(a)$ and $(b, c) \in g$ as $c = g(b)$. Then by substitution, $(a, c) \in g \circ f$ if and only if $c = g(f(a))$. Thus, if f and g are functions, for all $a \in \text{Dom}(f)$, $(g \circ f)(a) = g(f(a))$. For instance, if $f : \mathbf{R} \to \mathbf{R}$ is defined by $f(x) = x^2$ and $g : \mathbf{R} \to \mathbf{R}$ is defined by $g(x) = x + 1$, then $(g \circ f)(x) = g(f(x)) = g(x^2) = x^2 + 1$ for all $x \in \mathbf{R}$ and $(f \circ g)(x) = f(g(x)) = f(x+1) = (x+1)^2 = x^2 + 2x + 1$ for all $x \in \mathbf{R}$. Observe that $g \circ f \neq f \circ g$ on the domain $\mathbf{R}$. As a matter of fact, in this example, $(g \circ f)(x) = (f \circ g)(x)$ only for $x = 0$. Verify this. Earlier, we had seen that the composition of two relations is not commutative. This example shows that the composition of two functions in not commutative either.

The following theorem states that the composition of functions is associative.

Theorem 5.2 Let f be a function from A to B, let g be a function from B to C, and let h be a function from C to D, then $h \circ (g \circ f) = (h \circ g) \circ f$.

Proof: In the proof of theorem 5.1, we demonstrated that the domain of the composition of two functions, $g \circ f$, is the domain of the function which is on the right. Thus, $\text{Dom}(g \circ f) = \text{Dom}(f)$ and $\text{Dom}(h \circ (g \circ f)) = \text{Dom}(g \circ f) = \text{Dom}(f)$. Likewise, $\text{Dom}((h \circ g) \circ f) = \text{Dom}(f)$. Hence, both $h \circ (g \circ f)$ and $(h \circ g) \circ f$ have the same domain—namely, $\text{Dom}(f)$. Theorem 5.1 states that the composition of two functions is a function. Hence, by two applications of theorem 5.1, $g \circ f$ is a function and so is $h \circ (g \circ f)$. Likewise, $h \circ g$ is a function and so is $(h \circ g) \circ f$. Thus far, we have shown $h \circ (g \circ f)$ and $(h \circ g) \circ f$ are both functions with the same domain. To show $h \circ (g \circ f) = (h \circ g) \circ f$ we choose an arbitrary $x \in A$ and perform the following computation

$$(h \circ (g \circ f))(x) = h((g \circ f)(x)) = h(g(f(x))) = (h \circ g)(f(x)) = ((h \circ g) \circ f)(x)$$

Thus, for all $x \in A$, $(h \circ (g \circ f))(x) = ((h \circ g) \circ f)(x)$ and, consequently, $h \circ (g \circ f) = (h \circ g) \circ f$. That is, composition of functions is associative.

EXERCISES 5.1

1. Let f be the function which assigns to each month of the year its number of days in a non-leap year.

   a. What is the domain of f?

   b. What is the range of f?

   c. What is (i) f(January)? (ii) f(February)? (iii) f(April)?

   d. What is the pre-image of (i) 28? (ii) 30? (iii) 31? (iv) 36?

2. Let g be the function which assigns to each state in the United States the capital of that state.

   a. How many elements are there in the domain of g? Describe the domain of g in words, do not list the elements.

   b. How many distinct elements are there in the range of g?

   c. What is (i) g(Alaska)? (ii) g(Louisiana)? (iii) f(Vermont)?

   d. What is the pre-image of (i) Hartford? (ii) Jefferson City? (iii) Salem? (iv) Charleston?

   e. What state x satisfies g(x) = Frankfort?

3. Let $A = \{1, 2, 3, 4\}$ and $B = \mathbf{R}$. Which of the following relations from A to B are functions?

   a. $a = \{(1, 1), (2, 2)\}$

   b. $b = \{(1, 1), (2, 1), (3, 1), (4, 1)\}$

   c. $c = \{(1, 1), (2, 3), (3, 4), (4, 3)\}$

   d. $d = \{(1, 8), (2, 10), (3, 10), (3, 4), (4, 3)\}$

4. Draw arrow diagrams for the relations in exercise 3.

5. Which of the following relations are functions?

   a. $a = \{(x, y) \in \mathbf{Z} \times \mathbf{Z} \mid x^2 + y^2 = 9\}$

   b. $b = \{(x, y) \in \mathbf{Z} \times \mathbf{Z} \mid x^2 = y^2\}$

   c. $c = \{(x, y) \in \mathbf{Q} \times \mathbf{Q} \mid 3x + 4 = y\}$

   d. $d = \{(x, y) \in \mathbf{Q} \times \mathbf{Q} \mid x = 3y + 4\}$

   e. $e = \{(x, y) \in \mathbf{Q} \times \mathbf{Q} \mid x^3 = y\}$

   f. $f = \{(x, y) \in \mathbf{Q} \times \mathbf{Q} \mid x = y^3\}$

   g. $g = \{(x, y) \in \mathbf{R} \times \mathbf{R} \mid x = \cos y\}$

   h. $h = \{(x, y) \in \mathbf{R} \times \mathbf{R} \mid y = \cos x\}$

6. Prove that the following relations are not functions.

   a. $A = \{(x, y) \in \mathbf{R} \times \mathbf{R} \mid y > x\}$

   b. $B = \{(x, y) \in \mathbf{R} \times \mathbf{R} \mid y^2 - x^2 = 9\}$

   c. $C = \{(x, y) \in \mathbf{R} \times \mathbf{R} \mid |x| + |y| = 4\}$

   d. $D = \{(x, y) \in \mathbf{R} \times \mathbf{R} \mid x = \sin y\}$

7. Let the universe be $\mathbf{R}$ and let $A = \{-3\} \cup (-2, 1] \cup \{\sqrt{2}\}$. Graph the following sets.

   a. χ_A b. $\chi_{A'}$ c. $\chi_{\mathbf{Z}}$

8. Let $f : \mathbf{R} \to \mathbf{R}$ be defined by $f(x) = x^2 + 4x$.

   a. Calculate (i) $f(-3)$ (ii) $f(-2)$ (iii) $f(0)$ (iv) $f(1)$

   b. Find the pre-image(s) of (i) 0 (ii) -3 (iii) -4 (iv) -5

   c. What is $\text{Rng}(f)$?

9. Let $f(x) = \dfrac{x^2 - 4}{x - 2}$ and $g(x) = x + 2$.

   a. Does $f = g$ for all $x \in \mathbf{R}$?

   b. Does $f = g$ for all $x \in (\mathbf{R} - \{2\})$?

10. Let $f(x) = \sqrt{x^2}$, $g(x) = x$, and $h(x) = |x|$.

   a. Does $f = g$ for all $x \in [0, \infty)$?

   b. What is $f(-3)$? What is $g(-3)$?

   c. Does $f = g$ for all $x \in \mathbf{R}$?

   d. Does $f = h$ for all $x \in \mathbf{R}$?

11. Determine the domain and range of the following functions.

   a. $\{(m, n) \in \mathbf{N} \times \mathbf{N} \mid n = 9\}$

   b. $\{(m, n) \in \mathbf{N} \times \mathbf{N} \mid n = m^2 - 5\}$

   c. $\{(m, n) \in \mathbf{Z} \times \mathbf{Z} \mid n = 2m + 3\}$

   d. $\{(m, n) \in \mathbf{Z} \times \mathbf{Z} \mid n = \dfrac{9 - m^2}{m - 3}\}$

   e. $\{(r, s) \in \mathbf{Q} \times \mathbf{Q} \mid 3r - 2s = 4\}$

   f. $\{(r, s) \in \mathbf{Q} \times \mathbf{Q} \mid s = 4 - r^2\}$

   g. $\{(x, y) \in \mathbf{R} \times \mathbf{R} \mid y = e^x + e^{-x}\}$

   h. $\{(x, y) \in \mathbf{R} \times \mathbf{R} \mid y = \dfrac{3}{\sin(2x)}\}$

12. Assuming each of the following equations defines a real valued function of a real variable, determine the domain and range of the function.

   a. $f(x) = 3x - 4$

   b. $f(x) = \dfrac{x^2 + 2x - 8}{x + 4}$

   c. $f(x) = \sqrt{2x + 3}$

   d. $f(x) = \dfrac{-5}{\sqrt{e - x}}$

   e. $f(x) = \ln(x + 4) + \ln(3 - x)$

   f. $f(x) = \sqrt{x + 5} - \sqrt{-(x + 5)}$

   g. $f(x) = 2 - 3 \sin 5x$

   h. $f(x) = \dfrac{4}{\cos x}$

13. Let $X = \{x, y, z\}$ and $Y = \{1, 2\}$.

   a. How many different functions are there from X to Y?

   b. How many different functions are there from Y to X?

   c. If the number of elements in the nonempty set X is $n(X)$ and the number of elements in the nonempty set Y is $n(Y)$, how many different functions are there from X to Y?

14. For the given pairs of functions f and g, (i) write an expression for $f \circ g$, (ii) determine the domain and range of $f \circ g$, (iii) write an expression for $g \circ f$, and (iv) determine the domain and range of $g \circ f$.

    a. $f = \{(1,2),(2,3),(3,1)\}, \quad g = \{(1,3),(2,1),(3,3)\}$

    b. $f(x) = 3x + 4, \quad g(x) = 5 - 2x$

    c. $f(x) = 2 - 3x, \quad g(x) = x^2 + x$

    d. $f(x) = |x|, \quad g(x) = \cos x$

    e. $f(x) = \cos x, \quad g(x) = \tan x$

    f. $f(x) = 2x^2 - 1, \quad g(x) = \dfrac{x-2}{x-3}$

15. For a pair of functions f and g, $f \circ g$ is **well-defined** if and only if $\text{Rng}(g) \subseteq \text{Dom}(f)$. For the following pairs of functions (i) determine if $f \circ g$ is well-defined or not and (ii) if $f \circ g$ is well-defined, give an explicit expression for $f \circ g$.

    a. $f = \{(1,3),(2,4),(3,1)\}, \quad g = \{(1,2),(2,3),(3,2)\}$

    b. $f = \{(1,2),(2,3),(3,2)\}, \quad g = \{(1,3),(2,4),(3,1)\}$

    c. $f : \mathbf{Z} \to \mathbf{Z}$ defined by $f(m) = 2m + 3$

       $g : \mathbf{N} \to \mathbf{Z}$ defined by $g(n) = n - 10$

    d. $f : \mathbf{N} \to \mathbf{N}$ defined by $f(m) = m + 2$

       $g : \mathbf{Z} \to \mathbf{Z}$ defined by $g(n) = n^2 - 4$

    e. $f : \mathbf{N} \to \mathbf{N}$ defined by $f(m) = m + 2$

       $g : \mathbf{Z} \to \mathbf{Z}$ defined by $g(n) = n^2 + 4$

    f. $f : \mathbf{N} \to \mathbf{R}$ defined by $f(m) = 7 - \sqrt{m}$

       $g : \mathbf{N} \to \mathbf{Z}$ defined by $g(n) = n + n^2$

    g. $f : \mathbf{N} \to \mathbf{Z}$ defined by $f(m) = m - m^2$

       $g : \mathbf{N} \to \mathbf{R}$ defined by $g(n) = 7 + \sqrt{n}$

16. For $f(x) = x^3 + 1$, $g(x) = \sin x$, and $h(x) = x^2 - 4$, what is

    a. $f \circ (g \circ h)$? b. $(f \circ g) \circ h$?

17. Prove if f and g are functions, then $f \cap g$ is a function.

5.2 Onto Functions, One-to-One Functions, and One-to-One Correspondences

In this section we consider special classes of functions and general results which hold for these classes. First, we define an onto function.

DEFINITION Onto Function

Let A and B be sets. A function f from A to B is **onto** (or **surjective**) if and only if

$$(\forall y \in B)\ (\exists x \in A) \text{ such that } f(x) = y$$

We denote that a function f from A to B is onto by writing $f : A \overset{\text{onto}}{\longrightarrow} B$.

By definition, a function $f : A \to B$ is onto if and only if $\text{Rng}(f) = B$. Also by definition, a function f is onto if and only if every element in the codomain has an inverse image. Hence, whether or not a function is onto depends upon the choice of the codomain, B. When $\text{Rng}(f) = B$, f is onto; but, when $\text{Rng}(f) \subset B$, f is not onto. It is always possible to make a function onto by selecting its codomain to be its range. However, it may be difficult to actually specify the codomain explicitly. For example, it is not easy to specify the range of the function f defined on the natural numbers, $\mathbf{N}$, by $f(n) = (n!)^{2/n}$. It is more convenient to just write $f : \mathbf{N} \to \mathbf{R}$ and not try to specify the set $\text{Rng}(f)$.

It follows from the definition that in order to prove a function $f : A \to B$ is onto, we must show for arbitrary $y \in B$ how to select an $x \in A$ such that $f(x) = y$. It also follows from the negation of the definition of onto that to show a function f is not onto, we must show that there exists a $y \in b$ such that for all $x \in A$, $f(x) \neq y$. The following examples illustrate how to prove a function is onto or not onto.

The function $f : \mathbf{Z} \to \mathbf{Z}$ defined by $f(n) = 2n - 1$ is not onto, since there is no integer in the domain $\mathbf{Z}$ such that $f(n) = 2n - 1 = 2$. Notice that the solution of the equation $2n-1 = 2$ is $n = \dfrac{3}{2} \notin \mathbf{Z}$. That is, 2 is in the codomain $\mathbf{Z}$, but there is no n in the domain $\mathbf{Z}$ such that $f(n) = 2$. Let O be the set of odd integers and let $g : \mathbf{Z} \to O$ defined by $g(n) = 2n - 1$. The function g is onto. Let arbitrary $m \in O$. Since m is odd, there exists a $k \in \mathbf{Z}$ such that $m = 2k + 1$. Solving $2n - 1 = m = 2k + 1$ for n, we find $n = k + 1$. Since k is an integer $n = k + 1$ is an integer. In fact, it is the integer such that $g(n) = m$. Verify this.

EXAMPLE Proving a Function Is Onto

Prove the function $f : \mathbf{R} \rightarrow \mathbf{R}$ defined by $f(x) = 2x - 3$ is onto.

SOLUTION

Let y be in the codomain $\mathbf{R}$. << We must find an x in the domain $\mathbf{R}$ such that $f(x) = 2x - 3 = y$. Solving for x we find, $x = \frac{y+3}{2}$. >> Since $y, 3, 2 \in \mathbf{R}$ and the real numbers are closed under addition and nonzero division, $x = \frac{y+3}{2} \in \mathbf{R}$. Computing, we see

$$f(x) = f(\frac{y+3}{2}) - 3 = 2(\frac{y+3}{2}) - 3 = (y+3) - 3 = y.$$

Hence, f is onto. << Observe we have shown that given any y in the range $\mathbf{R}$, $x = \frac{y+3}{2} \in \mathbf{R}$ is the element in the domain which satisfies $f(x) = y$.>>

EXAMPLE Proving a Function Is Not Onto

Prove the function $g : \mathbf{R} \rightarrow \mathbf{R}$ defined by $g(x) = x^2 - 3$ is not onto.

SOLUTION

Since for all $x \in \mathbf{R}$, $x^2 \geq 0$, it follows $x^2 - 3 \geq -3$. Let $y = -4 \in \mathbf{R}$, the codomain of g. Since for all $x \in \mathbf{R}$, $x^2 - 3 \geq -3$, there is no x in the domain $\mathbf{R}$ such that $g(x) = -4$. Therefore, g is not onto.

The following **Horizontal Line Test**, which is similar in nature to the Vertical Line Test, can be used to indicate if a function is onto a particular set or not. The test, itself, does not constitute a proof that the function is onto or not. It only gives an indication that the function may be onto or may not be onto.

HORIZONTAL LINE TEST FOR ONTO

Let f be a function from A to B where A and B are subsets of $\mathbf{R}$.

If for every $b \in B$ the horizontal line $\{(a, b) \in f \mid a \in A\}$ intersects the graph of f in one or more points, then the function f is onto.

If for some $b \in B$ there is a horizontal line $\{(a, b) \in f \mid a \in A\}$ which does not intersect the graph of f in any point, then the function f is not onto.

Figure 5.7 is a graph of the function $f : \mathbf{R} \rightarrow \mathbf{R}$ defined by $f(x) = 2x - 3$. Since every horizontal line through Figure 5.7 intersects the graph of f exactly once, the function f appears to be onto.

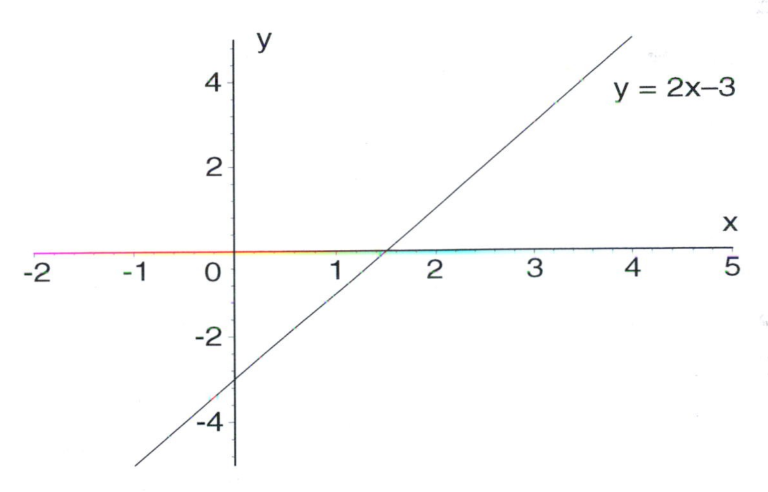

FIGURE 5.7: A Graph of $y = f(x) = 2x - 3$

Figure 5.8 is a graph of the function $g : \mathbf{R} \rightarrow \mathbf{R}$ defined by $g(x) = x^2 - 3$. Since the horizontal line through the point $(0, -4)$ of Figure 5.8 does not intersect the graph of g, the function g can be proven not to be onto $\mathbf{R}$. From the graph in Figure 5.8, it appears that if the range of g were changed from $\mathbf{R}$ to $[-3, \infty)$, then the newly defined function would be onto the new codomain.

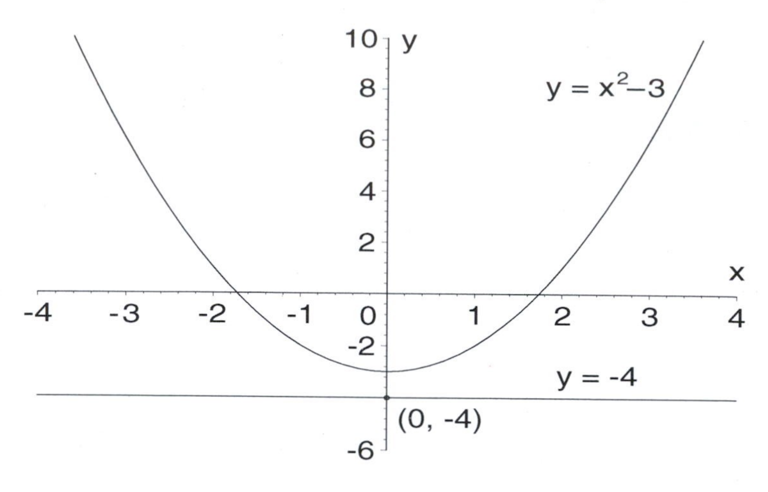

FIGURE 5.8: A Graph of $y = g(x) = x^2 - 3$

The next theorem states that the composition of two functions which are onto is a function which is onto.

Theorem 5.3 If $f : A \overset{\text{onto}}{\longrightarrow} B$ and $g : B \overset{\text{onto}}{\longrightarrow} C$, then $g \circ f : A \overset{\text{onto}}{\longrightarrow} C$.

Proof: Let $c \in C$. Since g is onto, there exists a $b \in B$ such that $g(b) = c$. Also since f is onto, there exists an $a \in A$ such that $f(a) = b$. By substitution, $g(f(a)) = (g \circ f)(a) = c$. Therefore, for every $c \in C$, there exists an $a \in A$ such that $(g \circ f)(a) = c$. Hence, the function $g \circ f$ is onto.

We state the following theorem and leave it for you to prove in the exercises.

Theorem 5.4 If $f : A \to B$, $g : B \to C$, and $g \circ f : A \overset{\text{onto}}{\longrightarrow} C$, then g is onto.

EXAMPLE Proving a Statement Is False

Prove the statement "If $f : A \to B$, $g : B \to C$, and $g \circ f : A \overset{\text{onto}}{\longrightarrow} C$, then $f : A \overset{\text{onto}}{\longrightarrow} B$." is false.

SOLUTION

Recall that to prove the statement $P \Rightarrow Q$ is false, we need to produce an example in which P is true and Q is false. In this instance, we need to exhibit three sets A, B, and C; a function f from A to B which is not onto; and a function g from B to C such that $g \circ f$ from A to C is onto. The best counterexamples are those which are as simple as possible. So we will choose A, B, and C to be small finite sets; f to be a subset of $A \times B$ which is not onto B; and g to be a subset of $B \times C$ such that $g \circ f$ is onto C. First, we select

$A = \{a\}$. Then so that f from A to B cannot be onto, we select $B = \{x, y\}$ and define $f = \{(a, x)\}$. So now we have a function f from set A to set B which is not onto, because the element $y \in B$ has no pre-image. Since we want $g \circ f$ to be onto C, we can have only one element in C. So we let $C = \{c\}$ and define $g = \{(x, c), (y, c)\}$ so that $g \circ f = \{(a, c)\}$ is onto. Our counterexample is complete. We have three sets $A = \{a\}$, $B = \{x, y\}$, $C = \{c\}$; a function $f = \{(x, c)\}$ from A to B which is not onto; and a function $g = \{(x, c), (y, c)\}$ from B to C such that $g \circ f = \{(a, c)\}$ is onto.

For a relation f from A to B to be a function it is necessary that every element of A appear *exactly once* as a first coordinate in the ordered pairs of f. For a function g from A to B to be onto, it is necessary that every element of B appear *at least once* as a second coordinate in the ordered pairs of g. A function h from A to B in which every element of B appears *at most once* as a second coordinate in the ordered pairs of h is called a **one-to-one** function. Thus, we have the following formal definition of a one-to-one function.

DEFINITION One-to-One Function

Let A and B be sets. A function $f : A \to B$ is **one-to-one** (or **injective**) if and only if $(x, y) \in f$ and $(z, y) \in f$ implies $x = z$. We denote that a function f from A to B is one-to-one by writing $f : A \xrightarrow{1\text{-}1} B$.

Using functional notation instead of ordered pair notation, a function $f : A \to B$ is **one-to-one** if and only if $f(x) = f(z)$ implies $x = z$. We can give a direct proof that f is one-to-one by assuming $f(x) = f(z)$ and showing $x = z$. Or, we can give an indirect proof by assuming $x \neq z$ and showing $f(x) \neq f(z)$.

EXAMPLE Proving a Function Is One-to-One

Prove the function $f : \mathbf{R} \to \mathbf{R}$ defined by $f(x) = 2x - 3$ is one-to-one.

SOLUTION

Assume $f(x) = f(z)$. That is, assume $2x - 3 = 2z - 3$. Hence, $2x = 2z$ and $x = z$, so the function $f(x) = 2x - 3$ is one-to-one.

EXAMPLE Proving a Function Is Not One-to-One

Prove the function $g : \mathbf{R} \to \mathbf{R}$ defined by $g(x) = x^2 - 3$ is not one-to-one.

SOLUTION

Let us see what happens if we attempt to prove g is one-to-one. We assume $g(x) = g(z)$. That is, we assume $x^2 - 3 = z^2 - 3$. Hence, (1) $x^2 = z^2$. Of course, it is wrong to conclude from equation (1) that $x = z$ and g is one-to-one. Observe that one solution of equation (1) is $x = 3$ and $z = -3$. Hence, we have the following counterexample to g being one-to-one: Let $x = 3$ and $z = -3$. Then $g(x) = g(3) = (3)^2 - 3 = 6$ and $g(z) = g(-3) = (-3)^2 - 3 = 6$ but $x \neq z$.

The following **Horizontal Line Test** can be used to indicate if a function is one-to-one or not. The test, itself, does not constitute a proof that the function is one-to-one or not. It only gives an indication that the function may be one-to-one or may not be one-to-one.

HORIZONTAL LINE TEST FOR ONE-TO-ONE

Let f be a function from A to B where A and B are subsets of $\mathbf{R}$.

If for every $b \in B$ the horizontal line $\{(a, b) \in f \mid a \in A\}$ intersects the graph of f in at most one point, then the function f is one-to-one.

If for some $b \in B$ there is a horizontal line $\{(a, b) \in f \mid a \in A\}$ which intersects the graph of f in two or more points, then the function f is not one-to-one.

Figure 5.7 is a graph of the function $f : \mathbf{R} \to \mathbf{R}$ defined by $f(x) = 2x-3$ and Figure 5.8 is a graph of the function $g : \mathbf{R} \to \mathbf{R}$ defined by $g(x) = x^2-3$. Since every horizontal line through Figure 5.7 intersects the graph of f exactly once, the function f appears to be one-to-one. Since the horizontal line through the point $(0, 6)$ of Figure 5.8 intersects the graph of g in two points, the function g can be proven not to be one-to-one, as we have done already.

The next theorem states that the composition of two functions which are one-to-one is a function which is one-to-one. We leave this theorem for you to prove in the exercises.

Theorem 5.5 If $f : A \xrightarrow{\text{1-1}} B$ and $g : B \xrightarrow{\text{1-1}} C$, then $g \circ f : A \xrightarrow{\text{1-1}} C$.

The following theorem says if the composition of two functions is one-to-one, then the right function of the composition must also be one-to-one.

Theorem 5.6 If $f : A \to B$, $g : B \to C$, and $g \circ f : A \xrightarrow{\text{1-1}} C$, then f is one-to-one.

Proof: Assume $x, y \in A$ and $f(x) = f(z)$. Since f is a function from A to B, $\text{Rng}(f) \subseteq B$. Since g is a function from B to C, $g(f(x)) = g(f(z))$. That is, $(g \circ f)(x) = (g \circ f)(z)$. By hypothesis, the function $g \circ f$ is one-to-one, so $x = z$; and, therefore, the function f is one-to-one.

Some functions are neither onto nor one-to-one. For example, $f : \mathbf{R} \to \mathbf{R}$ defined by $f(x) = x^2$ is not onto $\mathbf{R}$ and not one-to-one. Some functions are onto but not one-to-one. For instance, $g : \mathbf{R} \to [0, \infty)$ defined by $g(x) = x^2$ is onto $[0, \infty)$ but not one-to-one. Some functions are one-to-one but not onto. As an example, $h : [0, \infty) \to \mathbf{R}$ defined by $h(x) = \sqrt{x}$ is one-to-one but not onto $\mathbf{R}$. And some functions are onto and one-to-one. An example is the function $k : \mathbf{R} \to \mathbf{R}$ defined by $k(x) = x$ is onto $\mathbf{R}$ and is one-to-one.

DEFINITION One-to-One Correspondence

Let A and B be sets. A function $f : A \to B$ which is onto its codomain B and is one-to-one is called a **one-to-one correspondence** or **bijection**. We denote that a function f from A to B is a one-to-one correspondence by writing $f : A \xrightarrow[\text{onto}]{\text{1-1}} B$.

The following theorem is a direct consequence of theorems 5.3 and 5.5.

Theorem 5.7 If $f : A \xrightarrow[\text{onto}]{\text{1-1}} B$ and $g : B \xrightarrow[\text{onto}]{\text{1-1}} C$, then $g \circ f : A \xrightarrow[\text{onto}]{\text{1-1}} C$.

Theorem 5.8 follows directly from theorems 5.4 and 5.6.

Theorem 5.8 If $f : A \to B$, $g : B \to C$, and $g \circ f : A \xrightarrow[\text{onto}]{\text{1-1}} C$, then $g : B \xrightarrow{\text{onto}} C$ and $f : A \xrightarrow{\text{1-1}} B$.

EXERCISES 5.2

1. Let $A = \{a, b, c\}$. Specify a codomain B and a function $f : A \to B$ such that

   a. f is neither onto B nor one-to-one.

   b. f is onto B but not one-to-one.

   c. f is not onto B but is one-to-one.

   d. f is onto B and is one-to-one.

2. Let A be a finite set with m elements and let B be a finite set with n elements. Then there are 2^{mn} relations from A to B and n^m functions from A to B.

   a. How many functions are there from A onto B when (i) $m < n$, (ii) $m = n$, (iii) $m > n$?

   b. How many one-to-one functions are there from A to B when (i) $m < n$, (ii) $m = n$, (iii) $m > n$?

   c. How many one-to-one correspondences are there from A to B when (i) $m < n$, (ii) $m = n$, (iii) $m > n$?

3. Give examples of sets A, B, and C and functions $f : A \to B$ and $g : B \to C$ such that

   a. g is onto C but $g \circ f$ is not onto C.

   b. f is onto B but $g \circ f$ is not onto C.

   c. $g \circ f$ is onto C but f is not onto B.

   d. g is one-to-one but $g \circ f$ is not one-to-one.

   e. f is one-to-one but $g \circ f$ is not one-to-one.

   f. $g \circ f$ is one-to-one but g is not one-to-one.

4. Give an example of a function $f : \mathbf{N} \to \mathbf{N}$ such that

   a. f is neither onto nor one-to-one.

   b. f is onto but not one-to-one.

   c. f is not onto but is one-to-one.

   d. f is onto and is one-to-one.

5. Let $f : A \to \mathbf{R}$ be defined as given below. Determine a set $A \subseteq \mathbf{R}$ as large as possible so that f is one-to-one.

   a. $f(x) = (x - 2)^2 + 4$ b. $f(x) = \cos x$

   c. $f(x) = \tan x$ d. $f(x) = |3x + 2|$

6. Prove theorem 5.4

7. Prove theorem 5.5

Exercises 8-15 refer to the following functions.

a. $f : \mathbf{N} \to \mathbf{N}$ defined by $f(m) = m + 2$

b. $f : \mathbf{N} \to \mathbf{N}$ defined by $f(m) = m^3 + m$

c. $f : \mathbf{R} \to \mathbf{R}$ defined by $f(x) = e^x$

d. $f : \mathbf{R} \to (0, \infty)$ defined by $f(x) = e^x$

e. $f : \mathbf{R} \to \mathbf{R}$ defined by $f(x) = |x|$

f. $f : \mathbf{R} \to [-4, \infty)$ defined by $f(x) = |x| - 4$

g. $f : (-\infty, -\sqrt{3}) \to \mathbf{R}$ where $f(x) = \sqrt{x^2 - 3}$

h. $f : [\sqrt{3}, \infty) \to [0, \infty)$ where $f(x) = \sqrt{x^2 - 3}$

i. $f : \mathbf{R} \to \mathbf{R}$ defined by $f(x) = \sin x$

j. $f : \mathbf{R} \to [-1, 1]$ defined by $f(x) = \sin x$

k. $f : [-\pi/2, \pi/2] \to \mathbf{R}$ where $f(x) = \sin x$

l. $f : [-\pi/2, \pi/2] \to [-1, 1]$ where $f(x) = \sin x$

m. $f : \mathbf{R} - \{-3\} \to \mathbf{R} - \{1\}$ where $f(x) = \dfrac{x}{x + 3}$

n. $f : \mathbf{R} \times \mathbf{R} \to \mathbf{R}$ where $f((x, y)) = xy$

8. Which functions given above are onto the specified codomain? Prove they are onto.

9. Which functions given above are not onto the specified codomain? For each function which is not onto, give a counterexample which shows it is not onto.

10. Which functions given above are one-to-one? Prove they are one-to-one.

11. Which functions given above are not one-to-one? For each function which is not one-to-one, give a counterexample which shows it is not one-to-one.

12. Which functions are onto but not one-to-one?

13. Which functions are not onto but are one-to-one?

14. Which functions are one-to-one correspondences?

15. Which functions are neither onto nor one-to-one?

5.3 Inverse of a Function

Recall that a relation R from A to B is any set of ordered pairs in $A \times B$ and that the inverse relation R^{-1} from B to A is the relation defined by $R^{-1} = \{(b, a) \mid (a, b) \in R\}$. Since every function $f : A \to B$ is a relation, f^{-1} is a relation from B to A. However, f^{-1} may not be a function as the following two examples illustrate. First, consider the function $F : A \to B$ where $A = \{a\}$, $B = \{x, y\}$, and $F = \{(a, x)\}$. The inverse relation $F^{-1} = \{(x, a)\}$ is a relation from B to A; however, it is not a function since Dom $(F^{-1}) \neq B$—there is no ordered pair in F^{-1} with first element y. Consequently, for the inverse of a function to be a function it is necessary that the codomain of the function be the range of the function. That is, for the inverse of a function to be a function it is necessary that the function be onto. Next, consider the function $G : C \to D$ where $C = \{a, b\}$, $D = \{x\}$, and $G = \{(a, x), (b, x)\}$. The inverse relation $G^{-1} = \{(x, a), (x, b)\}$ is a relation from D to C; however, it is not a function since $(x, a), (x, b) \in G^{-1}$ but $a \neq b$. That is, for the inverse of a function to be a function it is necessary that the function be one-to-one. From these two examples, it is clear that for the inverse relation f^{-1} of a function f to be a function it is necessary that f be onto and one-to-one. The following theorem proves that these conditions are sufficient also.

Theorem 5.9 Let $f : A \to B$ be a function. Then the inverse relation f^{-1} from B to A is a function if and only if f is onto and one-to-one.

Proof: Assume f^{-1} is a function from B to A. First, we show f is onto B. Let $b \in B$. Since f^{-1} is assumed to be function, there exists a unique element $a \in A$ such that $f^{-1}(b) = a$—that is, $(b, a) \in f^{-1}$. By definition of the inverse relation, $(a, b) \in f$. Hence, for every $b \in B$ there exists an $a \in A$ such that $f(a) = b$ and f is onto B. Next, we show f is one-to-one. Assume $(a_1, b) \in f$ and $(a_2, b) \in f$. This implies $(b, a_1) \in f^{-1}$ and $(b, a_2) \in f^{-1}$. Since f^{-1} is assumed to be a function, $a_1 = a_2$ and, therefore, f is one-to-one. Hence, f^{-1} is a function from B to A implies f is onto B and one-to-one. Consequently, if f^{-1} is a function, f is a one-to-one correspondence from A onto B.

To prove the converse, we assume the function $f : A \to B$ is a one-to-one correspondence. To show that f^{-1} is a function from B to A, we must show

(1) $\text{Dom}(f^{-1}) = B$ and (2) if $(b, a_1) \in f^{-1}$ and $(b, a_2) \in f^{-1}$, then $a_1 = a_2$. To prove (1), we assume $b \in B$. Since f is onto B, there exists an $a \in A$ such that $f(a) = b$. Hence, $(a, b) \in f$ which implies $(b, a) \in f^{-1}$ for some $a \in A$—that is, $b \in \text{Dom}(f^{-1})$. To prove (2), we assume $(b, a_1) \in f^{-1}$ and $(b, a_2) \in f^{-1}$ which implies $(a_1, b) \in f$ and $(a_2, b) \in f$. Since f is one-to-one, $a_1 = a_2$. Since (1) and (2) are true, f^{-1} is a function from B to A.

EXAMPLE Determining an Inverse Function

Find an expression for the inverse of the function $f : \mathbf{R} \to \mathbf{R}$ defined by $f(x) = x^3 - 2$.

SOLUTION

First, we prove f is a one-to-one correspondence by showing f is onto and one-to-one. Let $y \in \mathbf{R}$. Solving the equation $y = x^3 - 2$ for x, we find $x = \sqrt[3]{y+2}$. Calculation then yields

$$f(\sqrt[3]{y+2}) = (\sqrt[3]{y+2})^3 - 2 = (y+2) - 2 = y$$

Since for all $y \in \mathbf{R}$, $x = \sqrt[3]{y+2}$ is such that $f(x) = y$, f is onto. Next, let $x, z \in \mathbf{R}$ and assume $f(x) = f(z)$. Thus, $x^3 - 2 = z^3 - 2$. Adding 2, we get $x^3 = z^3$ which implies $x = z$. Hence, f is one-to-one. By theorem 5.9, f^{-1} is the inverse function of f. Since f is a relation, $f^{-1} = \{(y, x) \mid (x, y) \in f\}$. That is, $f^{-1}(y) = x$ if and only if $y = f(x) = x^3 - 2$. We solved this last equation for x above when we showed f is onto. The solution is $x = \sqrt[3]{y+2}$ so $f^{-1}(y) = \sqrt[3]{y+2}$.

The next theorem says "If f is a one-to-one correspondence from A onto B, then f^{-1} is a one-to-one correspondence from B onto A."

Theorem 5.10 If $f : A \xrightarrow[\text{onto}]{\text{1-1}} B$, then $f^{-1} : B \xrightarrow[\text{onto}]{\text{1-1}} A$.

Proof: To show f^{-1} is one-to-one, we assume $f^{-1}(b_1) = f^{-1}(b_2) = a$. Hence, $(b_1, a) \in f^{-1}$ and $(b_2, a) \in f^{-1}$. By the definition of inverse relation, $(a, b_1) \in f$ and $(a, b_2) \in f$. Since f is a function, $b_1 = b_2$; and, therefore, f^{-1} is one-to-one. To show f^{-1} is onto, we let $a \in A$. Since f is a function from A to B, there exists a $b \in B$ such that $(a, b) \in f$ which implies $(b, a) \in f^{-1}$ for some $b \in B$. Hence, f^{-1} is onto A.

The following theorem states that "If f is a one-to-one correspondence from A to B and if g is a one-to-one correspondence from B to C, then the inverse function of the composition $g \circ f$ is the function $(g \circ f)^{-1} = f^{-1} \circ g^{-1}$."

Theorem 5.11 If $f : A \xrightarrow[\text{onto}]{\text{1-1}} B$ and $g : B \xrightarrow[\text{onto}]{\text{1-1}} C$, then the function $(g \circ f)^{-1} = f^{-1} \circ g^{-1}$.

Proof: By theorem 5.7 since f and g are one-to-one correspondences, $g \circ f$ is a one-to-one correspondence from A to C. By theorem 5.10, $(g \circ f)^{-1}$ is a one-to-one correspondence from C to A. And since f and g are relations, by theorem 4.2, $(g \circ f)^{-1} = f^{-1} \circ g^{-1}$.

Suppose $f : A \to B$ be a one-to-one correspondence. By theorem 5.9, f^{-1} is a function from B to A and by theorem 5.10, $f^{-1} : B \to A$ is a one-to-one correspondence. By theorem 5.7, if $f : A \to B$ is a one-to-one correspondence and $f^{-1} : B \to A$ is a one-to-one correspondence, then $f^{-1} \circ f$ is a one-to-one correspondence from A to A and $f \circ f^{-1}$ is a one-to-one correspondence from B to B. In fact, as the following theorem states $f^{-1} \circ f = I_A$ and $f \circ f^{-1} = I_B$ where I_A and I_B are the identity functions on the sets A and B respectively.

Theorem 5.12 If f is a one-to-one correspondence from A to B, then

$$(1)\ \ f^{-1} \circ f = I_A \quad \text{and} \quad (2)\ \ f \circ f^{-1} = I_B$$

Proof: << We prove (1). >> Since f is a one-to-one correspondence from A to B, by theorem 5.10, f^{-1} is a one-to-one correspondence from B to A. Therefore, by theorem 5.7, $f^{-1} \circ f$ is a one-to-one correspondence from A to A. Let $a \in A$ and suppose $f(a) = b$. Then $a = f^{-1}(b)$. Computing, we find

$$(f^{-1} \circ f)(a) = f^{-1}(f(a)) = f^{-1}(b) = a = I_A(a)$$

Hence, $(f^{-1} \circ f) = I_A$. Similarly, we can show $f \circ f^{-1} = I_B$.

Given two functions $f : A \to B$ and $g : B \to A$, we would like a test which will tell us if they are inverse functions of one another or not. The last of the following three theorems, Theorem 5.15, provides this test.

Theorem 5.13 If $f : A \to B$, $g : B \to A$, and $g \circ f = I_A$, then f is one-to-one.

Proof: Let $x, z \in A$ and assume that $f(x) = f(z)$. Applying g to both sides of this equation, we obtain $g(f(x)) = g(f(z))$—that is, $(g \circ f)(x) = (g \circ f)(z)$. Since $g \circ f = I_A$, $x = z$ and f is one-to-one.

Theorem 5.14 If $f : A \to B$, $g : B \to A$, and $f \circ g = I_B$, then f is onto B.

Proof: Let $b \in B$ and let $x = g(b) \in A$. Computing, we find $f(x) = f(g(b)) = (f \circ g)(b) = I_B(b) = b$. Hence, the function f is onto B, since for every $b \in B$ there is an $x = g(b) \in A$ such that $f(x) = b$.

Theorem 5.15 If $f : A \to B$, $g : B \to A$, $f \circ g = I_B$, and $g \circ f = I_A$, then f and g are one-to-one correspondences and $f^{-1} = g$.

Proof: It follows from theorems 5.13 and 5.14 that f and g are one-to-one correspondences. << The result for g follows by interchanging f and g and interchanging A and B in the statements of theorems 5.13 and 5.14. >> The following computation proves $f^{-1} = g$.

$$f^{-1} = f^{-1} \circ I_B = f^{-1} \circ (f \circ g) = (f^{-1} \circ f) \circ g = I_A \circ g = g$$

EXAMPLE Proving One Function is the Inverse of Another Function

Show that the inverse of the function $f : \mathbf{R} - \{2\} \to \mathbf{R} - \{3\}$ defined by $f(x) = \dfrac{3x}{x-2}$ is the function $g : \mathbf{R} - \{3\} \to \mathbf{R} - \{2\}$ defined by $g(y) = \dfrac{2y}{y-3}$.

SOLUTION

By theorem 5.15, the function $g = f^{-1}$ if (1) for all $x \in \mathbf{R} - \{2\}$, $g \circ f = I_{\mathbf{R}-\{2\}}$ and (2) for all $y \in \mathbf{R} - \{3\}$, $f \circ g = I_{\mathbf{R}-\{3\}}$.

Let $x \in \mathbf{R} - \{2\}$, then

$$(g \circ f)(x) = g(f(x)) = g\left(\frac{3x}{x-2}\right) = \frac{2\left(\dfrac{3x}{x-2}\right)}{\dfrac{3x}{x-2} - 3} = \frac{6x}{3x - 3(x-2)} = \frac{6x}{6} = x$$

Thus, $g \circ f = I_{\mathbf{R}-\{2\}}$.

Let $y \in \mathbf{R} - \{3\}$, then

$$(f \circ g)(y) = f(g(y)) = f\left(\frac{2y}{y-3}\right) = \frac{3\left(\dfrac{2y}{y-3}\right)}{\dfrac{2y}{y-3} - 2} = \frac{6y}{2y - 2(y-3)} = \frac{6y}{6} = y$$

Thus, $f \circ g = I_{\mathbf{R}-\{3\}}$. Consequently, $g = f^{-1}$. Furthermore, it follows from the computations above and theorem 5.15 that f and g are both one-to-one correspondences.

Often a function is not a one-to-one correspondence from its domain to its codomain and, therefore, the inverse of the function is a relation and not a function. In many such cases, we would like to modify the function and obtain a new function which is similar to the original function and whose inverse is a function. We can always modify a function and make it onto by changing its codomain to its range; although, as we noted earlier, it may be difficult to actually specify the range explicitly. In order to modify a function and obtain a new function which is one-to-one, it is often necessary to limit the domain of the original function. The following definition provides the means for doing this.

DEFINITION Restriction of a Function

Let function be a f from A to B and let $C \subseteq A$. The **restriction of** f **to** C, denoted by $f|_C$, is

$$f|_C = \{(x, y) \mid (x, y) \in f \text{ and } x \in C\}$$

Clearly, the restriction of f to C is a function whose domain is the set C. When g is a restriction of a function h, we may also say h is an **extension of** g.

For example, let $A = \{a, b, c, d\}$, $B = \{1, 2, 3, 4\}$, and define the function f from A to B by $f = \{(a, 1), (b, 1), (c, 3), (d, 4)\}$. The function f is not onto B because $\text{Rng}(f) = \{1, 3, 4\} \neq B$. If we let $g : A \to C$ where $C = \{1, 3, 4\}$ be defined by $g = \{(a, 1), (b, 1), (c, 3), (d, 4)\}$, then g is onto C; however, g is not one-to-one because $(a, 1), (b, 1) \in g$. Observe that the restriction of g to $D = \{a, c, d\}$,

$$g|_D = \{(a, 1), (c, 3), (d, 4)\} = h,$$

is a one-to-one correspondence from D onto C and $h^{-1} = \{(1, a), (3, c), (4, d)\}$. Also observe that the restriction of g to $E = \{b, c, d\}$,

$$g|_E = \{(b, 1), (c, 3), (d, 4)\} = k,$$

is a one-to-one correspondence from E onto C and $k^{-1} = \{(1, b), (3, c), (4, d)\}$. Thus, it is possible for a function which is not one-to-one to have more than one restriction which is one-to-one.

Recall from calculus that a real valued function of a real variable f that is defined on an interval I is **increasing on** I if and only if for all $x_1, x_2 \in I$ if $x_1 < x_2$, then $f(x_1) < f(x_2)$. Also the function f is **decreasing on** I if and only if for all $x_1, x_2 \in I$ if $x_1 < x_2$, then $f(x_1) > f(x_2)$. We leave the following fact as an exercise for you to prove: "If f is increasing or decreasing on an interval I, then f is one-to-one on the domain I." Consider the function f defined by $f(x) = |x + 2| + 3$. For $x \geq -2$, $|x + 2| = x + 2$ and for $x < -2$, $|x + 2| = -(x + 2) = -x - 2$. So

$$f(x) = |x + 2| + 3 = \begin{cases} x + 5, & \text{if } x \geq -2 \\ -x + 1, & \text{if } x < -2 \end{cases}$$

Notice that $f|_{[-2,\infty)}$ is an increasing function from the domain $[-2,\infty)$ onto the range $[3,\infty)$; and, therefore, a one-to-one correspondence whose inverse function is defined by $(f|_{[-2,\infty)})^{-1}(y) = y-5$. Also notice that $f|_{(-\infty,-2)}$ is a decreasing function from $(-\infty,-2]$ onto the range $[3,\infty)$ and, therefore, a one-to-one correspondence whose inverse function is defined by $(f|_{(-\infty,-2)})^{-1}(y) = 1-y$.

None of the six trigonometric functions sin, cos, tan, csc, sec, or cot have an inverse which is a function. A graph of $y = \sin x$ is displayed in Figure 5.9.

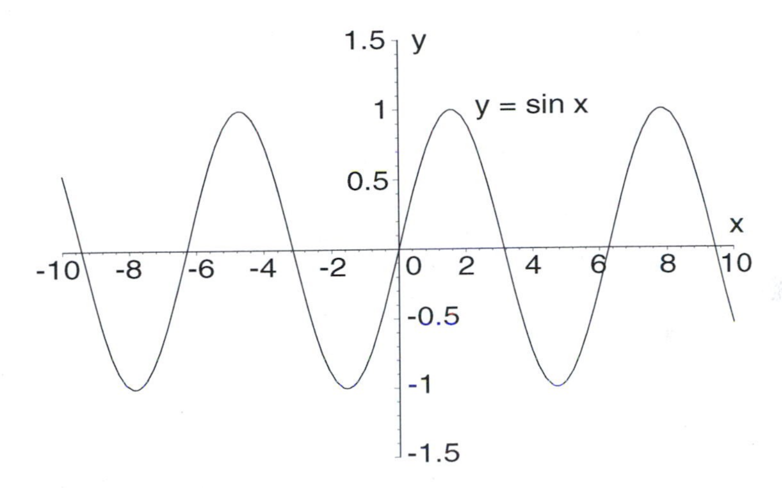

FIGURE 5.9: A Graph of the Function $y = \sin x$

As you well know, for $x \in \mathbf{R}$, $-1 \leq \sin x \leq 1$. So $\text{Rng}(\sin x) = [-1,1]$. The sine function increases on the intervals $\left[-\frac{\pi}{2} + 2n\pi, \frac{\pi}{2} + 2n\pi\right]$ for $n \in \mathbf{Z}$ and decreases on the intervals $\left[\frac{\pi}{2} + 2n\pi, \frac{3\pi}{2} + 2n\pi\right]$ for $n \in \mathbf{Z}$. Consequently, there are an infinite number of possible restrictions of the sine function which will result in a function whose inverse is also a function. The Sine function, $\text{Sin} : \left[-\frac{\pi}{2}, \frac{\pi}{2}\right] \to [-1,1]$ is defined by $\text{Sin}(x) = \sin(x)$. That is, $\text{Sin} = \sin|_{[-\frac{\pi}{2},\frac{\pi}{2}]}$. A graph of $y = \text{Sin}(x)$ is displayed in Figure 5.10 and a graph of its inverse function $y = \text{Sin}^{-1}(x)$ is displayed in Figure 5.11. A definition of the inverse Sine function is

$$\text{Sin}^{-1}(y) = x \quad \text{if and only if} \quad y = \sin x \quad \text{and} \quad -\frac{\pi}{2} \leq x \leq \frac{\pi}{2}$$

Furthermore,

$$\text{for all} \quad x \in \left[-\frac{\pi}{2}, \frac{\pi}{2}\right], \quad (\text{Sin}^{-1} \circ \text{Sin})(x) = (\text{Sin}^{-1}(\text{Sin}(x))) = x$$

and

$$\text{for all} \quad y \in [-1,1], \quad (\text{Sin} \circ \text{Sin}^{-1})(y) = (\text{Sin}(\text{Sin}^{-1}(y))) = y$$

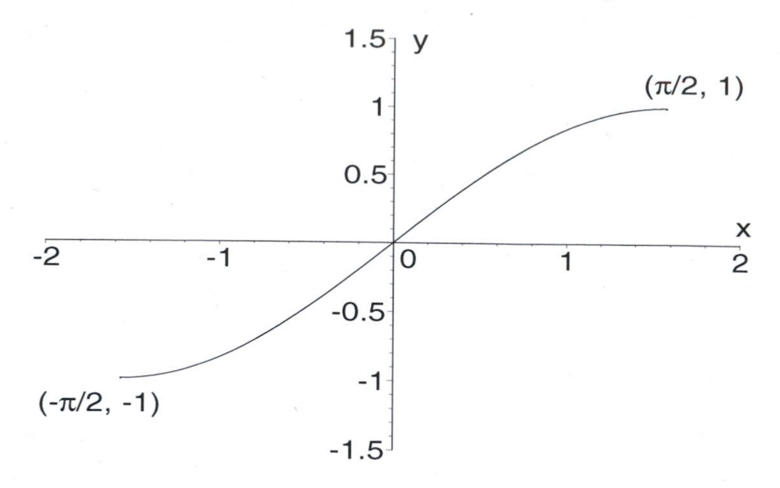

FIGURE 5.10: A Graph of the Function $y = \mathrm{Sin}(x)$

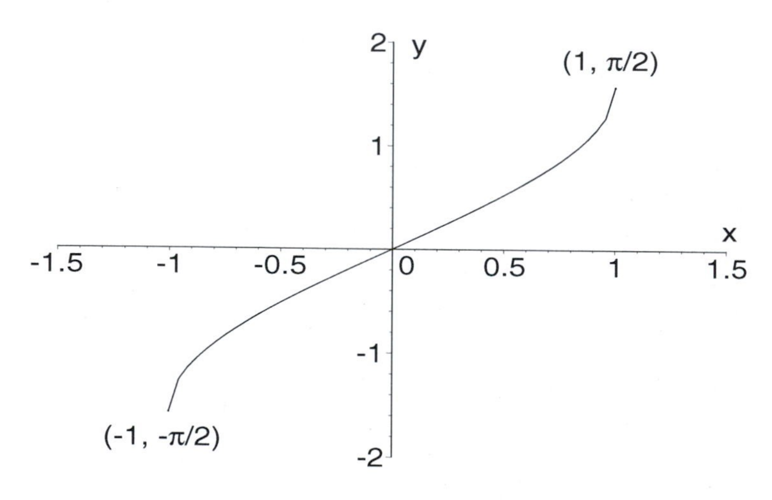

FIGURE 5.11: A Graph of the Function $y = \mathrm{Sin}^{-1}(x)$

EXERCISES 5.3

1. Prove that $f : \mathbf{R} \to \mathbf{R}$ defined by $f(x) = 3x - 2$ is a one-to-one correspondence and find f^{-1}.

2. Prove that $f : \mathbf{R} \to \mathbf{R}$ defined by $f(x) = ax + b$ where $a \neq 0$ is a one-to-one correspondence and find its inverse function.

3. Prove that $f : (-\infty, 0] \to [1, \infty)$ defined by $f(x) = x^2 + 1$ is a one-to-one correspondence and find f^{-1}.

4. Let $f : \mathbf{Z} \to \mathbf{N}$ be defined by

$$f(m) = \begin{cases} -2m, & \text{if } m < 0 \\ 2m + 1, & \text{if } m \geq 0 \end{cases}$$

   Prove that f is a one-to-one correspondence and find f^{-1}.

5. Let $f : \mathbf{R} \to \mathbf{R}$ defined by $f(x) = -2x + 3$ and let $g : \mathbf{R} \to \mathbf{R}$ defined by $g(x) = 4x - 5$.
   a. Write expressions for f^{-1} and g^{-1}.
   b. Write expressions for $g \circ f$, $(g \circ f)^{-1}$ and $f^{-1} \circ g^{-1}$.
   c. Does $(g \circ f)^{-1} = f^{-1} \circ g^{-1}$?

6. Let $f(x) = \dfrac{1}{1-x}$ and $\text{Dom}(f) = [0, 1)$.
   a. Find $\text{Rng}(f)$.
   b. Show that f is a one-to-one correspondence from $[0, 1)$ onto $\text{Rng}(f)$.
   c. Write an expression for $f^{-1} : \text{Rng}(f) \to [0, 1)$.
   d. Verify $f^{-1} \circ f = I_{[0,1)}$ and $f \circ f^{-1} = I_{\text{Rng}(f)}$.

7. Let $f(x) = \dfrac{x}{1-x}$ and $\text{Dom}(f) = (0, 1)$.
   a. Find $\text{Rng}(f)$.
   b. Show that f is a one-to-one correspondence from $(0, 1)$ onto $\text{Rng}(f)$.
   c. Write an expression for $f^{-1} : \text{Rng}(f) \to (0, 1)$.
   d. Verify $f^{-1} \circ f = I_{(0,1)}$ and $f \circ f^{-1} = I_{\text{Rng}(f)}$.

8. Show that the following pairs of functions are inverses by showing $g \circ f = I_A$ and $f \circ g = I_B$.
   a. $A = (-\infty, 0]$, $B = [0, \infty)$, $f : A \to B$ is defined by $f(x) = x^2$, and $g : B \to A$ is defined by $g(y) = -\sqrt{y}$.
   b. $A = [0, \infty)$, $B = [-2, \infty)$, $f : A \to B$ is defined by $f(x) = \sqrt{x} - 2$, and $g : B \to A$ is defined by $g(y) = y^2 + 4y + 4$.
   c. $A = \mathbf{R}$, $B = (0, \infty)$, $f : A \to B$ is defined by $f(x) = 3e^{2x-1}$, and $g : B \to A$ is defined by $g(y) = \dfrac{1}{2} \ln\left(\dfrac{y}{3}\right) + \dfrac{1}{2}$.

9. Give an example of two functions $f : \mathbf{R} \to \mathbf{R}$ and $g : \mathbf{R} \to \mathbf{R}$ which are both one-to-one correspondences, but the functions $f + g$ and fg are both not one-to-one correspondences.

10. Give an example of two functions $f : A \to B$ and $g : B \to A$ such that $f \circ g = I_B$ but f^{-1} is not a function.

11. Let f be a one-to-one correspondence from A to B and let g be a one-to-one correspondence from B to A. Prove: If $g \circ f = I_A$ or $f \circ g = I_B$, then $f^{-1} = g$.

12. Let $A = \{2, 3, 5, 7, 11\}$, $B = \{1, 2, 3, 5, 8, 13\}$, and let f from A to B be defined by $f = \{(2, 2), (3, 3), (5, 8), (7, 1), (11, 8)\}$.

    a. Find $\text{Rng}(f)$ b. Is f onto B? c. What is $f|_{\{5,11\}}$?

    d. Find two sets C with four elements such that $f|_C$ is one-to-one.

13. Let $f : \mathbf{R} \to \mathbf{R}$ be defined by $f(x) = 2 - 5x$ and let $A = \{-2, 0, 1, 3\}$. Find a. $f|_A$ and b. $\text{Rng}(f|_{[-2,3]})$.

14. Let $A \subseteq \mathbf{R}$ and let $f : A \to \mathbf{R}$ be as defined below. Determine A in two different ways so that $f|_A$ is a one-to-one function. Choose the set A to be as large as possible. For both choices of A determine $\text{Rng}(f|_A)$ and write the expression for $(f|_A)^{-1} : \text{Rng}(f|_A) \to A$.

    a. $f(x) = |3x + 4| - 6$ b. $f(x) = -x^2 + 4x - 8$

15. Determine the ranges of the following restricted trigonometric functions.

    a. $\text{Rng}(\cos|_{[0,\frac{\pi}{2}]})$ b. $\text{Rng}(\cos|_{[\frac{\pi}{2},\pi]})$

    c. $\text{Rng}(\tan|_{[0,\frac{\pi}{4}]})$ d. $\text{Rng}(\tan|_{[\frac{\pi}{4},\frac{\pi}{2})})$

16. Prove that if $f : \mathbf{R} \to \mathbf{R}$ is an increasing function on the interval I, then $f|_I$ is one-to-one.

17. We saw that the Sine function is the restriction of the sine function to the interval $\left[-\frac{\pi}{2}, \frac{\pi}{2}\right]$—that is, $\text{Sin} = \sin|_{[-\frac{\pi}{2},\frac{\pi}{2}]}$—and that $\text{Rng}(\text{Sin}) = [-1, 1]$. Associated with each trigonometric function sin, cos, tan, csc, sec, cot there is a restricted trigonometric function Sin, Cos, Tan, Csc, Sec, Cot whose inverse is a function. Complete the following table.

Function	Domain	Range	Definition by Restriction
Sin	$\left[-\frac{\pi}{2}, \frac{\pi}{2}\right]$	$[-1, 1]$	$\text{Sin} = \sin|_{[-\frac{\pi}{2},\frac{\pi}{2}]}$
Cos			
Tan			
Csc			
Sec			
Cot			

5.4 Images and Inverse Images of Sets

Earlier, we noted that if $f : A \to B$ is a function and $y = f(x)$, then y is the image of x under f and x is the pre-image of y under f. These concepts extend naturally from elements to sets as follows.

DEFINITION Image of a Set and Inverse Image of a Set

Let f be a function from A to B, $f : A \to B$.

If $X \subseteq A$, then the **image of** X or **image of the set** X, denoted by $f(X)$, is

$$f(X) = \{y \in B \mid y = f(x) \text{ for some } x \in X\}$$

If $Y \subseteq B$, then the **inverse image of** Y, denoted by $f^{-1}(Y)$, is

$$f^{-1}(Y) = \{x \in A \mid f(x) \in Y\}$$

By definition, the image of the set X under the function f, $f(X)$, is the set of all images of the elements of X. The set $f(X)$ is a subset of $\text{Rng}(f) \subseteq B$. The inverse image of the set Y under the function f, $f^{-1}(Y)$, is the set of all pre-images of the elements of Y. The set $f^{-1}(Y) \subseteq A$. Figure 5.12 graphically illustrates this situation.

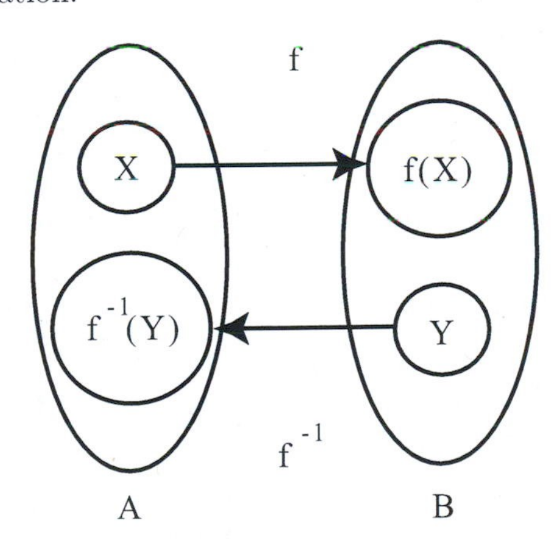

FIGURE 5.12: Image of X under f, $f(X)$, and the Inverse Image of Y under f^{-1}, $f^{-1}(Y)$

The next theorem follows easily from the definitions of image and inverse image of sets.

Theorem 5.16 Let $f : A \to B$, $C \subseteq A$, $D \subseteq B$, and $x \in A$. Then

a. $f(\emptyset) = \emptyset$.

b. $f^{-1}(\emptyset) = \emptyset$.

c. $f(\{x\}) = \{f(x)\}$.

d. $x \in C \Rightarrow f(x) \in f(C)$.

e. $x \in f^{-1}(D) \Rightarrow f(x) \in D$

f. $f(x) \in D \Rightarrow x \in f^{-1}(D)$

EXAMPLE Determining Images and Inverse Images

Let $A = \{1, 2, 3, 4, 5\}$, let $B = \{a, b, c, d\}$, and let $f : A \to B$ be defined by $f = \{(1, a), (2, a), (3, b), (4, b), (5, c)\}$. Find a. $f(C)$ for $C = \{2, 4\}$, b. $f(A)$, c. $f^{-1}(D)$ for $D = \{b, c\}$, d. $f^{-1}(\{d\})$, and e. $f^{-1}(B)$.

SOLUTION

a. $f(C) = \{y \in B \mid y = f(x) \text{ for some } x \in C\} = \{f(x) \mid x \in C\} = \{f(2), f(4)\} = \{a, b\}$

b. $f(A) = \{f(x) \mid x \in A\} = \{f(1), f(2), f(3), f(4), f(5)\} = \{a, a, b, b, c\} = \{a, b, c\} = \text{Rng}(f)$

c. $f^{-1}(D) = \{x \in A \mid f(x) \in D\} = \{x \in A \mid f(x) = b \text{ or } f(x) = c\} = \{3, 4, 5\}$

d. $f^{-1}(\{d\}) = \{x \in A \mid f(x) = d\} = \emptyset$

e. $f^{-1}(B) = \{x \in A \mid f(x) \in B\} = \{x \in A \mid f(x) = a \text{ or } f(x) = b \text{ or } f(x) = c \text{ or } f(x) = d\} = \{1, 2, 3, 4, 5\} = A = \text{Dom}(f)$

EXAMPLE Determining Images and Inverse Images

Let $f : \mathbf{R} \to \mathbf{R}$ be defined by $y = f(x) = |x| - 1$. Find a. $f((2, 3])$ and b. $f^{-1}((1, 2])$.

SOLUTION

a. A graph of $y = f(x) = |x| - 1$ is displayed in Figure 5.13. First, we marked the interval $(3, 5]$ on the x-axis. Since $3 \notin (3, 5]$, we drew a dashed vertical line from the point $(3, 0)$ to the graph of $y = f(x)$. The vertical line $x = 3$ intersected the graph at the point $(3, 2)$. We projected the point $(3, 2)$ onto the y-axis by drawing the dashed horizontal line $y = 2$. Next, since $5 \in (3, 5]$, we drew the solid vertical line from the point $(5, 0)$ to the graph of

$y = f(x)$. The line $x = 5$ intersected the graph at the point $(5, 4)$. We then projected the point $(5, 4)$ onto the y-axis by drawing the line $y = 4$. From Figure 5.13, it appears that $f((3, 5]) = (2, 4]$. To prove this we must show that (1) $f((3, 5]) \subseteq (2, 4]$ and (2) $(2, 4] \subseteq f((3, 5])$.

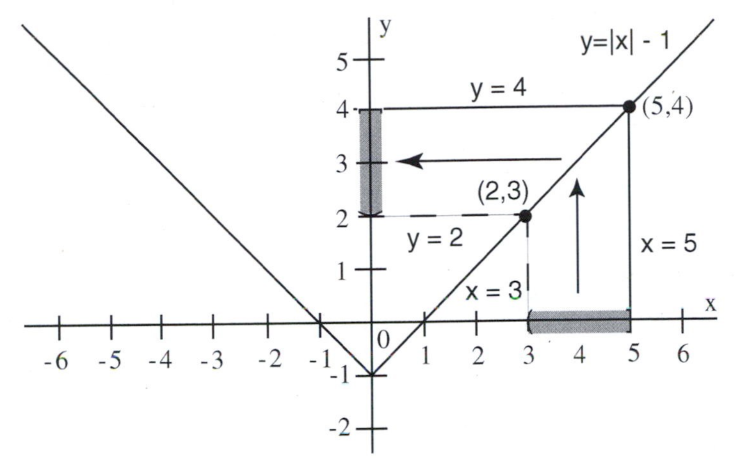

FIGURE 5.13: A Graph of $y = f(x) = |x| - 1$ and $f((3, 5]) = (2, 4]$

(1) Let $y \in f((3, 5])$. By definition, there exists an $x \in (3, 5]$ such that $f(x) = y$ or $|x| - 1 = y$. For $x \in (3, 5]$, $|x| = x$, so there exists an $x \in (3, 5]$ such that $x - 1 = y$. Since $x \in (3, 5]$, $3 < x \leq 5$ and, therefore, $3 - 1 < x - 1 \leq 5 - 1$ or $2 < y \leq 4$—that is, $y \in (2, 4]$.

(2) Let $y \in (2, 4]$. We must find an $x \in (3, 5]$ such that $f(x) = y = |x| - 1$. Since for $x \in (3, 5]$, $|x| = x$, we must find an $x \in (3, 5]$ such that $y = x - 1$. Solving for x, we see $x = y + 1$. Since $y \in (2, 4]$, $2 < y \leq 4$, and, therefore, $2 + 1 < y + 1 \leq 4 + 1$ or $3 < x \leq 5$. That is, if $y \in (2, 4]$, then $x = y + 1 \in (3, 5]$.

Since by (1) $f((3, 5]) \subseteq (2, 4]$ and by (2) $(2, 4] \subseteq f((3, 5])$, it follows that $f((3, 5]) = (2, 4]$

b. A second graph of $y = f(x) = |x| - 1$ is shown in Figure 5.14. We marked the interval $(2, 4]$ on the y-axis and drew horizontal lines through the points $(0, 2)$ and $(0, 4)$ to the graph of $y = f(x)$. These lines intersected the graph at the points $(-5, 4)$, $(-3, 2)$, $(3, 2)$, and $(5, 4)$. Projecting these points onto the x-axis it appears from Figure 5.14, that $f^{-1}((2, 4]) = [-5, -3) \cup (3, 5]$.

The proof that $f^{-1}((2, 4]) = [-5, -3) \cup (3, 5]$ is as follows:

$$x \in f^{-1}((2, 4]) \Leftrightarrow f(x) \in (2, 4] \Leftrightarrow 2 < |x| - 1 \leq 4$$

$$\Leftrightarrow 3 < |x| \leq 5 \Leftrightarrow 3 < x \leq 5 \text{ or } 3 < -x \leq 5$$

$$\Leftrightarrow 3 < x \leq 5 \text{ or } -5 \leq x < -3 \Leftrightarrow x \in (3, 5] \text{ or } x \in [-5, -3)$$

$$\Leftrightarrow x \in [-5, -3) \cup (3, 5]$$

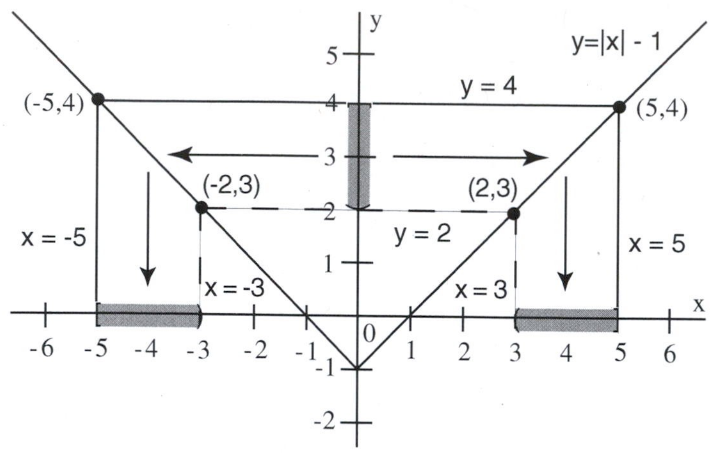

FIGURE 5.14: A Graph of $y = f(x) = |x| - 1$ and $f^{-1}((2, 4]) = [-5, -3) \cup (3, 5]$

This example proves that the following statement is false: (*) "If $f : A \to B$, if $C \subseteq A$, and if $D \subseteq B$, then $f(C) = D \Rightarrow C = f^{-1}(D)$." That is, this example is a counterexample to the statement (*).

The next example examines the relationship between the images and inverse images of unions and intersections of sets and the unions and intersections of sets of images and inverse images.

EXAMPLE Images, Inverse Images, Unions, and Intersections

Let $A = \{a, b, c, d, e\}$, $B = \{1, 2, 3, 4\}$, $C = \{a, b\}$, $D = \{b, c\}$, $E = \{1, 2\}$, $F = \{2, 4\}$, and let $f : A \to B$ be defined by $f = \{(a, 2), (b, 1), (c, 2), (d, 3), (e, 2)\}$.

a. Find $f(C \cup D)$ and $f(C) \cup f(D)$. How are they related?

b. Find $f(C \cap D)$ and $f(C) \cap f(D)$. How are they related?

c. Find $f^{-1}(E \cup F)$ and $f^{-1}(E) \cup f^{-1}(F)$. How are they related?

d. Find $f^{-1}(E \cap F)$ and $f^{-1}(E) \cap f^{-1}(F)$. How are they related?

SOLUTION

a. $C \cup D = \{a, b\} \cup \{b, c\} = \{a, b, c\}$. So $f(C \cup D) = f(\{a, b, c\}) = \{f(a), f(b), f(c)\} = \{2, 1\}$.

$f(C) = f(\{a,b\}) = \{f(a), f(b)\} = \{2,1\}$ and $f(D) = f(\{b,c\}) = \{f(b), f(c)\} = \{1,2\}$.

So $f(C) \cup f(D) = \{1,2\}$. Thus, in this instance, $f(C \cup D) = f(C) \cup f(D)$.

b. $C \cap D = \{a,b\} \cap \{b,c\} = \{b\}$. So $f(C \cap D) = f(\{b\}) = \{f(b)\} = \{1\}$.

From part a., $f(C) = \{2,1\}$ and $f(D) = \{1,2\}$, so $f(C) \cap f(D) = \{1,2\}$.

In this example, $f(C \cap D) = \{1\} \subset \{1,2\} = f(C) \cap f(D)$.

c. $E \cup F = \{1,2\} \cup \{2,4\} = \{1,2,4\}$. And $f^{-1}(E \cup F) = f^{-1}(\{1,2,4\}) = \{x \in A \mid f(x) = 1 \text{ or } f(x) = 2, \text{ or } f(x) = 4\} = \{b, a, c, e\}$.

$f^{-1}(E) = f^{-1}(\{1,2\}) = \{x \in A \mid f(x) = 1 \text{ or } f(x) = 2\} = \{b, a, c, e\}$.

$f^{-1}(F) = f^{-1}(\{2,4\}) = \{x \in A \mid f(x) = 2 \text{ or } f(x) = 4\} = \{a, c, e\}$.

So $f^{-1}(E) \cup f^{-1}(F) = \{b, a, c, e\}$.

In this case, $f^{-1}(E \cup F) = f^{-1}(E) \cup f^{-1}(F)$.

d. $E \cap F = \{1,2\} \cap \{2,4\} = \{2\}$ and $f^{-1}(E \cap F) = f^{-1}(\{2\}) = \{x \in A \mid f(x) = 2\} = \{a, c, e\}$.

From part c., $f^{-1}(E) = \{b, a, c, e\}$ and $f^{-1}(F) = \{a, c, e\}$, so $f^{-1}(E \cap F) = \{a, c, e\}$.

For this example, $f^{-1}(E \cap F) = f^{-1}(E) \cap f^{-1}(F)$.

The following theorem, which applies to the previous example, indicates how to calculate the image and inverse image of the unions and intersections of sets.

Theorem 5.17 Let $f : A \to B$, let $C, D \subseteq A$, and let $E, F \subseteq B$. Then

a. $f(C \cup D) = f(C) \cup f(D)$

b. $f(C \cap D) \subseteq f(C) \cap f(D)$

c. $f^{-1}(E \cup F) = f^{-1}(E) \cup f^{-1}(F)$

d. $f^{-1}(E \cap F) = f^{-1}(E) \cap f^{-1}(F)$

Proof: << We will prove part b. and d. >>

b. Let $y \in f(C \cap D)$. By definition, $y = f(x)$ for some $x \in C \cap D$. Thus, $y = f(x)$ for some $x \in C$ and $x \in D$. Since $x \in C$, $f(x) \in f(C)$ and since $x \in D$, $f(x) \in f(D)$. Consequently, $y = f(x) \in f(C) \cap f(D)$.

d. For every $x \in A$,

$$x \in f^{-1}(E \cap F) \Leftrightarrow f(x) \in E \cap F \Leftrightarrow f(x) \in E \text{ and } f(x) \in F$$

$$\Leftrightarrow x \in f(E) \text{ and } x \in f(F) \Leftrightarrow x \in f(E) \cap f(F)$$

The next theorem states the relationship between the inverse image of the difference of two sets and the difference of their inverse images.

Theorem 5.18 Let $f : A \to B$ and let $E, F \subseteq B$. Then

$$f^{-1}(E - F) = f^{-1}(E) - f^{-1}(F).$$

Proof: $x \in f^{-1}(E - F) \Leftrightarrow f(x) \in E - F \Leftrightarrow f(x) \in E$ and $f(x) \notin F$

$$\Leftrightarrow x \in f^{-1}(E) \text{ and } x \notin f^{-1}(F)$$

$$\Leftrightarrow x \in f^{-1}(E) - f^{-1}(F)$$

EXERCISES 5.4

1. Let $A = \{v, w, x, y, z\}$, $B = \{1, 2, 3, 4, 5\}$, and $f : A \to B$ be defined by $f = \{(v, 2), (w, 5), (x, 3), (y, 3), (z, 4)\}$. Find

   a. $f(\{v\})$ b. $f(\{x, y\})$ c. $f^{-1}(\{3\})$

   d. $f^{-1}(\{1\})$ e. $f^{-1}(f(\{v, x, z\}))$ f. $f(f^{-1}(\{3, 4, 5\}))$

2. Let $f : \mathbf{R} \to \mathbf{R}$ be defined by $f(x) = 2 - 3x$. Find

   a. $f(\{-2, -1, 0, 1\})$ b. $f(\mathbf{N})$

   c. $f(\mathbf{Z})$ d. $f([1, 4))$

   e. $f([-2, 0) \cup (1, 2])$ f. $f^{-1}((5, 10))$

   g. $f^{-1}((2, 4) \cap [3, 5])$ h. $f^{-1}(\mathbf{R})$

   i. $f(f^{-1}([-2, 4)))$ j. $f^{-1}(f((1, 5]))$

3. Let $f : \mathbf{N} \times \mathbf{N} \to \mathbf{N}$ be defined by $f((m, n)) = 3^m 5^n$. Find

   a. $f(A \times B)$ where $A = \{1, 2, 3\}$ and $B = \{1, 2\}$.

   b. $f^{-1}(\{10, 15, 25, 45, 65, 75\})$

4. Let $f : \mathbf{R} \to \mathbf{R}$ be defined by $f(x) = -8x - x^2$. Find

   a. $f([-5, 0))$
   b. $f([2, 5])$
   c. $f(\mathbf{R})$
   d. $f^{-1}((-20, 0))$
   e. $f^{-1}([-20, 20])$
   f. $f^{-1}(\mathbf{R})$
   g. $f^{-1}([-5, 0])$
   h. $f(f^{-1}((-20, 0)))$

5. Prove theorem 5.16

6. Prove parts a. and c. of theorem 5.17.

In exercises 7-23 let $f : A \to B$, let $C, D \subseteq A$, and let $E, F \subseteq B$. Prove the following theorems.

7. If $C \subseteq D$, then $f(C) \subseteq f(D)$.
8. If $E \subseteq F$, then $f^{-1}(E) \subseteq f^{-1}(F)$.
9. $C \subseteq f^{-1}(f(C))$.
10. $f(f^{-1}(E)) \subseteq E$.
11. $f(f^{-1}(E)) = f(A) \cap E$.
12. $f(C \cap f^{-1}(E)) = f(C) \cap E$.
13. $f^{-1}(B - E) = A - f^{-1}(E)$.
14. If f is onto, then $f(f^{-1}(E)) = E$.
15. If $f(f^{-1}(E)) = E$ for all $E \subseteq B$, then f is onto.
16. If f is onto and $f^{-1}(E) = f^{-1}(F)$, then $E = F$.
17. If f is one-to-one, then $f^{-1}(f(C)) = C$.
18. If $f^{-1}(f(C)) = C$ for all $C \subseteq A$, then f is one-to-one.
19. If f is one-to-one, then $f(C \cap D) = f(C) \cap f(D)$.
20. If f is one-to-one, then $f(C - D) = f(C) - f(D)$.
21. If f is one-to-one, then $f(x) \in f(C)$ if and only if $x \in C$.
22. If f is one-to-one correspondence, then $f(C) = E$ if and only if $C = f^{-1}(E)$.
23. If f is one-to-one correspondence, then $f(A - C) = B - f(C)$.

In exercises 24-30 let $f : A \to B$, let $C, D \subseteq A$, and let $E, F \subseteq B$. Provide counterexamples to the following statements.

24. If $f(C) \subseteq f(D)$, then $C \subseteq D$.
25. $f(C - D) = f(C) - f(D)$.
26. If $f^{-1}(E) \subseteq f^{-1}(F)$, then $E \subseteq F$.
27. $f^{-1}(f(C)) = C$.
28. $f(f^{-1}(E)) = E$.
29. If $E \neq \emptyset$, then $f^{-1}(E) \neq \emptyset$.
30. $x \in C$ if and only if $f(x) \in f(C)$.

5.5 Chapter Review

Definitions

Let A and B be sets. A **function f from A to B**, denoted by $f : A \to B$, is a relation from A to B such that

(1) $\text{Dom}(f) = A$

(2) If $(a, b) \in f$ and $(a, c) \in f$, then $b = c$.

The **domain** of a function f from A to B is the set

$$\text{Dom}(f) = \{x \in A \mid (\exists y \in B)((x, y) \in f)\}$$

The **codomain** of the function f from A to B is the set B.

The **range** of a function f from A to B is the set

$$\text{Rng}(f) = \{y \in B \mid (\exists x \in A)((x, y) \in f)\}$$

Let f be a function from A to B, $f : A \to B$, and let (x, y) denote the elements of f.

The **image of x under f** is $f(x)$. The image of x under f is also called the **value of f at x**.

The **pre-image of y under f** is x.

Two functions f and g are **equal**, written $f = g$, if and only if $\text{Dom}(f) = \text{Dom}(g)$ and for all $x \in \text{Dom}(f)$, $f(x) = g(x)$.

Let A and B be sets. A function f from A to B is **onto** (or **surjective**) if and only if $(\forall y \in B)\,(\exists x \in A)$ such that $f(x) = y$.

Let A and B be sets. A function $f : A \to B$ is **one-to-one** (or **injective**) if and only if $(x, y) \in f$ and $(z, y) \in f$ implies $x = z$.

Let A and B be sets. A function $f : A \to B$ which is onto its codomain B and is one-to-one is called a **one-to-one correspondence** or **bijection**.

Let function be a f from A to B and let $C \subseteq A$. The **restriction of f to** C is $f|_C = \{(x, y) \mid (x, y) \in f \text{ and } x \in C\}$.

Let f be a function from A to B, $f : A \to B$.

If $X \subseteq A$, then the **image of** X or **image of the set** X is $f(X) = \{y \in B \mid y = f(x) \text{ for some } x \in A\}$.

If $Y \subseteq B$, then the **inverse image of** Y is $f^{-1}(Y) = \{x \in A \mid f(x) \in Y\}$.

Review Exercises

1. Which of the following relations on the set $A = \{a, b, c, d\}$ are functions? For those relations which are not functions, indicate why they are not functions.

   a. $f = \{(a, b), (c, d)\}$

   b. $g = \{(a, a), (b, c), (c, a), (d, b)\}$

   c. $h = \{(a, b), (b, c), (c, d), (d, a), (a, c)\}$

   d. $k = \{(a, b), (b, c), (c, d), (d, d)\}$

   e. $\ell = \{(a, b), (b, c), (c, d), (d, a)\}$

2. a. Which functions in exercise 1 are onto?

   b. Which functions in exercise 1 are one-to-one?

   c. Which functions in exercise 1 are one-to-one correspondences?

3. Assuming the equation $f(x) = \dfrac{x^2 + 9}{x^2 - 9}$ defines a real valued function of a real variable, determine the domain and range of f.

4. Prove that the relation $g = \{(x, y) \in \mathbf{R} \times \mathbf{R} \mid x^2 = y^2\}$ is not a function on $\mathbf{R}$.

5. Let $f : \mathbf{R} \to \mathbf{R}$ defined by $f(x) = -x^2 + 12$. What are the pre-images of 4?

6. For the given pairs of functions f and g, determine

   (i) $\mathrm{Dom}(g)$ (ii) $\mathrm{Rng}(f)$ (iii) $f \circ g$ (iv) $\mathrm{Dom}(f \circ g)$ (v) $\mathrm{Rng}(f \circ g)$.

   a. $f = \{(a, b), (b, a), (c, c)\}$, $g = \{(a, c), (b, a)\}$

   b. $f : \mathbf{R} \to \mathbf{R}$ defined by $f(x) = -x^2 + 1$,
      $g : \mathbf{R} \to \mathbf{R}$ defined by $g(x) = 2x - 1$

7. Let A be a nonempty set. Suppose R is a function and an equivalence relation on the set A. What is the function R?

8. a. Give an example of two functions f and g such that $f \circ g = g \circ f$.

   b. Give an example of two functions h and k such that $h \circ k \neq k \circ h$.

9. Give two different pairs of functions f and g such that

   a. $(f \circ g) = \sqrt{2x - 5}$ b. $(f \circ g) = \cos|3x + 2|$

10. Classify each of the following functions as (i) onto, (ii) one-to-one, (iii) a one-to-one correspondence, or (iv) none of these.

    a. $f : \mathbf{R} \to \mathbf{R}$ defined by $f(x) = 3x - 4$

    b. $f : \mathbf{R} \to \mathbf{R}$ defined by $f(x) = 2x^2 + 1$

    c. $f : \mathbf{R} \to [1, \infty)$ defined by $f(x) = 2x^2 + 1$

    d. $f : (-\infty, 0] \to [1, \infty)$ defined by $f(x) = 2x^2 + 1$

    e. $f : \mathbf{R} \to \mathbf{R}$ defined by $f(x) = \tan x$

    f. $f : (-\pi/2, \pi/2) \to \mathbf{R}$ defined by $f(x) = \tan x$

    g. $f : (-\infty, 0) \to (0, \infty)$ defined by $f(x) = 1/x^2$

    h. $f : (-\infty, 0) \to \mathbf{R}$ defined by $f(x) = 1/x^2$

    i. $f : \mathbf{R} \to \mathbf{R}$ defined by $f(x) = e^{2x}$

    j. $f : \mathbf{R} \to (0, 0)$ defined by $f(x) = e^{2x}$

11. For each one-to-one correspondence found in exercise 10, write an expression that defines the inverse function, f^{-1}.

12. Give an example of functions f and g such that $f \circ g$ is a one-to-one correspondence, but neither f nor g is a one-to-one correspondence.

13. Let $A \subset \mathbf{R}$ and let $f : A \to \mathbf{R}$ be as defined below. Determine two different sets A so that $f_{|A}$ is one-to-one. Choose the sets to be as large as possible.

    a. $f(x) = \dfrac{2}{(x-3)^2}$ b. $f(x) = \cot x$

14. Let $f : (0, \infty) \to [-1, 1]$ be defined by $f(x) = \sin(\ln x)$. Write an expression for $f^{-1}(x)$.

15. Let $A = \{1, 2, 3, 4\}$, let $B = \{a, b, c, d, e\}$, and let $f : A \to B$ be defined by $f = \{(1, e), (2, c), (3, a), (4, e)\}$. Determine
    a. $f(\{1, 3\})$ b. $f^{-1}(\{a, b, c\})$ c. $f^{-1}(f(\{2, 4\}))$ d. $f(f^{-1}(\{b, d, e\}))$

16. Let $f : \mathbf{R} \to \mathbf{R}$ be defined by $f(x) = |x - 2| + 3$. Calculate

    a. $f((-2, 5])$ b. $f((-2, -1) \cup (3, 6))$ c. $f^{-1}([0, 2))$

    d. $f^{-1}([4, 7))$ e. $f^{-1}(f((-1, 3)))$ f. $f(f^{-1}([-1, 5]))$

17. Let h be a function on some set S. Prove that if $h \circ h$ is a one-to-one correspondence, then h is a one-to-one correspondence.

Chapter 6

Mathematical Induction

In this chapter, we state Peano's fifth axiom, also known as the Axiom of Induction, and explore the consequences of this axiom. Among the consequences are four versions of the Principle of Mathematical Induction, the Well-Ordering Principle for set of the natural numbers, and the Fundamental Theorem of Arithmetic. It is difficult, if not impossible, to determine when the very first proof by induction was employed. In 1202, Leonardo Pisano Bigollo (1180-1250), also known as Leonardo of Pisa and after his death called Fibonacci, used induction in his *Book of the Abacus* to prove $6(1^2 + 2^2 + 3^2 + \cdots + n^2) = n(n+1)(n+n+1)$. In 1321, Levi Ben Gershon (1288-1344) completed *The Art of the Calculator* in which several propositions were proven using mathematical induction. In 1654, the French mathematician Blaise Pascal (1623-1662) gave the first definitive explanation of the method of mathematical induction. However, the name "mathematical induction" was not associated with this method of proof until August De Morgan published his article on "Induction (Mathematics)" in the *Penny Cyclopaedia* of 1838.

6.1 Mathematical Induction

In chapter 2, we discussed the first four axioms, P1-P4, which the Italian mathematician and logician Giuseppe Peano (1858-1932) used in his axiomatic development of the set of natural numbers. The following axiom, called the Axiom of Induction, is a rewording of Peano's fifth axiom, P5.

Axiom of Induction If $S \subseteq \mathbf{N}$ satisfying the two properties:

(1) $1 \in S$

(2) for all $n \in \mathbf{N}$, $n \in S \Rightarrow n + 1 \in S$,

then $S = \mathbf{N}$.

The Fundamental Theorem of Mathematical Induction stated below follows immediately from the Axiom of Induction.

The Fundamental Theorem of Mathematical Induction

For each $n \in \mathbf{N}$, let $P(n)$ be a statement which is either true or false but not both.

Let $S = \{n \in \mathbf{N} \mid P(n) \text{ is true}\}$.

If (i) $1 \in S$ and

if (ii) for all $n \in \mathbf{N}$, $n \in S \Rightarrow n+1 \in S$,

then $S = \mathbf{N}$. (That is, for all $n \in \mathbf{N}$ the statement $P(n)$ is true.)

The next example illustrates how to use the Fundamental Theorem of Mathematical Induction to prove for all $n \in \mathbf{N}$ that the statement $P(n)$ is true. Notice in this example, we first prove that (i) $1 \in S$ and then we prove that (ii) for any $n \in \mathbf{N}$, if $n \in S$, then $n+1 \in S$.

EXAMPLE Using the Fundamental Theorem of Mathematical Induction To Prove: For All $n \in \mathbf{N}$, the Statement $P(n)$ Is True

Prove that for all $n \in \mathbf{N}$, $1 + 2 + \cdots + n = \dfrac{n(n+1)}{2}$.

SOLUTION

Let $P(n)$ be the statement $1 + 2 + \cdots + n = \dfrac{n(n+1)}{2}$ and let

$$S = \{n \in \mathbf{N} \mid P(n) \text{ is true}\}.$$

(i) $<<$ First we prove that $1 \in S$. $>>$ For $n = 1$, the left-hand side of the statement $P(1)$ is 1 and the right-hand side is $\dfrac{1(1+1)}{2} = \dfrac{2}{2} = 1$. Consequently, $P(1)$ is true and $1 \in S$.

(ii) $<<$ Next, we prove that for $n \in \mathbf{N}$, $n \in S \Rightarrow n+1 \in S$. $>>$ Let $n \in \mathbf{N}$ and assume that $n \in S$. Hence, $P(n)$, which is

$$(1) \qquad 1 + 2 + \cdots + n = \frac{n(n+1)}{2}$$

is assumed to be true. The statement $P(n+1)$, which we must prove to be true using (1), is

$$(2) \qquad 1+2+\cdots+n+(n+1)=\frac{(n+1)((n+1)+1)}{2}=\frac{(n+1)(n+2)}{2}$$

Grouping the summands on the left-hand side of (2) as shown below and then substituting from (1), we find

$$(1+2+\cdots+n)+(n+1)=\frac{n(n+1)}{2}+(n+1)$$

$$=(n+1)\left(\frac{n}{2}+1\right)=\frac{(n+1)(n+2)}{2}$$

That is, if (1) is true, it follows that (2) is true and, therefore, $n+1 \in S$.

Since the hypotheses (i) and (ii) of the Fundamental Theorem of Mathematical Induction are true, the conclusion $S = \mathbf{N}$ is true. Consequently, the statement $P(n)$ is true for all $n \in \mathbf{N}$—that is, $1+2+\cdots+n=\dfrac{n(n+1)}{2}$ for all $n \in \mathbf{N}$.

Instead of defining S to be the set of all natural numbers n such that the statement $P(n)$ is true, proofs by mathematical induction are usually presented in the following logically equivalent, but less formal, form.

The Principle of Mathematical Induction

For each $n \in \mathbf{N}$, let $P(n)$ be a statement which is either true or false but not both.

If (i) $P(1)$ is true and

if (ii) for all $n \in \mathbf{N}$, $P(n)$ is true $\Rightarrow P(n+1)$ is true,

then the statement $P(n)$ is true for all $n \in \mathbf{N}$.

Step (i) above is called the **basis step** or **initial step** and step (ii) is called the **inductive step**. The hypothesis of the inductive step, "$P(n)$ is true," is known as the **inductive hypothesis**. The next example shows how to use the Principle of Mathematical Induction to prove for all $n \in \mathbf{N}$ that the statement $P(n)$ is true.

EXAMPLE Using the Fundamental Theorem of Mathematical Induction To Prove: For All $n \in \mathbf{N}$, the statement $P(n)$ Is True

Prove for all $n \in \mathbf{N}$ that $1^2 + 3^2 + \cdots + (2n-1)^2 = \dfrac{n(4n^2-1)}{3}$.

SOLUTION

Let $P(n)$ be the statement

$$(1) \qquad 1^2 + 3^2 + \cdots + (2n-1)^2 = \frac{n(4n^2-1)}{3}$$

(i) For $n = 1$, the left-hand side of the statement $P(1)$ is 1 and the right-hand side is

$$\frac{1(4(1)^2-1)}{3} = \frac{1 \cdot 3}{3} = 1.$$

So, $P(1)$ is true.

(ii) Let $n \in \mathbf{N}$ and assume that the statement $P(n)$ is true. That is, for $n \in \mathbf{N}$ assume that (1) is true. The statement $P(n+1)$ is the statement

$$1^2 + 3^2 + \cdots + (2n-1)^2 + (2(n+1)-1)^2 = \frac{(n+1)(4(n+1)^2-1)}{3}.$$

Substituting (1) into the left-hand side of $P(n+1)$, we obtain

$$(2) \quad 1^2 + 3^2 + \cdots + (2n-1)^2 + (2(n+1)-1)^2 =$$

$$[1^2 + 3^2 + \cdots + (2n-1)^2] + (2n+1)^2 = \frac{n(4n^2-1)}{3} + \frac{3(2n+1)^2}{3} =$$

$$\frac{4n^3 - n + 12n^2 + 12n + 3}{3} = \frac{4n^3 + 12n^2 + 11n + 3}{3}$$

Expanding the right-hand side of $P(n+1)$, we find

$$(3) \qquad \frac{(n+1)(4(n+1)^2-1)}{3} = \frac{(n+1)(4n^2+8n+3)}{3}$$

$$= \frac{4n^3 + 12n^2 + 11n + 3}{3}$$

Because the right-hand sides of (2) and (3) are equal, it follows from the Principle of Mathematical Induction that for all $n \in \mathbf{N}$,

$$1^2 + 3^2 + \cdots + (2n-1)^2 = \frac{n(4n^2-1)}{3}.$$

Some statements $Q(n)$ are false for some or all $n = 1, 2, \ldots, n_0 - 1$ but are true for all $n = n_0, n_0 + 1, \ldots$. The generalized principle of mathematical induction stated below can be used to prove statements of the form: "For all $n \in \mathbf{N}$ with $n \geq n_0$, the statement $Q(n)$ is true".

The Generalized Principle of Mathematical Induction

For each $n \in \mathbf{N}$ let $Q(n)$ be a statement which is either true or false but not both.

If (i) $Q(n_0)$ is true and

if (ii) for all $n \in \mathbf{N}$ with $n \geq n_0$, $Q(n)$ is true $\Rightarrow Q(n+1)$ is true,

then for all natural numbers $n \geq n_0$, the statement $Q(n)$ is true.

It is easy to see that the principle of mathematical induction and the generalized principle of mathematical induction are logically equivalent by setting $P(n) = Q(n + n_0 - 1)$ for all $n \in \mathbf{N}$. In the next example, we demonstrate how to use the generalized principle of mathematical induction.

EXAMPLE Using the Generalized Principle of Mathematical Induction To Prove: For All $n \in \mathbf{N}$ with $n \geq n_0$, the Statement $Q(n)$ Is True

Prove for all $n \in \mathbf{N}$ with $n \geq 5$ that $2^n > n^2$.

SOLUTION

Let $Q(n)$ be the statement $2^n > n^2$.

(i) For $n = 5$, we have $2^5 = 32 > 25 = 5^2$, so the statement $Q(5)$ is true.

(ii) Assume for n a natural number greater than or equal to 5 that the statement $Q(n)$ is true—that is, for $n \in \mathbf{N}$ and $n \geq 5$ assume that (1) $2^n > n^2$. We must now prove the statement $Q(n+1)$ is true. Thus, given (1) is true, we must prove that (2) $2^{n+1} > (n+1)^2$. Multiplying equation (1) by 2, we find

$$2 \cdot 2^n = 2^{n+1} > 2n^2 \tag{3}$$

If we can show for $n \geq 5$ that $2n^2 \geq (n+1)^2$, we are done. Assume that (4) $n \geq 5$. Multiplying (4) by n, we find $n^2 = n \cdot n \geq 5n$. And multiplying (4) by 3, we find $3n \geq 15$. Consequently, for $n \geq 5$ we have

$$2n^2 = n^2 + n^2 \geq n^2 + 5n = n^2 + 2n + 3n \geq n^2 + 2n + 15 > n^2 + 2n + 1 = (n+1)^2.$$

EXERCISES 6.1

In exercises 1-12 use the Principle of Mathematical Induction to prove the given formula.

1. $1 + 3 + 5 + \cdots + (2n - 1) = n^2$

2. $1 + 5 + 9 + \cdots + (4n - 3) = 2n^2 - n$

3. For $r \in \mathbf{R}$ and $r \neq 1$, $1 + r + r^2 + \cdots r^n = \dfrac{1 - r^{n+1}}{1 - r}$

4. $1^2 + 2^2 + 3^2 + \cdots + n^2 = \dfrac{n(n+1)(2n+1)}{6}$

5. $1 + 2^1 + 2^2 + \cdots + 2^{n-1} = 2^n - 1$

6. $1^3 + 2^3 + 3^3 + \cdots + n^3 = \dfrac{n^2(n+1)^2}{4} = (1 + 2 + \cdots + n)^2$

7. $1 \cdot 2 + 2 \cdot 3 + 3 \cdot 4 + \cdots + n(n+1) = \dfrac{n(n+1)(n+2)}{3}$

8. $\dfrac{1}{1 \cdot 2} + \dfrac{1}{2 \cdot 3} + \dfrac{1}{3 \cdot 4} + \cdots + \dfrac{1}{n(n+1)} = \dfrac{n}{n+1}$

9. $\dfrac{1}{3 \cdot 4} + \dfrac{1}{4 \cdot 5} + \dfrac{1}{5 \cdot 6} + \cdots + \dfrac{1}{(n+2)(n+3)} = \dfrac{n}{3(n+3)}$

10. $\dfrac{1}{2!} + \dfrac{2}{3!} + \dfrac{3}{4!} + \cdots + \dfrac{n}{(n+1)!} = 1 - \dfrac{1}{(n+1)!}$

11. $2 \cdot 6 \cdot 10 \cdots (4n - 2) = \dfrac{(2n)!}{n!}$

12. For all $x, y \in \mathbf{R}$, $x^{n+1} - y^{n+1} = (x - y)(x^n + x^{n-1}y + \cdots + xy^{n-1} + y^n)$
    Hint: $x^{n+2} - y^{n+2} = x^{n+1}(x - y) + y(x^{n+1} - y^{n+1})$

In exercises 13-21 use the Generalized Principle of Mathematical Induction to prove the given statement.

13. $3^n > n^2$

14. $4^n > n^3$

15. $1^3 + 2^3 + 3^3 + \cdots + n^3 < n^4$ for $n \geq 2$

16. $1^3 + 2^3 + 3^3 + \cdots + n^3 < \dfrac{n^4}{2}$ for $n \geq 3$

17. $n! > 2^n$ for $n \geq 4$

18. $n! > n^2$ for $n \geq 4$

19. $(n+1)! > 2^{n+3}$ for $n \geq 5$

20. $n^3 < n!$ for $n \geq 6$

21. $2^n > n^3$ for $n \geq 10$

22. Prove De Moivre's Theorem: For all $n \in \mathbf{N}$ and for θ any real number, $(\cos\theta + i\sin\theta)^n = \cos n\theta + i\sin n\theta$.

    Hint: You will need to use these facts: $i^2 = -1$, $\cos(A+B) = \cos A\cos B - \sin A\sin B$, and $\sin(A+B) = \sin A\cos B + \cos A\sin B$.

6.2 The Well-Ordering Principle and the Fundamental Theorem of Arithmetic

Three consequences of the Axiom of Induction are the Fundamental Theorem of Mathematical Induction, The Principle of Mathematical Induction, and the Generalized Principle of Mathematical Induction. Another important consequence is the Well-Ordering Principle for the natural numbers which we now prove by contradiction using mathematical induction.

The Well-Ordering Principle: Every nonempty subset of natural numbers has a least element.

Proof: Let $T \neq \emptyset$ be a subset of $\mathbf{N}$. Suppose T does not have a least element and let $S = \mathbf{N} - T$.

(i) Since 1 is the least element of $\mathbf{N}$ and since T is a subset of $\mathbf{N}$ with no least element, $1 \notin T$. Consequently, $1 \in S$.

(ii) Suppose $n \in S$. It follows from the definition of S that $n \notin T$. Also, no natural number k less than n is in T. For if any natural number or numbers less than n were in T, then T would have a least element. Since $n \notin T$ and no k less than n is in T, $n+1 \notin T$; otherwise, $n+1$ would be the least element in T. Hence, $n+1 \in S$. By the Principle of Mathematical Induction, $S = \mathbf{N}$ and, consequently, $T = \emptyset$. Contradiction.

We devote the remainder of this section to proving the Fundamental Theorem of Arithmetic. First, we restate the definitions of a divides b, prime number, and composite number. Then, we state and prove some facts we need in order to prove the Fundamental Theorem of Arithmetic.

DEFINITIONS Divides, Prime, and Composite

Let a and b be integers. The number a **divides** b if and only if there exists an integer c such that $ac = b$. If a divides b, then we also say a **is a factor of** b and b **is divisible by** a.

A **prime number** (or, simply, a **prime**) is a natural number greater than one which is divisible only by itself and one.

A **composite number** (or, simply, a **composite**) is a natural number greater than one which is not a prime number.

The natural number 1 has special status among the natural numbers, it is not a prime number and it is not a composite number. All other natural numbers are either prime or composite but not both. Observe that 1 and a divide the natural number a, since $1a = a$ and that the only natural number which is a factor of 1 is 1 itself. Notice that if the natural number $a \neq b$ divides the natural number b, then $1 < a < b$. For if $ac = b$ and $a \neq b$, then the natural number $c > 1$. Multiplying this inequality by a, we have $b = ac > a$. It follows directly from the definition of composite that the natural number n is composite if and only if there exist natural numbers a and b such that $n = ab$ where $1 < a < n$ and $1 < b < n$.

The greatest common divisor of two nonzero integers is defined as follows.

DEFINITION The Greatest Common Divisor

Let a and b be two nonzero integers. The natural number d is the **greatest common divisor of** a **and** b, denoted by $\gcd(a, b)$, if

(i) d divides both a and b and

(ii) every divisor of both a and b is a divisor of d.

In theorem 6.1 we prove that two nonzero integers a and b always have a greatest common divisor. It is easy to prove that the greatest common divisor, if it exists, is unique. Suppose, to the contrary, there exist two distinct natural numbers d and d' satisfying the definition of the greatest common divisor. By part (ii) of the definition, d divides d' and also d' divides d. Therefore, since d and d' are natural numbers, $d = d'$.

In order to prove the existence of the $\gcd(a, b)$, we need the concept of a **linear combination of a and b.**

DEFINITION A Linear Combination of Integers a and b

Let a and b be two integers, then any integer of the form $ax + by$ where x and y are integers is a **linear combination of a and b.**

The following theorem establishes the existence of the $\gcd(a, b)$ and provides a representation for it.

Theorem 6.1 If a and b are nonzero integers, then the least natural number d which can be expressed as a linear combination of a and b is the $\gcd(a, b)$. That is, if $d = \gcd(a, b)$, then there exist integers x_1 and y_1 such that $d = ax_1 + by_1$ and d is the least natural number which is expressible in this form.

Proof: Let $S = \{ax + by \mid x, y \in \mathbf{Z} \text{ and } ax + by > 0\}$. Since a and b are nonzero integers, $a^2 + b^2 > 0$ and $S \neq \emptyset$. By the Well-Ordering Principle, S has a least element, say, $d = as + bt$. We claim $d = \gcd(a, b)$. By the division algorithm, there exist integers q and r such that (1) $a = dq + r$ where $0 \le r < d$. Assume that $0 < r < d$. Then from (1),

$$r = a - dq = a - (as + bt)q = a - asq - btq = a(1 - sq) + b(-tq) \in S$$

This contradicts d is the least element of S. Consequently, $r = 0$ and from equation (1), $a = dq$—that is, d divides a. A similar argument shows d divides b also. By definition and our previous uniqueness proof, d is *the* $\gcd(a, b)$.

The next lemma appeared in Euclid's *Elements* and is known as Euclid's Lemma.

Euclid's Lemma: If a and b are two nonzero integers and if p is a prime which divides ab, then p divides a or p divides b.

Proof: Suppose p does not divide a. Since the only factors of p are 1 and p and since p does not divide a, the greatest common divisor of p and a is 1. By theorem 6.1 there exist integers x and y such that $1 = ax + by$. Multiplying this equation by b, we obtain $b = abx + bpy$. Since p divides ab, p divides the right-hand side of this equation. Hence, p divides b.

The following lemma is a generalization of Euclid's Lemma.

Lemma 6.1 For a natural number $m \ge 2$, if p is a prime, if $a_1, a_2, \ldots,$ a_m are all nonzero integers, and if p divides the product $a_1 a_2 \cdots a_m$, then p divides a_i for some $i \in \{1, 2, \ldots, m\}$.

Proof: << We prove this lemma using the generalized principle of mathematical induction on m. >>

(i) For $m = 2$, this lemma is true, since it is Euclid's Lemma.

(ii) Suppose this lemma is true for some $k \geq 2$. That is, assume if p is a prime which divides $a_1 a_2 \cdots a_k$, then p divides a_i for some $i \in \{1, 2, \ldots, k\}$. Let $a_1, a_2, \ldots, a_{k+1}$ be nonzero integers and let p divide the product $a_1 a_2 \cdots a_{k+1}$. Define $b = a_1 a_2 \cdots a_k$. By Euclid's Lemma p divides b or p divides a_{k+1}. If p divides b, then by the induction hypothesis p divides a_i for some $i \in \{1, 2, \ldots, k\}$. Hence, p divides a_i for some $i \in \{1, 2, \ldots, k+1\}$. Consequently, the lemma is true for $k + 1$.

Hence, by the generalized principle of mathematical induction, Lemma 6.1 is true for all natural numbers $m \geq 2$.

In many cases, the assumption that a statement $P(n)$ is true for a natural number n does not easily lead to a proof that the statement $P(n+1)$ is true. In this instance, it is prudent to use one of the following two forms of induction.

The Second Principle of Mathematical Induction

For each $n \in \mathbf{N}$ let $P(n)$ be a statement which is either true or false but not both.

If (i) $P(1)$ is true and

if (ii) for all $n \in \mathbf{N}$, $P(1), P(2), \ldots, P(n)$ are true $\Rightarrow P(n+1)$ is true,

then the statement $P(n)$ is true for all $n \in \mathbf{N}$.

Notice in the second principle of mathematical induction the induction hypothesis of the principle of mathematical induction "$P(n)$ is true" has been replaced by the "stronger" hypothesis "$P(1), P(2), \ldots, P(n)$ are true." We obtain the following generalized second principle of mathematical induction from the generalized principle of mathematical induction by making a similar change in the induction hypothesis.

The Generalized Second Principle of Mathematical Induction

For each $n \in \mathbf{N}$ let $Q(n)$ be a statement which is either true or false but not both.

If (i) $Q(n_0)$ is true and

if (ii) for all $n \in \mathbf{N}$ with $n \geq n_0$,

$Q(n_0), Q(n_0+1), \ldots, Q(n)$ are true $\Rightarrow Q(n+1)$ is true,

then for all natural numbers $n \geq n_0$, the statement $Q(n)$ is true.

We will use the generalized second principle of mathematical induction in proving the Fundamental Theorem of Arithmetic, which is also known as the Unique Factorization Theorem.

The Fundamental Theorem of Arithmetic Every natural number greater than one is a prime or can be written uniquely as a product of primes except for the order in which the prime factors are written.

Proof: Let $Q(n)$ be the statement: "For $n \in \mathbf{N}$ and $n \geq 2$, n is a prime or a product of primes."

(i) For $n = 2$, the statement $Q(2)$ is true, because 2 is a prime.

(ii) Assume that for $n \in \mathbf{N}$ and $n \geq 2$, $Q(2)$, $Q(3)$, ..., $Q(n)$ are true. If $n+1$ is a prime, then the statement $Q(n+1)$ is true. On the other hand, if $n+1$ is a composite, then there exist natural numbers a and b such that $n+1 = ab$ where $1 < a < n+1$ and $1 < b < n+1$—that is, $n+1 = ab$ where $2 \leq a \leq n$ and $2 \leq b \leq n$. By the induction hypothesis, $Q(a)$ and $Q(b)$ are true. Therefore, $n+1$ is a product of primes and $Q(n+1)$ is true.

Consequently, by the generalized second principle of mathematical induction, every natural number greater than one is a prime or can be written as a product of primes.

We prove the uniqueness part of this theorem using the generalized second principle of mathematical induction as well. For each $n \in \mathbf{N}$ and $n \geq 2$, let $S(n)$ be the statement "If $n = p_1 p_2 \cdots p_r = q_1 q_2 \cdots q_s$ where $p_1, p_2, \ldots, p_r$ and $q_1, q_2, \ldots, q_s$ are primes with $p_1 \leq p_2 \leq \cdots \leq p_r$ and $q_1 \leq q_2 \leq \cdots \leq q_s$, then $r = s$ and $p_i = q_i$ for $i = 1, 2, \ldots, r$."

(i) Since 2 is a prime, its only factorization is $2 = 2$. Hence, $S(2)$ is true.

(ii) For $n \in \mathbf{N}$ and $n \geq 2$ we assume the statements $S(2)$, $S(3)$, ... , $S(n)$ are true and we assume $n+1$ has two different factorizations into a product of primes. Thus, we assume $n+1 = p_1 p_2 \cdots p_r = q_1 q_2 \cdots q_s$ where $p_1, p_2, \ldots, p_r$ and $q_1, q_2, \ldots, q_s$ are primes with $p_1 \leq p_2 \leq \cdots \leq p_r$ and $q_1 \leq q_2 \leq \cdots \leq q_s$. By the trichotomy law either $p_1 \leq q_1$ or $p_1 \geq q_1$. We assume $p_1 \leq q_1$. (The proof for the case $p_1 \geq q_1$ is the same as the proof which follows if p and q are interchanged and r and s are interchanged throughout.) By lemma 6.1, since p_1 is a prime and $n+1 = p_1(p_2 \cdots p_r) = q_1 q_2 \cdots q_s$, the prime p_1 divides q_j for some $j \in \{1, 2, \ldots, s\}$. Because p_1 and q_j are both primes, we have $p_1 = q_j$.

Since $q_1 \leq q_j$ for all $j \in \{1, 2, \ldots, s\}$, the prime $q_1 \leq p_1$. Since we assumed that $p_1 \leq q_1$, we conclude that $p_1 = q_1$ and $n + 1 = p_1 p_2 \cdots p_r = q_1 q_2 \cdots q_s = p_1 q_2 \cdots q_s$. Cancelling p_1, we have $m = p_2 \cdots p_r = q_2 \cdots q_s < n + 1$. The statement $S(m)$ is assumed to be true by the induction hypothesis; therefore, $r = s$ and $p_i = q_i$ for $i \in \{2, 3, \ldots, r\}$.

By the second general principle of mathematical induction, $S(n)$ is true for all $n \in \mathbf{N}$ with $n \geq 2$.

We now examine a class of problems for which the generalized second principal of mathematical induction is an appropriate method of proof.

A **sequence of real numbers** is a function from $\mathbf{N}$ into $\mathbf{R}$. For each $n \in \mathbf{N}$, let $x_n = f(n)$. The value x_n is called the **nth term** of the sequence f. We will denote the sequence by $< x_n >_{n=1}^{\infty}$, or simply $< x_n >$. Often a formula is given for the nth term—for instance,

$$x_n = 2n + 3, \qquad x_n = \frac{(-1)^n}{n}, \qquad \text{or} \qquad x_n = \frac{\sin n}{n}.$$

Sometimes a sequence is defined recursively. For example, $x_1 = 1$, $x_2 = 2$, and $x_n = 3x_{n-1} - 2x_{n-2}$ for $n \geq 3$. In this case, the first two terms are given explicitly and the remaining terms are defined in terms of the two immediately preceeding terms. The following example illustrates how to prove a result when a sequence is defined recursively.

EXAMPLE Finding and Proving a Formula for a Sequence Which is Defined Recursively

Let $< x_n >$ be the sequence defined by $x_1 = 1$; $x_2 = 2$; and for $n \geq 3$, defined recursively by $x_n = 3x_{n-1} - 2x_{n-2}$. Find a formula for the nth term, x_n, and prove that the formula is correct.

SOLUTION

The first few terms of the sequence are $x_1 = 1$, $x_2 = 2$,

$$\begin{aligned} x_3 &= 3x_2 - 2x_1 = 3(2) - 2(1) = 6 - 2 = 4 \\ x_4 &= 3x_3 - 2x_2 = 3(4) - 2(2) = 12 - 4 = 8 \\ x_5 &= 3x_4 - 2x_3 = 3(8) - 2(4) = 24 - 8 = 16 \end{aligned}$$

From the first five terms of the sequence, it appears the formula for the nth term may be $x_n = 2^{n-1}$.

(i) Using this formula, for $n = 1$ we find, $x_1 = 2^{1-1} = 2^0 = 1$ and for $n = 2$, we obtain $x_2 = 2^{2-1} = 2$. So the formula is correct for $n = 1$ and $n = 2$.

(ii) For $k = 1, 2, \ldots, n$, we assume that $x_k = 2^{k-1}$. We must now show that $x_{n+1} = 2^{(n+1)-1} = 2^n$. From the recursive definition for x_n and the inductive hypothesis, we find

$$x_{n+1} = 3x_n - 2x_{n-1} = 3{\cdot}2^{n-1} - 2{\cdot}2^{(n-1)-1} = 3{\cdot}2^{n-1} - 2^{n-1} = 2^{n-1}(3-1) = 2^n.$$

Hence, by induction, $x_n = 2^{n-1}$ for all natural numbers.

In this chapter, we stated the Axiom of Induction, a version of Peano's fifth axiom. Then we stated and discussed the following theorems.

1. The Fundamental Theorem of Mathematical Induction
2. The Principle of Mathematical Induction
3. The Generalized Principle of Mathematical Induction
4. The Well-Ordering Principle
5. The Second Principle of Mathematical Induction
6. The Generalized Second Principle of Mathematical Induction

Theorems 1 through 6 are all logically equivalent. Although we did not prove this fact, we did show that (1) $\Rightarrow$ (2) $\Leftrightarrow$ (3) and (2) $\Rightarrow$ (4). The proof of (5) $\Leftrightarrow$ (6) is similar to the proof of (2) $\Leftrightarrow$ (3).

EXERCISES 6.2

1. For the given pairs of integers a and b, (i) express a and b as a product of primes, (ii) find the $\gcd(a, b)$, and (iii) find integers x and y such that $\gcd(a, b) = ax + by$.

   (a) 112 and 320 (b) 2387 and 7469 (c) -4174 and 10672

   (d) $2^3 \cdot 3^4 \cdot 5^2 \cdot 7$ and $2^4 \cdot 3^2 \cdot 5 \cdot 7^2$

2. Two integers a and b are **relatively prime** if and only if $\gcd(a, b) = 1$.

   a. Show that if a and b are relatively prime there exist integers x and y such that $ax + by = 1$.

   b. Show that 4 and 7 are relatively prime.

   c. Find two distinct pairs of integers x and y such that $4x + 7y = 1$.

   d. Prove that every integer n can be written as $n = 4x' + 7y'$ for appropriately chosen x' and y'.

3. a. Let a and b be nonzero integers and let $D = gcd(a, b)$. Prove that if $a = da'$ and $b = db'$, then d divides D.

   b. Let a and b be nonzero integers and let $d = gcd(a, b)$. Prove that if $a = da'$ and $b = db'$, then $gcd(a', b') = 1$.

   c. Let $n \in \mathbf{N}$. Prove that gcd $(na, nb) = n \cdot \text{gcd}(a, b)$.

4. Let the sequence $< x_n >$ be defined by $x_1 = 1$; $x_2 = 3$; and for $n \geq 3$ defined recursively by $x_n = 2x_{n-1} - x_{n-2}$. Find a formula for the nth term, x_n, and prove that the formula is correct.

5. Let the sequence $< x_n >$ be defined by $x_1 = 1$; $x_2 = 4$; and for $n \geq 3$ defined recursively by $x_n = 2x_{n-1} - x_{n-2} + 2$. Find a formula for the nth term, x_n, and prove that the formula is correct.

6. Let the sequence $< x_n >$ be defined by $x_1 = 1$; $x_2 = 3$; and for $n \geq 3$ defined recursively by $x_n = 3x_{n-1} - 2x_{n-2}$. Find a formula for the nth term, x_n, and prove that the formula is correct.

7. Let the sequence $< x_n >$ be defined by $x_1 = 1$ and for $n \geq 2$ defined recursively by $x_n = \sqrt{5 + x_{n-1}}$. Prove that $2 < x_n < 3$ for all $n \in \mathbf{N}$.

8. Let the sequence $< x_n >$ be defined by $x_1 = 1$; $x_2 = 2$; and for $n \geq 3$ defined recursively by $x_n = (x_{n-1} + x_{n-2})/2$. Prove that $1 \leq x_n \leq 3$ for all $n \in \mathbf{N}$.

9. The first ten numbers in the Fibonacci sequence are $F_1 = 1$, $F_2 = 1$, $F_3 = 2$, $F_4 = 3$, $F_5 = 5$, $F_6 = 8$, $F_7 = 13$, $F_8 = 21$, $F_9 = 34$, $F_{10} = 55$. The recursive definition for this sequence is $F_1 = 1$, $F_2 = 1$, and $F_n = F_{n-1} + F_{n-2}$ for $n \geq 3$.

   a. For all $n \in \mathbf{N}$ prove that $F_n < 2^n$.

   b. For all $n \in \mathbf{N}$ prove simultaneously that F_{3n} is even, F_{3n+1} is odd, and F_{3n+2} is odd. (**Hint:** For all $n \in \mathbf{N}$, let $P(n)$ be the statement "F_{3n} is even and both F_{3n+1} and F_{3n+2} are odd.")

   c. Prove for all $n \in \mathbf{N}$ that $F_1 + F_2 + \cdots + F_n = F_{n+2} - 1$.

   d. Prove for all $n \in \mathbf{N}$ that $F_{n+6} = 4F_{n+3} + F_n$.

   e. Prove for all $n \in \mathbf{N}$ that $F_1^2 + F_2^2 + \cdots + F_n^2 = F_n F_{n+1}$.

10. Let $< E_n >$ be the sequence of equations

    $E_1 : \ 2 + 3 + 4 = 1 + 8$

    $E_2 : \ 5 + 6 + 7 + 8 + 9 = 8 + 27$

    $E_3 : \ 10 + 11 + 12 + 13 + 14 + 15 + 16 = 27 + 64$

    a. What is the equation E_4? E_5?

    b. Write a formula for E_n?

    c. Prove that E_n is true for all $n \in \mathbf{N}$.

Chapter 7

Cardinalities of Sets

Study of the concept of infinity dates back to the Greek philosopher Zeno of Elea (c. 450 B. C.). However, the modern era of the study of infinity was begun by the Italian physicist and mathematician Galileo Galilei (1564-1642). In 1638, Galileo published a *Dialog Concerning Two New Sciences.* His primary goal in *Dialog* was to establish the Copernican heliocentric theory of the solar system over the accepted and church supported Ptolemaic geocentric theory. For this, Galileo was charged with heresy by the Inquisition, forced to recant, and spent the last eight years of his life under house arrest. In *Dialog*, Galileo also discussed the concepts of infinite and infinitesimal. He observed that the set of natural numbers, **N**, properly contains the set of perfect squares, $S = \{1, 4, 9, 16, \ldots\}$, and that there are an infinite number of elements in both **N** and S. He concluded that there were as many elements in S as in **N**. However, Galileo believed this conclusion was absurd, since this would mean it was possible to apply the terms "equal to," "greater than," and "less than" to infinite quantities. In 1691, the English mathematician Edmond Halley (1656-1742) published an article in the *Philosophical Transactions of the Royal Society* titled "On the Several Species of Infinite Quantity and the Proportions They Bear to One Another" in which he suggested that the phrases "twice as infinite" and "one-fourth as infinite" might be meaningful. In other words, Halley was suggesting that infinite quantities might have different "sizes" and be "comparable."

On November 29, 1873, Cantor wrote a letter to Dedekind in which he stated he had proved there is a one-to-one correspondence between the set of natural numbers and the set of rational numbers and in which he speculated on whether there was a one-to-one correspondence between the set of natural numbers and the set of real numbers. A few days later, on December 7, Cantor wrote to Dedekind stating there was no one-to-one correspondence between the natural numbers and the real numbers. In 1874, Cantor published these results in *Crelle's Journal* in the article "On a Property of the Real Algebraic Numbers." Prior to this publication, all infinite collections were thought to be "the same size." That is, prior to Cantor's work orders of infinity did not exist.

7.1 Finite Sets

What property do the sets $A = \{e, \pi, i, \sqrt{2}\}$ and $B = \{\#, *, \$, @\}$ have in common? Clearly, they are not equal, and neither is a subset of the other. However, they do have the same number of elements. To verify this fact, you would probably count the number of elements in A by associating e with the number "1", associating π with the number "2", i with "3", and $\sqrt{2}$ with "4," and conclude "A has four elements." In the same manner, you would count the elements in B and conclude "B has four elements." Then since A and B both have four elements, you conclude A and B have the same number of elements. When two sets are finite and the number of elements in both sets is "small," counting the number of elements in each set and deciding if the two numbers are the same or not is an appropriate technique for determining if the two sets have the same number of elements or not. However, when the number of elements is "large" (and you are apt to loose count) or the sets are infinite, a different technique is needed to decide if the two sets have the same number of elements or not. Let f be the one-to-one correspondence from the set A onto the set B defined by $f = \{(e, \#), (\pi, *), (i, \$), (\sqrt{2}, @)\}$. This one-to-one correspondence pairs each element of the set A with exactly one element of the set B. Thus, to determine if sets A and B have the same number of elements or not, we do not need to know the number of elements in the sets. All we need do is determine if there is or is not a one-to-one correspondence from A onto B. (Observe that there are 24 distinct one-to-one correspondences from A onto B, so a one-to-one correspondence which shows two sets have the same number of elements is not unique.) The formal definition of set equivalence follows.

DEFINITION Set Equivalence

Two sets A and B are **equivalent**, denoted by $A \sim B$ if and only if there is a one-to-one correspondence from A onto B. When set A is not equivalent to set B, we write $A \not\sim B$.

Instead of saying A is equivalent to B, we may say (i) A and B **have the same cardinality**, (ii) A and B are **numerically equivalent**, (iii) A and B are **equinumerous**, or (iv) A and B are **equipotent**.

Theorems 7.1, 7.2, and 7.3 are general theorems about set equivalence and apply to both finite and infinite sets. As you might easily anticipate, the next theorem states that set equivalence is an equivalence relation.

Theorem 7.1 Set equivalence is an equivalence relation.

Proof: Reflexive: For any set A, the identity function I_A is a one-to-one correspondence from A onto A, so $A \sim A$.

Symmetry: Suppose $A \sim B$. By definition, there exists a one-to-one correspondence from A onto B, call it f. By theorem 5.10, the function f^{-1} is a one-to-one correspondence from B onto A, so $B \sim A$.

Transitivity: Let A, B, and C be sets such that $A \sim B$ and $B \sim C$. By the definition of set equivalence, there exists a function $f : A \xrightarrow[\text{onto}]{\text{1-1}} B$ and there exists a function $g : B \xrightarrow[\text{onto}]{\text{1-1}} C$. By theorem 5.7, the function $g \circ f : A \xrightarrow[\text{onto}]{\text{1-1}} C$, so $A \sim C$.

Since $\sim$ is reflexive, symmetric, and transitive, $\sim$ is an equivalence relation.

The next theorem states if sets A and C are equivalent and sets B and D are equivalent, then the Cartesian products $A \times B$ and $C \times D$ are equivalent.

Theorem 7.2 Let A, B, C, and D be sets such that $A \sim C$ and $B \sim D$, then $A \times B \sim C \times D$.

Proof: Since $A \sim C$, there exists a function $f : A \xrightarrow[\text{onto}]{\text{1-1}} C$, and since $B \sim D$ there exists a function $g : B \xrightarrow[\text{onto}]{\text{1-1}} D$. Define $(f \times g) : A \times B \to C \times D$ by $(f \times g)(a, b) = (f(a), g(b))$. We claim $f \times g$ is a one-to-one correspondence from $A \times B$ onto $C \times D$. To prove the function $f \times g$ is one-to-one, we assume (a_1, b_1), $(a_2, b_2) \in A \times B$ and $(f \times g)(a_1, b_1) = (f \times g)(a_2, b_2)$. Thus, by definition of $f \times g$, we have $(f(a_1), g(b_1)) = (f(a_2), g(b_2))$. Hence, $f(a_1) = f(a_2)$ and $g(b_1) = g(b_2)$. Since f and g are both one-to-one functions, $a_1 = a_2$ and $b_1 = b_2$. Consequently, $(a_1, b_1) = (a_2, b_2)$ and the function $f \times g$ is one-to-one. To prove the function $f \times g$ is onto, suppose $(c, d) \in C \times D$. Since the functions f and g are both onto, there exists an $a \in A$ such that $f(a) = c$ and there exists a $b \in B$ such that $g(b) = d$. The ordered pair (a, b) is an element of $A \times B$ and by definition, $(f \times g)(a, b) = (f(a), g(b)) = (c, d)$. Thus, the function $f \times g$ is onto and $f \times g$ is a one-to-one correspondence from $A \times B$ onto $C \times D$.

The following theorem states: "If A and B are equivalent sets, if C and D are equivalent sets, if A and C are disjoint sets, and if B and D are disjoint sets, then the sets $A \cup C$ and $B \cup D$ are equivalent sets."

Theorem 7.3 If A, B, C, and D are sets with $A \sim B$, $C \sim D$, $A \cap C = \emptyset$, and $B \cap D = \emptyset$, then $A \cup C \sim B \cup D$.

Proof: Since $A \sim B$ there exists a one-to-one correspondence f from A onto B. And since $C \sim D$ there exists a one-to-one correspondence g from C onto

D. Define $f \cup g : A \cup C \to B \cup D$ by

$$(f \cup g)(x) = \begin{cases} f(x), & \text{if } x \in A \\ g(x), & \text{if } x \in C \end{cases}$$

<< To prove $f \cup g$ is a one-to-one correspondence from $A \cup C$ onto $B \cup D$, we must prove $f \cup g$ is a function with domain $A \cup C$ and range $B \cup D$. Then we must prove $f \cup g$ is one-to-one.>>

Since f and g are functions, f, g, and $f \cup g$ are relations. To prove $f \cup g$ is a function, suppose $(x, y) \in f \cup g$ and $(x, z) \in f \cup g$. Since $A \cap C = \emptyset$, either (i) $x \in A$ and $x \notin C$ or (ii) $x \notin A$ and $x \in C$. If $x \in A$ and $x \notin C$, then $y = (f \cup g)(x) = f(x) = z$, because f is a function. If $x \notin A$ and $x \in C$, then $y = (f \cup g)(x) = g(x) = z$, because g is a function. In both cases, $y = z$ and, therefore, $f \cup g$ is a function.

Suppose $x \in A \cup C$. Since $A \cap C = \emptyset$, either (i) $x \in A$ and $x \notin C$, and, consequently, $(f \cup g)(x) = f(x)$ or (ii) $x \notin A$ and $x \in C$, and, consequently, $(f \cup g)(x) = g(x)$. In both cases, $x \in A \cup C$ implies $x \in \text{Dom}(f \cup g)$—that is, (1) $A \cup C \subseteq \text{Dom}(f \cup g)$. Now suppose $x \in \text{Dom}(f \cup g)$. By definition of $f \cup g$, we have $x \in \text{Dom}(f) = A$ or $x \in \text{Dom}(g) = C$. Hence, $x \in A \cup C$. Therefore, (2) $\text{Dom}(f \cup g) \subseteq A \cup C$. It follows from (1) and (2) that $\text{Dom}(f \cup g) = A \cup C$.

Next suppose $y \in B \cup D$. Since $B \cap D = \emptyset$, either (i) $y \in B$ and $y \notin D$ or (ii) $y \notin B$ and $y \in D$. In case (i), since $y \in B$ and since f is a function from A onto B, there exists an $x \in A$ such that $y = f(x) = (f \cup g)(x)$. In case (ii), since $y \in D$ and since g is a function from C onto D, there exists an $x \in C$ such that $y = g(x) = (f \cup g)(x)$. Hence, for every $y \in B \cup D$ there exists an $x \in A \cup C$ such that $(f \cup g)(x) = y$. Therefore, the function $f \cup g$ is onto.

To prove the function $f \cup g$ is one-to-one, we assume $x, y \in A \cup C$ and $(f \cup g)(x) = (f \cup g)(y)$. There are four cases to consider. (1) If $x, y \in A$, then $(f \cup g)(x) = f(x) = (f \cup g)(y) = f(y)$, and because f is a one-to-one function, $x = y$. (2) If $x, y \in C$, then $(f \cup g)(x) = g(x) = (f \cup g)(y) = g(y)$, and because g is a one-to-one function, $x = y$. (3) If $x \in A$ and $y \in C$, then $(f \cup g)(x) = f(x) = (f \cup g)(y) = g(y)$. But $f(x) \in B$ and $g(y) \in D$ which contradicts the hypothesis $B \cap D = \emptyset$. (4) As in case (3), the case $y \in A$ and $x \in C$ contradicts the hypothesis $B \cap D = \emptyset$. It follows from cases (1)-(4) that $x, y \in A \cup C$ and $(f \cup g)(x) = (f \cup g)(y)$ implies $x = y$. Hence, the function $f \cup g$ is one-to-one; and, therefore, $f \cup g$ is a one-to-one correspondence from $A \cup C$ onto $B \cup D$.

For each $k \in \mathbf{N}$, we define $\mathbf{N}_k = \{1, 2, 3, \ldots, k\}$. The elements of the set $\mathbf{N}_k$ are the numbers we would normally use to count the elements of any set with exactly k elements. Consequently, the collection of sets $\{\mathbf{N}_k \mid k \in \mathbf{N}\}$ serves as our stand of measure for finite sets as indicated by the following definitions.

DEFINITION Cardinality of Finite Sets

A set A is **finite** if and only if $A = \emptyset$ or $A \sim \mathbf{N}_k$ for some $k \in \mathbf{N}$.

A set is **infinite**, if it is not finite.

The **cardinality** of a finite set A is denoted by $|A|$.

The cardinality of the empty set is $|\emptyset| = 0$.

If $A \sim \mathbf{N}_k$ for some $k \in \mathbf{N}$, then the cardinality of A is $|A| = k$.

By definition, the empty set is finite. Since the identity function $I_{\mathbf{N}_k}$ is a one-to-one function from $\mathbf{N}_k$ onto $\mathbf{N}_k$, the cardinality of the set $\mathbf{N}_k$ is k—that is, $|\mathbf{N}_k| = k$.

Theorems 7.4 through 7.11, which concern finite sets, may appear to be obvious; however, we want to state and prove these theorems from the definitions given above to convince ourselves that the concept of cardinality for a finite set coincides with our idea of the number of elements in the set. Also, this will allow us to compare and contrast the results we obtain for finite sets with the results we obtain for infinite sets in the next two sections. Usually, those results are very different.

Theorem 7.4 Any set B which is equivalent to a finite set A is a finite set and $|B| = |A|$.

Proof: Let A be a finite set and let $B \sim A$.

Case 1 If $A = \emptyset$, then $|A| = 0$ and there exists a one-to-one correspondence $f : B \to \emptyset$. Since the codomain of f is the empty set, $\text{Dom}(f) = \emptyset = B$. Hence, B is finite and $|B| = 0 = |A|$.

Case 2 If $A \neq \emptyset$, then $A \sim \mathbf{N}_k$ for some $k \in \mathbf{N}$ and $|A| = k$. Since $B \sim A$, $A \sim \mathbf{N}_k$, and $\sim$ is transitive, $B \sim \mathbf{N}_k$. Therefore, B is finite and $|B| = k = |A|$.

Theorem 7.5 The cardinality of a finite set A is unique.

Proof: If $A = \emptyset$, then by definition $|\emptyset| = 0$. By way of contradiction, suppose there exists a nonempty, finite set A such that $|A| = m$, $|A| = n$, and $m \neq n$. Since $|A| = m$, we have $A \sim \mathbf{N}_m$ and since $|A| = n$, we have $A \sim \mathbf{N}_n$. Since $\sim$ is an equivalence relation, $\mathbf{N}_m \sim \mathbf{N}_n$ which implies by theorem 7.4 that $|\mathbf{N}_m| = |\mathbf{N}_n|$. Hence, $m = n$ which is a contradiction.

Theorem 7.6 If A is a finite set and $x \notin A$, then $A \cup \{x\}$ is a finite set and $|A \cup \{x\}| = |A| + 1$.

Proof: Case 1 If $A = \emptyset$, then $|A| = 0$ and $A \cup \{x\} = \{x\}$. Since $\{x\} \sim \mathbf{N}_1$, $|\{x\}| = 1$ and

$$|A \cup \{x\}| = |\{x\}| = 1 = 0 + 1 = |A| + 1.$$

Case 2 If $A \neq \emptyset$, then $A \sim \mathbf{N}_k$ for some $k \in \mathbf{N}$ and, therefore, $|A| = k$. Since $A \sim \mathbf{N}_k$ there exists a one-to-one correspondence $f : A \to \mathbf{N}_k$. Let g be the function from $A \cup \{x\}$ to $\mathbf{N}_{k+1}$ defined by

$$g(t) = \begin{cases} f(t), & \text{if } t \in A \\ \\ k+1, & \text{if } t = x \end{cases}$$

<< We now prove g is a one-to-one correspondence. First, we prove g is a one-to-one function.>> Let $t_1, t_2 \in A \cup \{x\}$ and assume $t_1 \neq t_2$. There are two cases to consider.

Case (i) If t_1 and t_2 are both elements of A, then since f is a one-to-one function $f(t_1) \neq f(t_2)$ which implies $g(t_1) \neq g(t_2)$.

Case (ii) If $t_2 = x$, then $g(t_2) = k + 1$. Since $t_1 \neq t_2 = x$, we conclude $t_1 \in A$. Hence, $g(t_1) = f(t_1) \in \mathbf{N}_k$. Since $g(t_1) \in \mathbf{N}_k$ and $g(t_2) = k + 1$, $g(t_1) \neq g(t_2)$.

In cases (i) and (ii), $t_1 \neq t_2$ implies $g(t_1) \neq g(t_2)$. Consequently, g is a one-to-one function.

<< Next, we prove g is onto.>> Suppose $m \in \mathbf{N}_{k+1}$. Either $m \in \mathbf{N}_k$ or $m = k + 1$. If $m \in \mathbf{N}_k$, then since $f : A \to \mathbf{N}_k$ is onto, there exists a $t \in A$ such that $f(t) = m = g(t)$. If $m = k + 1$, then $t = x$ and $g(x) = k + 1 = m$. Thus, g is onto.

Consequently, g is a one-to-one correspondence from $A \cup \{x\}$ onto $\mathbf{N}_{k+1}$, $A \cup \{x\} \sim \mathbf{N}_{k+1}$, and

$$|A \cup \{x\}| = |\mathbf{N}_{k+1}| = k + 1 = |A| + 1.$$

Theorem 7.6 says adjoining one "new" element to a finite set increases the cardinality by one. The following theorem says any subset of $\mathbf{N}_n$ is a finite set and has cardinality less than or equal to n.

Theorem 7.7 If $A \subseteq \mathbf{N}_n$, then A is a finite set and $|A| \leq n$.

Proof: << We prove this theorem by induction.>> Let $P(n)$ be the statement "If $A \subseteq \mathbf{N}_n$, then A is a finite set and $|A| \leq n$."
(1) If $A \subseteq \mathbf{N}_1$, then either $A = \emptyset$ or $A = \mathbf{N}_1$. If $A = \emptyset$, then A is finite and $|A| = 0 < 1$. If $A = \mathbf{N}_1$, then A is finite and $|A| = 1$. Consequently, $P(1)$ is true.
(2) Let $k \in \mathbf{N}$ and assume $P(k)$ is true. Suppose $A \subseteq \mathbf{N}_{k+1}$. Then $A - \{k + 1\} \subseteq \mathbf{N}_k$ and by the induction hypothesis, (*) $A - \{k + 1\}$ is finite

and $|A - \{k+1\}| \leq k$. There are two cases to consider: (i) $k+1 \notin A$ and (ii) $k+1 \in A$. (i) If $k+1 \notin A$, then $A - \{k+1\} = A$ and from (*) A is finite and $|A| \leq k < k+1$. (ii) If $k+1 \in A$, then $A = (A - \{k+1\}) \cup \{k+1\}$. By theorem 7.6, since $A - \{k+1\}$ is a finite set and since $\{k+1\} \notin (A - \{k+1\})$, $(A - \{k+1\}) \cup \{k+1\} = A$ is a finite set and $|A| = |A - \{k+1\}| + 1 \leq k+1$. In both cases, $P(n+1)$ is true. Hence, by mathematical induction, for all $n \in \mathbf{N}$, if $A \subseteq \mathbf{N}_n$, then A is a finite set and $|A| \leq n$.

Theorem 7.8 If $A \subseteq B$ where B is a finite set, then A is a finite set and $|A| \leq |B|$.

Proof: Let B be a finite set and let $A \subseteq B$.

Case 1 If $A = \emptyset$, then A is a finite set and $|A| = 0 \leq |B|$.

Case 2 Suppose $A \neq \emptyset$. Since $A \subseteq B$, the set $B \neq \emptyset$ and there exists a $k \in \mathbf{N}$ such that $B \sim \mathbf{N}_k$. That is, there exists a $k \in \mathbf{N}$ and there exists a one-to-one correspondence $f : B \to \mathbf{N}_k$. The restriction of f to the set A, $f|_A$, is a one-to-one function from A onto $f(A)$. Therefore, $A \sim f(A)$. By theorem 7.7, since $f(A)$ is a subset of $\mathbf{N}_k$, $f(A)$ is a finite set and $|f(A)| \leq k$. By theorem 7.4, because $A \sim f(A)$ which is finite, A is finite and $|A| = |f(A)| \leq k$.

Theorem 7.8 says if A is a subset of a finite set B, then A is finite and the cardinality of A is less than or equal to the cardinality of B. The following theorem says if A and B are finite sets which are disjoint, then $A \cup B$ is a finite set whose cardinality is equal to the sum of the cardinality of A and the cardinality of B.

Theorem 7.9 If A and B are finite sets and if $A \cap B = \emptyset$, then $A \cup B$ is a finite set and $|A \cup B| = |A| + |B|$.

Proof: Suppose A and B are finite sets and $A \cap B = \emptyset$. If $A = \emptyset$, then $A \cup B = B$ is finite and $|A \cup B| = |B| = 0 + |B| = |A| + |B|$. If $B = \emptyset$, then $A \cup B = A$ is finite and $|A \cup B| = |A| = |A| + 0 = |A| + |B|$. So if A or B is the empty set, $A \cup B$ is a finite set and $|A \cup B| = |A| + |B|$. Now suppose $A \neq \emptyset$ and $B \neq \emptyset$. Since A and B are finite sets, there exist $m, n \in \mathbf{N}$ such that $A \sim \mathbf{N}_m$ and $B \sim \mathbf{N}_n$. Let $M = \{m+1, m+2, \ldots, m+n\}$. Define $h : \mathbf{N}_n \to M$ by $h(x) = m + x$. The function h is a one-to-one correspondence. (Verify this.) Hence, $\mathbf{N}_n \sim M$ and by transitivity of set equivalence $B \sim M$. By theorem 7.3, since $A \sim \mathbf{N}_m$, $B \sim M$, $A \cap B = \emptyset$, and $\mathbf{N}_m \cap M = \emptyset$, the set $A \cup B \sim \mathbf{N}_m \cup M = \mathbf{N}_{m+n}$. Furthermore, since $A \sim \mathbf{N}_m$, since $B \sim \mathbf{N}_n$, since $A \cup B \sim \mathbf{N}_{m+n}$, since $|A| = m$, and since $|B| = n$, it follows that $|A \cup B| = m + n = |A| + |B|$.

Theorem 7.6 says adjoining one "new" element to a finite set increases the cardinality by one. The following theorem says removing one element from a nonempty, finite set decreases the cardinality by one.

Theorem 7.10 If B is a finite set and if $x \in B$, then $B - \{x\}$ is a finite set and $|B - \{x\}| = |B| - 1$.

Proof: Because $x \in B$, the set $B - \{x\}$ is a proper subset of B. By theorem 7.8, since B is finite, the set $B - \{x\}$ is finite. Let $A = B - \{x\}$. By theorem 7.6, since $x \notin (B - \{x\})$, the set $A \cup \{x\} = (B - \{x\}) \cup \{x\} = B$ has cardinality $|A \cup \{x\}| = |A| + 1$—that is, $|B| = |B - \{x\}| + 1$. Hence, $|B - \{x\}| = |B| - 1$.

The following theorem characterizes finite sets.

Theorem 7.11 A finite set is not equivalent to any of its proper subsets.

Proof: Let B be a finite set and assume A is a proper subset of B. Since $A \subset B$, there exists an element $x \in B - A$. Hence, $x \in B$ and by theorem 7.10, the set $B - \{x\}$ is finite and $|B - \{x\}| = |B| - 1$. Since $x \in B - A$, the set $A \subseteq B - \{x\}$ and by theorem 7.8, A is finite and $|A| \leq |B - \{x\}| = |B| - 1$. Consequently, $|A| < |B|$. By the contrapositive of theorem 7.4, we see that $A \not\sim B$.

The contrapositive of theorem 7.11 is "If a set is equivalent to one of its proper subsets, then it is an infinite set." This statement is a characterization of infinite sets.

EXERCISES 7.1

1. Which of the following sets are finite?

   a. The set of all hairs on your head.

   b. The set of all hairs on all the heads of all of the people on Earth.

   c. The set of natural numbers.

   d. The set of all prime numbers.

   e. The set of all composite numbers.

   f. The set of all words in the English language.

   g. The set of all words in all languages ever used on Earth.

   h. The set of all integers which satisfy the equation $x^2 + 4 = 0$.

2. Give examples of sets A, B, C, and D such that

   a. $A \sim B$, $C \sim D$, $A \cap C = \emptyset$ and $A \cup C \not\sim B \cup D$.

   b. $A \sim B$, $C \sim D$, $B \cap D = \emptyset$ and $A \cup C \not\sim B \cup D$.

   c. How do these examples relate to theorem 7.3?

3. Let A and B be sets. Prove that if A is an infinite set and if $A \subseteq B$, then B is an infinite set.

4. Let A and B be sets. Prove the following statements.

   a. If A is a finite set, then $A \cap B$ is a finite set.

   b. If $A \cup B$ is a finite set, then A and B are finite sets.

   c. If $A \cap B$ is an infinite set, then A and B are infinite sets.

   d. If A is an infinite set, then $A \cup B$ is an infinite set.

5. Let E be the set of even natural numbers. Prove that $E \sim \mathbf{N}$.

6. Prove that if A is a finite set and B is an infinite set, then $B - A$ is an infinite set.

7. Give an example of finite sets A and B such that $|A \cup B| \neq |A| + |B|$.

8. Let A and B be finite sets. Prove that $A \cup B$ is a finite set and that $|A \cup B| = |A| + |B| - |A \cap B|$.

9. Prove by mathematical induction that if $A_1, A_2, \ldots, A_n$ are finite sets, then $\bigcup_{i=1}^{n} A_i$ is a finite set.

10. Let A be a set and let x be an object. Prove that $A \times \{x\} \sim A$.

11. Prove that if A and B are two sets, then $A \times B \sim B \times A$.

12. Prove that if $(A - B) \sim (B - A)$, then $A \sim B$.

13. a. Prove that for all $m, n \in \mathbf{N}$, the set $\mathbf{N}_m \times \mathbf{N}_n$ is finite.

    b. Prove that if A and B are finite sets, then $A \times B$ is a finite set.

7.2 Denumerable and Countable Sets

Theorem 7.11 states "A finite set is not equivalent to any of its proper subsets." The contrapositive of this statement is a theorem as well and provides us with a method for proving a set is infinite. Theorem 7.12 is the contrapositive of theorem 7.11.

Theorem 7.12 If a set is equivalent to one of its proper subsets, then it is an infinite set.

The first infinite set we will discuss is the set of natural numbers. Let E denote the set of even natural numbers. That is, let $E = \{2, 4, 6, \ldots\}$. Clearly, E is a proper subset of the natural numbers and the function $f : \mathbf{N} \to E$

defined by $f(n) = 2n$ is a one-to-one correspondence from $\mathbf{N}$ onto E. (Verify this fact.) That is, $\mathbf{N} \sim E$. By theorem 7.12, since $E \subset \mathbf{N}$ and since $\mathbf{N} \sim E$, the set $\mathbf{N}$ is an infinite set.

To "count" the number of elements in a nonempty, finite set, we find an equivalence with $\mathbf{N}_k$ for some $k \in \mathbf{N}$. We call the cardinality of a finite set a **finite cardinal number**. Because $\emptyset \subset \mathbf{N}_1 \subset \mathbf{N}_2 \subset \cdots \subset \mathbf{N}_n \subset \mathbf{N}_{n+1} \subset \cdots$. The finite cardinal numbers have the same order relation as the natural numbers—namely, $|\emptyset| < |\mathbf{N}_1| < |\mathbf{N}_2| < \cdots < |\mathbf{N}_n| < |\mathbf{N}_{n+1}| < \cdots$ or $0 < 1 < 2 < \cdots < n < n+1 < \cdots$. Infinite sets which are equivalent to the set of natural numbers, and thereby "counted" by the natural numbers, Cantor called countable or denumerably infinite. Because finite sets are countable and can be assigned a number, it is appropriate to assign a number to the set of natural numbers (and to all sets equivalent to the set of natural numbers). No natural number can be used for this purpose, because each natural number is a finite cardinal and is assigned to finite sets. To represent the cardinality of infinite sets it was necessary to create a new collection of cardinal numbers called **transfinite cardinal numbers**. Because Cantor realized additional transfinite numbers other than the one for the set of natural numbers would be required and because he suspected an infinite number of transfinite cardinal numbers would be required, Cantor designated the number $\aleph_0$ (read "aleph-null," "aleph-naught," or "aleph-zero") to be the cardinality of the natural numbers. That is, by definition $|\mathbf{N}| = \aleph_0$. Thus, the order of the transfinite cardinals anticipated by Cantor was $\aleph_0, \aleph_1, \aleph_2, \ldots$. The letter $\aleph$ is the first letter of the Hebrew alphabet. We now present the following formal definitions.

DEFINITION Cardinality of N, Denumerable, Countable, and Uncountable Sets

The **cardinality of the set of natural numbers**, $\mathbf{N}$, is $|\mathbf{N}| = \aleph_0$.

A set A is **denumerable** or **countably infinite** if and only if $A \sim \mathbf{N}$.

A set is **countable** provided it is finite or denumerable.

A set which is not countable is **uncountable**.

Theorem 7.13 The set of integers, $\mathbf{Z}$, is denumerable and $|\mathbf{Z}| = \aleph_0$.

Proof: Let $f : \mathbf{Z} \to \mathbf{N}$ be the function defined by

$$f(n) = \begin{cases} 2n, & \text{if } n > 0 \\ 1 - 2n, & \text{if } n \leq 0 \end{cases}$$

In exercise 1, you are asked to prove f is a one-to-one correspondence from $\mathbf{Z}$ onto $\mathbf{N}$. Once proved, $\mathbf{Z} \sim \mathbf{N}$ and $|\mathbf{Z}| = \aleph_0$. That is, the set of integers is denumerable and has cardinality $\aleph_0$.

Recall that theorem 7.6 says "If A is a finite set and $x \notin A$, then $A \cup \{x\}$ is a finite set and $|A \cup \{x\}| = |A| + 1$." That is, if an element which does not belong to a finite set is adjoined to the set, then the cardinality of the new set is equal to the cardinality of the original set plus one. Contrast theorem 7.6 for finite sets with the following theorem for denumerable sets. This theorem says if an element is adjoined to a denumerable set, the cardinality of the new set is the same as the cardinality of the original set—namely, $\aleph_0$.

Theorem 7.14 If A is a denumerable set, then $A \cup \{x\}$ is a denumerable set.

Proof: Let A be a denumerable set. Then there exists a one-to-one correspondence $f : \mathbf{N} \to A$. There are two cases to consider. (1) If $x \in A$, then $A \cup \{x\} = A$ which is denumerable. (2) If $x \notin A$, define $g : \mathbf{N} \to A \cup \{x\}$ by

$$g(n) = \begin{cases} x, & \text{if } n = 1 \\ f(n-1), & \text{if } n \neq 1 \end{cases}$$

The function g is a one-to-one correspondence from $\mathbf{N}$ onto $A \cup \{x\}$. (You are asked to prove this in exercise 2.) Since $\mathbf{N} \sim A \cup \{x\}$, the set $A \cup \{x\}$ is denumerable.

The previous theorem shows that one property of cardinal arithmetic is $\aleph_0 + 1 = \aleph_0$. The next theorem states that for any $n \in \mathbf{N}$, $\aleph_0 + n = \aleph_0$.

Theorem 7.15 If A is a denumerable set and B is a finite set, then $A \cup B$ is a denumerable set.

Proof: The proof of this theorem is by mathematical induction and you are asked to provide the proof in exercise 3.

The following theorem states that the union of two disjoint denumerable sets is a denumerable set. The associated cardinal arithmetic is $\aleph_0 + \aleph_0 = \aleph_0$.

Theorem 7.16 If A and B are denumerable sets and $A \cap B = \emptyset$, then $A \cup B$ is a denumerable set.

Proof: Let A and B be disjoint denumerable sets and let $f : \mathbf{N} \to A$ and $g : \mathbf{N} \to B$ be one-to-one correspondences. Define $h : \mathbf{N} \to A \cup B$ by

$$h(n) = \begin{cases} f\left(\dfrac{n+1}{2}\right), & \text{if } n \text{ is odd} \\ g\left(\dfrac{n}{2}\right), & \text{if } n \text{ is even} \end{cases}$$

In exercise 4, you are asked to prove h is a one-to-one correspondence from $\mathbf{N}$ onto $A \cup B$, which proves that $A \cup B$ is a denumerable set.

Compare theorem 7.16 and theorem 7.9 and observe the difference in the result obtained for finite sets (theorem 7.9) versus the result obtained for denumerable sets (theorem 7.16).

Theorem 7.17 The Cartesian product $\mathbf{N} \times \mathbf{N}$ is denumerable.

Proof: Claim: The function $f : \mathbf{N} \times \mathbf{N} \to \mathbf{N}$ defined by $f((m,n)) = 2^{m-1}(2n-1)$ is a one-to-one correspondence from $\mathbf{N} \times \mathbf{N}$ onto $\mathbf{N}$. To prove f is one-to-one, we assume $f((m,n)) = f((r,s))$. Hence, $2^{m-1}(2n-1) = 2^{r-1}(2s-1)$. The natural numbers $2n-1$ and $2s-1$ are both odd and it follows from the Fundamental Theorem of Arithmetic that $2^{m-1} = 2^{r-1}$ and $2n-1 = 2s-1$. Therefore, $m = r$ and $n = s$. Consequently, $(m,n) = (r,s)$ and f is one-to-one. To prove f is onto, we assume $x \in \mathbf{N}$. It follows from the Fundamental Theorem of Arithmetic that $x = 2^{m-1}q$ where $m \in \mathbf{N}$ and q is an odd natural number. Since q is odd, there exists an $n \in \mathbf{N}$ such that $q = 2n-1$. If x is odd, then $x = 2^0(2n-1) = f((1,n))$. Whereas, if x is even, $x = 2^{m-1}(2n-1) = f((m,n))$. Hence, f is onto and a one-to-one correspondence from $\mathbf{N} \times \mathbf{N}$ onto $\mathbf{N}$. That is, $\mathbf{N} \times \mathbf{N} \sim \mathbf{N}$ and so $\mathbf{N} \times \mathbf{N}$ is denumerable.

The previous theorem illustrates that in cardinal arithmetic $\aleph_0\aleph_0 = \aleph_0$.

Theorem 7.18 If A and B are denumerable sets, then $A \times B$ is a denumerable set.

Proof: By theorem 7.2, since $A \sim \mathbf{N}$ and $B \sim \mathbf{N}$, the set $A \times B \sim \mathbf{N} \times \mathbf{N}$. From theorem 7.17, we have $\mathbf{N} \times \mathbf{N} \sim \mathbf{N}$, so by transitivity of set equivalence $A \times B \sim \mathbf{N}$ and $A \times B$ is denumerable.

Theorem 7.19 Every infinite subset of a denumerable set is denumerable.

Proof: Let B be an infinite subset of a denumerable set A. Since $A \sim \mathbf{N}$, there exists a function $f : A \to \mathbf{N}$ which is a one-to-one correspondence from A onto $\mathbf{N}$. The restriction of f to the set B, $f|_B$, is a one-to-one correspondence from B onto $f(B)$. Thus, $B \sim f(B)$. We now define a function $g : \mathbf{N} \to f(B)$ inductively. The set $f(B)$ is a subset of the natural numbers. Since B is an infinite set and $f(B) \sim B$, the set $f(B)$ is infinite and, therefore, nonempty. By the Well-Ordering Principle, $f(B)$ has a least element, call it $g(1)$. For each $n \in \mathbf{N}$, the set $f(B) - \{g(1), g(2), \ldots, g(n)\}$ is nonempty, because $f(B)$ is infinite. Let $g(n+1)$ be the least element of the set $f(B) - \{g(1), g(2), \ldots, g(n)\}$. To prove g is one-to-one, we assume $p, q \in \mathbf{N}$ and $p < q$. Then $g(p) \in \{g(1), g(2), \ldots, g(q-1)\}$, but $g(p)$ is not. Hence, $g(p) \neq g(q)$ and g is one-to-one. To prove g is onto, we assume $r \in f(B)$ and there are m natural numbers less than r in $f(B)$. By definition of g, $r = g(m+1)$. Therefore, g is onto and a one-to-one correspondence from $\mathbf{N}$

onto $f(B)$. By transitivity, $\mathbf{N} \sim f(B) \sim B$ and B is denumerable.

Our immediate goal is to prove the set of rational numbers is denumerable. First, however, we prove that the set of positive rational numbers is denumerable.

Theorem 7.20 The set of positive rational numbers, denoted by $\mathbf{Q}^+$, is denumerable.

Proof: We define the set of positive rational numbers by

$$\mathbf{Q}^+ = \{\frac{p}{q} \mid p, q \in \mathbf{N} \text{ and } \gcd(p, q) = 1\}.$$

Consider the function $f : \mathbf{Q}^+ \to \mathbf{N} \times \mathbf{N}$ defined by $f(\frac{p}{q}) = (p, q)$. The function f is one-to-one but not onto, since $(2, 2) \in \mathbf{N} \times \mathbf{N}$ but $\frac{2}{2} \notin \mathbf{Q}^+$. However, the restriction of f to $\mathbf{Q}^+$ is a one-to-one function from $\mathbf{Q}^+$ onto $f(\mathbf{Q}^+)$. That is, $\mathbf{Q}^+ \sim f(\mathbf{Q}^+)$. Since $\mathbf{N} \subset \mathbf{Q}^+$, the set $\mathbf{Q}^+$ is an infinite set and by set equivalence so is $f(\mathbf{Q}^+)$. By theorem 7.19 because $f(\mathbf{Q}^+)$ is an infinite subset of $\mathbf{N} \times \mathbf{N}$, a denumerable set, $f(\mathbf{Q}^+)$ is denumerable and by set equivalence $\mathbf{Q}^+$ is denumerable also.

Theorem 7.21 The set of rational numbers is denumerable.

Proof: Let the negative rational numbers be defined by

$$\mathbf{Q}^- = \{-\frac{p}{q} \mid \frac{p}{q} \in \mathbf{Q}^+\}.$$

By definition, $\mathbf{Q}^- \sim \mathbf{Q}^+$ and, therefore, $\mathbf{Q}^-$ is denumerable. By theorem 7.14, the set $\mathbf{Q}^- \cup \{0\}$ is denumerable. By theorem 7.16, since $\mathbf{Q}^-$ is denumerable and $(\mathbf{Q}^- \cup \{0\}) \cap \mathbf{Q}^+ = \emptyset$, the set $\mathbf{Q} = \mathbf{Q}^- \cup \{0\} \cup \mathbf{Q}^+$ is denumerable.

The relationship between countable sets, finite sets, and denumerable sets, and several specific sets are shown in the Venn diagram of Figure 7.1.

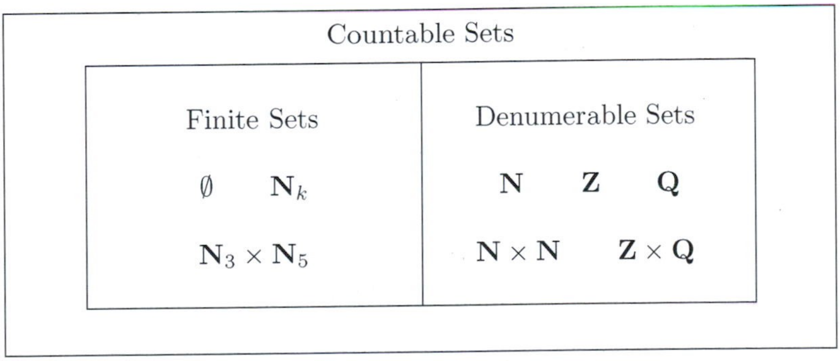

Figure 7.1 Finite, Denumerable, and Countable Sets

EXERCISES 7.2

1. Prove that the function f defined in the proof of theorem 7.13 is a one-to-one correspondence from $\mathbf{Z}$ onto $\mathbf{N}$.

2. Prove that the function g defined in the proof of theorem 7.14 is a one-to-one correspondence from $\mathbf{N}$ onto $A \cup \{x\}$.

3. Prove theorem 7.15 by mathematical induction. (**HINT:** Let $P(n)$ be the statement: "If A is a denumerable set and if B is a finite set with $|B| = n$, then $A \cup B$ is a denumerable set." The statement $P(1)$ is true by theorem 7.14.)

4. Prove that the function h defined in the proof of theorem 7.16 is a one-to-one correspondence from $\mathbf{N}$ onto $A \cup B$.

5. Prove that the following sets are denumerable.

   a. $F^+ = \{4, 8, 12, 16, \ldots\}$

   b. $F = \{\ldots, -16, -12, -8, -4, 0, 4, 8, 12, 16, \ldots\}$

   c. $G = \{n \mid n \in \mathbf{N} \text{ and } n > 100\}$

   d. $H = \{m \mid m \in \mathbf{Z} \text{ and } m < -50\}$

   e. $\mathbf{N} - \{1, 3, 5, 7, 9\}$

6. Give an example of two denumerable sets A and B such that $A \neq B$ and

   a. $A \cup B$ and $A \cap B$ are both denumerable.

   b. $A \cup B$ and $A - B$ are both denumerable.

   c. $A \cup B$ is denumerable and $A - B$ is finite but not empty.

7. Prove that if A is a denumerable set and B is a finite subset of A, then $A - B$ is denumerable. Compare this theorem with theorem 7.10.

8. Let A_1 and A_2 be denumerable sets. Prove that $A_1 \cup A_2$ is denumerable.

9. Let $A_1, A_2, \ldots, A_n$ be denumerable sets. Prove that the set $\bigcup_{i=1}^{n} A_i$ is denumerable.

7.3 Uncountable Sets

Every real number can be written as a decimal expansion with an infinite number of digits to the right of the decimal point. Every irrational number has a unique decimal expansion which is nonrepeating. Every rational number has a repeating expansion. However, some rational numbers have two different decimal expansions. For example, $1/4 = 0.250000\cdots$ and $1/4 = 0.249999\cdots$. If a rational number has two different decimal expansions, one ends with an infinite string of zeros and the other ends with an infinite string of nines. We will say a real number is in **normalized decimal form**, if it does not end with an infinite string of nines. Therefore, $1/4 = 0.250000\cdots$ in normalized decimal form. By making this choice for the decimal representation of real numbers, we have guaranteed the following statement is true: "Every real number can be written uniquely in normalized decimal form." Hence, **two real numbers written in normalized decimal form are equal if and only if in each position of their decimal expansion their digits are identical**.

Thus far, we have seen examples of finite sets, denumerable sets, countable sets, and infinite sets, but we have not exhibited an uncountable set. In the following theorem, we prove the open unit interval $(0, 1)$ is an uncountable set. The method of proof employed in this theorem is due to Georg Cantor and is called **Cantor's diagonal argument**.

Theorem 7.22 The open unit interval $(0, 1)$ is an uncountable set.

Proof: The set $S = \{\frac{1}{2}, \frac{1}{3}, \frac{1}{4}, \ldots\}$ is an infinite subset of $(0, 1)$; therefore, $(0, 1)$ is an infinite set. (See exercise 8 of section 7.2.) Consequently, $(0, 1)$ is denumerable or uncountable. We will show by contradiction that $(0, 1)$ is uncountable. Assume $(0, 1)$ is denumerable. Then there exists a one-to-one correspondence $f : \mathbf{N} \to (0, 1)$. We list all of the real numbers in $(0, 1)$ in normalized decimal form as follows:

$$\begin{aligned}
f(1) &= 0.a_{11}a_{12}a_{13}a_{14}a_{15}\cdots \\
f(2) &= 0.a_{21}a_{22}a_{23}a_{24}a_{25}\cdots \\
f(3) &= 0.a_{31}a_{32}a_{33}a_{34}a_{35}\cdots \\
f(4) &= 0.a_{41}a_{42}a_{43}a_{44}a_{45}\cdots \\
&\vdots \\
f(n) &= 0.a_{n1}a_{n2}a_{n3}a_{n4}a_{n5}\cdots \\
&\vdots
\end{aligned}$$

Observe that the jth decimal digit of the ith real number $f(i)$ is a_{ij}. We will now show how to construct a real number $b = 0.b_1b_2b_3b_4b_5\cdots$ in $(0,1)$ which does not appear in the list $f(1), f(2), f(3), \ldots$. We want to choose the digits b_k so that $b_k \neq a_{kk}$ and $b_k \neq 9$. We choose $b_k \neq a_{kk}$ to ensure b differs from $f(k)$ in the kth decimal digit. We choose $b_k \neq 9$ to ensure we do not create a decimal number which ends with an infinite string of nines. For each $k \in \mathbf{N}$, define

$$b_k = \begin{cases} 2, & \text{if } a_{kk} \neq 2 \\ \\ 4, & \text{if } a_{kk} = 2 \end{cases}$$

By our choice of the digits b_k, the number $b = 0.b_1b_2b_3b_4b_5\cdots$ is a real number expressed in normalized decimal form which is in the interval $(0,1)$ but which does not appear in the list $f(1), f(2), f(3), \ldots$. Therefore, the function f is not onto which contradicts f being a one-to-one correspondence from $\mathbf{N}$ onto $(0,1)$. << The choice of the digits 2 and 4 in the definition of b_k is arbitrary. Any other two distinct digits neither of which is 9 would suffice.>>

The open unit interval $(0,1)$ is our first example of an uncountable set. We will define the cardinality of $(0,1)$ to be $\mathbf{c}$. Just as the sets $\mathbf{N}_k$ for $k \in \mathbf{N}$ are our standards for measuring finite sets and $\mathbf{N}$ is our standard for measuring denumerable sets, the interval $(0,1)$ is one standard for measuring uncountable sets.

DEFINITION Cardinality of the Set $(0,1)$ and Cardinality c

The **cardinality of the set** $(0,1)$ is $|(0,1)| = \mathbf{c}$.

A set A is **uncountable** and has **cardinality c** if and only if $A \sim (0,1)$.

The next theorem states that every finite interval is uncountable and has cardinality $\mathbf{c}$.

Theorem 7.23 For $a, b \in \mathbf{R}$ with $a < b$, the open interval $(a,b) \sim (0,1)$ and $|(a,b)| = \mathbf{c}$.

Proof: The function $f : (a,b) \to (0,1)$ defined by $f(x) = \dfrac{x-a}{b-a}$ is a linear function from $\mathbf{R}$ onto $\mathbf{R}$ restricted to the domain (a,b); therefore, f is a one-to-one correspondence from (a,b) onto $(0,1)$.

Since $\sim$ is an equivalence relation, it follows that for $a, b, c, d \in \mathbf{R}$ with $a < b$ and $c < d$ the open intervals (a,b) and (c,d) are equivalent and have cardinality $\mathbf{c}$. The following theorem states that the set of real numbers is an uncountable set and has cardinality $\mathbf{c}$.

Theorem 7.24 The set of real numbers $\mathbf{R}$ is uncountable and $|\mathbf{R}| = \mathbf{c}$.

Proof: The function $g : (-\frac{\pi}{2}, \frac{\pi}{2}) \to \mathbf{R}$ defined by $g(x) = \tan x$ is a one-to-one correspondence from the interval $(-\frac{\pi}{2}, \frac{\pi}{2})$ onto $\mathbf{R}$. Since $(0,1) \sim (-\frac{\pi}{2}, \frac{\pi}{2})$ by the transitivity of $\sim$, we have $(0,1) \sim \mathbf{R}$. Consequently, $\mathbf{R}$ is countable and by definition $|\mathbf{R}| = \mathbf{c}$.

The set of real numbers is referred to as the **continuum**. For this reason, the cardinal number $\mathbf{c}$ is called the **cardinality of the continuum**.

By theorem 7.21 the set of rational numbers is denumerable and by theorem 7.24 the set of real numbers is uncountable. The following theorem proves that the set of irrational numbers is uncountable.

Theorem 7.25 The set of irrational numbers is uncountable.

Proof: By definition, the set of irrational numbers is the set $H = \mathbf{R} - \mathbf{Q}$. The set $\{n + \sqrt{2} \mid n \in \mathbf{N}\}$ is an infinite subset of H. Hence, H is an infinite set; and, consequently, H is either denumerable or uncountable. Suppose H is denumerable. By theorem 7.21, $\mathbf{Q}$ is denumerable. Since $\mathbf{Q}$ is denumerable, since $H = \mathbf{R} - \mathbf{Q}$ is assumed to be denumerable, and since $\mathbf{Q} \cap (\mathbf{R} - \mathbf{Q}) = \emptyset$, $\mathbf{Q} \cup (\mathbf{R} - \mathbf{Q}) = \mathbf{R}$ is denumerable by theorem 7.16. But this contradicts theorem 7.24. Hence, the set of irrational numbers is uncountable.

The relationship between countable sets, finite sets, denumerable sets, infinite sets, and uncountable sets and several specific sets are shown in the Venn diagram of Figure 7.2.

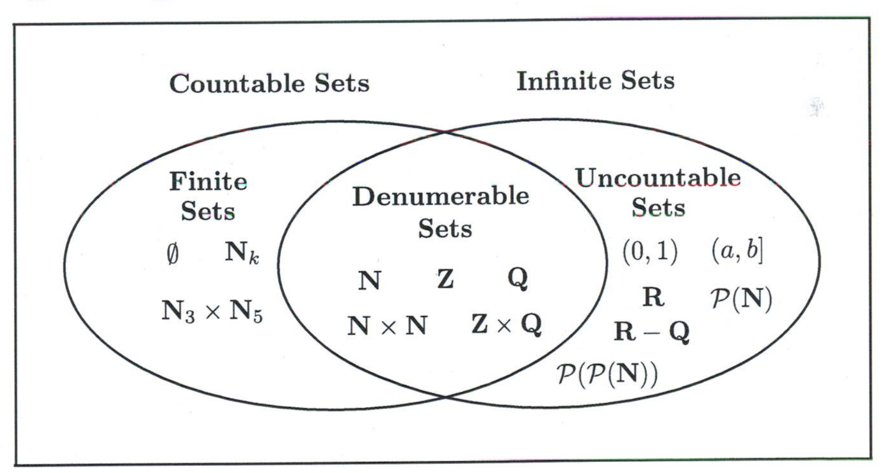

FIGURE 7.2: Finite, Denumerable, Countable, Infinite, and Uncountable Sets

We may compare cardinal numbers using the following definitions.

DEFINITION $=, \neq, \leq, <$ **for Cardinal Numbers**

Let A and B be sets.

$|A| = |B|$ if and only if $A \sim B$.

$|A| \neq |B|$ if and only if $A \not\sim B$.

$|A| \leq |B|$ if and only if there exists a one-to-one function f from A to B.

$|A| < |B|$ if and only if $|A| \leq |B|$ and $|A| \neq |B|$.

By definition, $|A| < |B|$ if there exists a one-to-one function from A to B but no one-to-one function which is onto B. For example, for all $k \in \mathbf{N}$, $|\mathbf{N}_k| < |\mathbf{N}|$, since the function $f : \mathbf{N}_k \to \mathbf{N}$ defined by $f(x) = x$ is one-to-one from $\mathbf{N}_k$ to $\mathbf{N}$ but there is no one-to-one function from $\mathbf{N}_k$ onto $\mathbf{N}$. Furthermore, $|\mathbf{N}| < |\mathbf{R}|$ or $\aleph_0 < \mathbf{c}$, because the function $g : \mathbf{N} \to \mathbf{R}$ defined by $g(x) = x$ is one-to-one from $\mathbf{N}$ to $\mathbf{R}$ but there is no one-to-one function from $\mathbf{N}$ onto $\mathbf{R}$.

So far, we have seen only two different transfinite cardinal numbers, $\aleph_0$ and $\mathbf{c}$. Some questions which arise quite naturally are "Are there more transfinite cardinal numbers?", "If so, what are they?", and "How many are there?" The next theorem, which Georg Cantor proved, is known as Cantor's Theorem and answers all of these questions. Recall that the power set $\mathcal{P}(A)$ of a set A is the set of all subsets of A.

Cantor's Theorem If A is a set, then $|A| < |\mathcal{P}(A)|$.

Proof: Let A be a set. If $A = \emptyset$, then $\mathcal{P}(A) = \{\emptyset\}$, and $|A| = 0 < 1 = |\mathcal{P}(A)|$. For $A \neq \emptyset$, we must show (i) $|A| \leq |\mathcal{P}(A)|$ and (ii) $|A| \neq |\mathcal{P}(A)|$. To prove (i) we must show there exists a function $f : A \to \mathcal{P}(A)$ which is one-to-one. We define $f : A \to \mathcal{P}(A)$ by $f(x) = \{x\}$. Suppose $f(x_1) = f(x_2)$. Then $\{x_1\} = \{x_2\}$ which implies $x_1 = x_2$. Consequently f is one-to-one. We prove (ii) by contradiction. Suppose there exists a function g from A onto $\mathcal{P}(A)$. Let $S = \{x \in A \mid x \notin g(x)\}$. Since $S \subseteq A$, we have $S \in \mathcal{P}(A)$ and since g is onto $\mathcal{P}(A)$ there exists some $y \in A$ such that $S = g(y)$. Either (1) $y \in S$ or (2) $y \notin S$.

(1) If $y \in S = \{x \in A \mid x \notin g(x)\}$, then $y \notin g(y) = S$. Contradiction.

(2) If $y \notin S = \{x \in A \mid x \notin g(x)\}$, then $y \in g(y) = S$. Contradiction.

In either case, there does not exist a $y \in A$ such that $g(y) = S$. Therefore, there is no function g from A to $\mathcal{P}(A)$ which is onto. Consequently, there is no one-to-one correspondence from A onto $\mathcal{P}(A)$ and $A \not\sim \mathcal{P}(A)$. Hence, $|A| \neq |\mathcal{P}(A)|$ and $|A| < |\mathcal{P}(A)|$.

One consequence of Cantor's theorem is $|\mathbf{N}| < |\mathcal{P}(\mathbf{N})|$. Since $\mathcal{P}(\mathbf{N})$ contains the infinite subset $\{\{1\}, \{2\}, \{3\}, \ldots\}$, the set $\mathcal{P}(\mathbf{N})$ is an infinite set. And since $|\mathbf{N}| < |\mathcal{P}(\mathbf{N})|$, the set $\mathcal{P}(\mathbf{N})$ is uncountable. Because $\mathcal{P}(\mathbf{N})$ is a set, $\mathcal{P}(\mathcal{P}(\mathbf{N}))$ is a set. A second consequence of Cantor's theorem is $|\mathcal{P}(\mathbf{N})| < |\mathcal{P}(\mathcal{P}(\mathbf{N}))|$. In this manner, we can generate a denumerable collection of cardinal numbers with the properties $\aleph_0 = |\mathbf{N}| < |\mathcal{P}(\mathbf{N})| < |\mathcal{P}(\mathcal{P}(\mathbf{N}))| < |\mathcal{P}(\mathcal{P}(\mathcal{P}(\mathbf{N})))| < \cdots$. Thus, there are infinitely many transfinite cardinal numbers and there is no largest cardinal number.

About 1880, Georg Cantor asked the question: "Is there a cardinal number which lies strictly between $\aleph_0$ and $\mathbf{c}$?" Stated in terms of the set of real numbers, this question is "Does there exist an infinite set of real numbers which is not equivalent to the set of natural numbers and which is not equivalent to the set of real numbers?" Cantor conjectured the answer to both of these questions is no. This conjecture became known as the **continuum hypothesis**, which can be stated as follows.

Continuum Hypothesis: There is no set X such that $\aleph_0 < |X| < \mathbf{c}$.

Cantor and other leading mathematicians tried to prove or disprove this conjecture with no success. A generalization of this conjecture is called the **generalized continuum hypothesis**. This conjecture states that there is no cardinal number which lies strictly between a transfinite cardinal number a and 2^a. Stated more formally, the conjecture is

Generalized Continuum Hypothesis: For any transfinite cardinal number a there is no cardinal number x such that $a < x < 2^a$.

At the International Congress of Mathematicians held in Paris in 1900, the famous German mathematician David Hilbert (1862-1943) presented a list of twenty-three important unsolved problems. At the top of the list was the continuum hypothesis. In 1931, the Austrian mathematician Kurt Gödel (1906-1978) proved that any consistent axiom system includes true, unprovable statements. In 1940, Gödel proved if a contradiction were to arise by assuming the axioms of set theory and the continuum hypothesis, then the contradiction could be deduced from the axioms of set theory alone. Thus, Gödel showed the continuum hypothesis was independent from the axioms of set theory and assuming either the continuum hypothesis or its negation would result in a consistent axiom system provided the axioms of set theory, themselves, are consistent. In this sense, the continuum hypothesis is an undecidable statement. In 1963, the American mathematician Paul J. Cohen (1934-) proved that neither the continuum hypothesis nor the axiom of choice can be proved from the axioms of set theory.

In 1895 and 1897, Cantor published his last two works on set theory. In 1895, he used the axiom of choice to prove if A and B are sets such that $|A| \leq |B|$ and $|B| \leq |A|$, then $|A| = |B|$. At that time, the axiom of choice was somewhat controversial. So in 1896, Ernest Schröder (1841-1902) proved the same result without using the axiom of choice. Independently, while attending Cantor's seminar at Halle in 1897, Felix Bernstein (1878-1956) also proved this result without employing the axiom of choice. The following theorem, which we will not prove, is called the **Schröder-Bernstein Theorem** or the **Cantor-Schröder-Bernstein Theorem**.

Schröder-Bernstein Theorem Let A and B be sets. If $|A| \leq |B|$ and $|B| \leq |A|$, then $|A| = |B|$.

The Schröder-Bernstein Theorem can be used to prove two sets are equivalent when it is difficult to produce a one-to-one correspondence between the two sets. For example, to prove $|(0,1)| = |[0,1)|$ and, therefore, $(0,1) \sim [0,1)$. We first note that the function $f : (0,1) \to [0,1)$ defined by $f(x) = x$ is a one-to-one function, so (1) $|(0,1)| \leq |[0,1)|$. Then, we observe that the function $g : [0,1) \to (-1,2)$ defined by $g(x) = x$ is one-to-one. Hence, $|[0,1)| \leq |(-1,2)|$. But since $(-1,2) \sim (0,1)$, we have $|(-1,2)| = |(0,1)|$. Thus, (2) $|[0,1)| \leq |(-1,2)| = |(0,1)|$. Consequently, from (1) and (2) by the Schröder-Bernstein Theorem $|(0,1)| = |[0,1)|$ and, therefore, $(0,1) \sim [0,1)$.

Recall for a finite set A with $|A| = n$ elements, the number of elements in the power set of A is $|\mathcal{P}(A)| = 2^n$. For any infinite set A with cardinal number $|A|$, we define the number of elements in the power set of A to be $|\mathcal{P}(A)| = 2^{|A|}$. Since $|\mathbf{N}| = \aleph_0$, by definition the cardinality of the power set of $\mathbf{N}$ is $|\mathcal{P}(\mathbf{N})| = 2^{\aleph_0}$. We now use the Schröder-Bernstein Theorem to prove $|\mathcal{P}(\mathbf{N})| = 2^{\aleph_0} = \mathbf{c} = |\mathbf{R}|$.

Theorem 7.26 $\mathbf{c} = 2^{\aleph_0}$

Proof: Let $I = [0,1)$. We have just shown that $|I| = \mathbf{c}$. First, we define a one-to-one function f from $\mathcal{P}(\mathbf{N})$ to I. For each set $A \in \mathcal{P}(\mathbf{N})$, let $x = f(A) \in I$ be the real number whose decimal expansion $x = 0.d_1d_2d_3\ldots$ is defined by

$$d_k = \begin{cases} 0, & \text{if } n \notin A \\ 1, & \text{if } n \in A \end{cases}$$

(You should verify f is one-to-one.) Since $f : \mathcal{P}(\mathbf{N}) \to I$ is one-to-one, (1) $|\mathcal{P}(\mathbf{N})| \leq |I|$. Next, we define a one-to-one function g from I to $\mathcal{P}(\mathbf{N})$. If $x \in I$ and x has binary expansion $x = 0.b_1b_2b_3\ldots$, then $g(x) \in \mathcal{P}(\mathbf{N})$ is defined to be the set $g(x) = \{n \in \mathbf{N} \mid b_n = 1\}$. (You should verify g is one-to-one.) Since $g : I \to \mathcal{P}(\mathbf{N})$ is one-to-one, (2) $|I| \leq |\mathcal{P}(\mathbf{N})|$. It follows from (1) and (2) by the Schröder-Bernstein Theorem that $|I| = |\mathcal{P}(\mathbf{N})|$ or $\mathbf{c} = 2^{\aleph_0}$.

EXERCISES 7.3

1. Prove that both of the following functions are one-to-one correspondences from $(0,1)$ onto $\mathbf{R}$. Consequently, each of these functions can be used to provide an alternate proof of theorem 7.24.

   a. $h(x) = \tan\left(\pi\left(x - \frac{1}{2}\right)\right)$ b. $h(x) = \dfrac{2x-1}{x^2 - x}$

2. Prove that if A is an uncountable set and if $B \sim A$, then B is uncountable.

3. Prove that if A is an uncountable subset of B, then B is an uncountable set.

4. For each of the given sets find a one-to-one correspondence from the set onto $\mathbf{R}$. This proves that each of the given sets has cardinality $\mathbf{c}$.

   a. $(0, \infty)$ b. (a, ∞) for any $a \in \mathbf{R}$ c. $\mathbf{R} - \{0\}$

   d. $\mathbf{R} - \{a\}$ for any $a \in \mathbf{R}$

5. Let A, B, and C be nonempty sets. Prove that

   a. If $|A| = |B|$ and $|B| = |C|$, then $|A| = |C|$.
   (Transitivity of $=$ for cardinal numbers)

   b. $|A| \leq |A|$ (Reflexivity for $\leq$ for cardinal numbers)

   c. If $|A| \leq |B|$ and $|B| \leq |C|$, then $|A| \leq |C|$.
   (Transitivity of $\leq$ for cardinal numbers)

   d. If $A \subseteq B$, then $|A| \leq |B|$.

   e. $|A \cap B| \leq |B|$

   f. $|A| \leq |A \times B|$

   g. If $A \subseteq B \subseteq C$ and $|A| = |C|$, then $|A| = |B|$.

6. Use $=$ and $<$ to order the following lists of cardinal numbers.

   a. $|\mathcal{P}(\mathbf{R})|$, $|\{\emptyset\}|$, $|\mathbf{R}|$, $|\emptyset|$, $|\mathcal{P}(\mathcal{P}(\mathbf{R}))|$, $|\mathbf{Q}|$, $|\mathbf{R} - \mathbf{Q}|$, $|\mathbf{N}|$, $|(0,3)|$, $|\{0,3\}|$, $|[0,3]|$, $|(0,\infty)|$

   b. $\mathbf{c}$, $|\mathbf{R} - \{0\}|$, $|\mathbf{Q} - \{0\}|$, $|\mathbf{Z} - \{0\}|$, $|\mathbf{R} - \mathbf{Q}|$, $|\mathbf{R} - \mathbf{Z}|$, $|\mathbf{Q} - \mathbf{Z}|$, $|\mathbf{R}|$, $|\mathbf{Z}|$, $|\mathbf{Q}|$, $\aleph_0$

7. Use the Schröder-Bernstein Theorem to prove the following.

   a. If A, B, and C are sets with $|A| \leq |B|$, $|B| \leq |C|$, and $|C| \leq |A|$, then $A \sim B \sim C$.

   b. $\mathbf{R} \times \mathbf{R} \sim \mathbf{R}$

Chapter 8

Proofs from Real Analysis

Real analysis is the study of real numbers, sets of real numbers, and functions on sets of real numbers. In this chapter, we study sequences of real numbers and their properties.

8.1 Sequences

Sequences play an important role in mathematics. In calculus, you undoubtedly discussed sequences of real numbers prior to studying series. In the broadest sense, a **sequence** is a function whose domain is the set of natural numbers. For example, a sequence can be a sequence of sets, a sequence of intervals, a sequence of functions, and so forth. In this chapter, we devote our attention to sequences of real numbers.

DEFINITIONS A Sequence of Real Numbers and the General Term of a Sequence

A **sequence of real numbers** is a function from the set of natural numbers to the set of real numbers.

Thus, a sequence of real numbers $a : \mathbf{N} \to \mathbf{R}$ is a function which assigns to each natural number, n, a unique real number $a(n)$. It is customary to denote the **general term** or **nth term** of the sequence by a_n instead of $a(n)$.

We will represent a sequence by $a_1, a_2, \cdots, a_n, \cdots$ or $< a_n >_{n=1}^{\infty}$ or, simply, $< a_n >$.

Sequences are usually defined (i) by listing the first few terms and assuming the pattern continues indefinitely, (ii) by giving an explicit formula for the general term, or (iii) by recursion. The following is an example of defining the same sequence by these three different methods.

$$(i)\ \ 0,1,0,1,0,1,\cdots \qquad (ii)\ \ a_n = \frac{1+(-1)^n}{2}$$

$$(iii)\ \ a_1 = 0 \ \text{ and } \ a_n = a_{n-1} + (-1)^n \ \text{ for } \ n \geq 2$$

It is important to understand the notational difference between the sequence $< a_n >_{n=1}^{\infty}$ and the range of the sequence $\{a_n\}_{n=1}^{\infty}$. For example, if $a_n = \cos(n\pi/2)$, then $< a_n >_{n=1}^{\infty} = 0,-1,0,1,0,-1,0,1,\cdots$ while $\{a_n\}_{n=1}^{\infty} = \{0,-1,1\}$. In this instance, the sequence $< a_n >$ is a countably infinite ordered set, while the range $\{a_n\}$ is a finite set with three elements.

Consider the sequence $\left\langle \frac{1}{n} \right\rangle = 1, \frac{1}{2}, \frac{1}{3}, \frac{1}{4}, \frac{1}{5}, \frac{1}{6}, \cdots$. It appears that as n gets very large, the general term $a_n = \frac{1}{n}$ approaches the value 0, although no term in the sequence is ever 0. On the other hand, the sequence $\left\langle \frac{1+(-1)^n}{2} \right\rangle = 0,1,0,1,0,1,\cdots$ does not approach any value as n gets very large, because successive terms continue to alternate between 0 and 1. The terms in the sequence $\left\langle \frac{1}{n} \right\rangle$ get smaller with each successive term. That is, for all $n \in \mathbf{N}$, we have $a_{n+1} < a_n$, because $0 < 1$ and $n \in \mathbf{N}$ implies $0 < n < n+1$ which implies $0 < \frac{1}{n+1} < \frac{1}{n}$. If we choose a positive value, say $\epsilon = .05$, how far out in the sequence $\left\langle \frac{1}{n} \right\rangle$ must we go before $\frac{1}{n} < .05$? Solving the last inequality for n, we see for $n > 20$ we have $\frac{1}{n} < .05$. Since $0 < a_{n+1} < a_n$ for all $n \in \mathbf{N}$ and since $\frac{1}{n} < .05$ for $n \geq 21$, we have $0 < \frac{1}{n} < .05$ for all $n \geq 21$. Notice that for $n \geq 21$, the set $\left\{ \frac{1}{n} \right\}_{n=21}^{\infty}$ is a subset of the interval $(0, .05)$—that is, all of the elements of the set $\left\{ \frac{1}{n} \right\}_{n=21}^{\infty}$ are within .05 of 0.

As we have seen, as n gets large some sequences approach a real value while others do not. We now give a formal definition of convergence for a sequence of real numbers.

DEFINITIONS Convergence and Divergence of a Sequence of Real Numbers

A sequence $< a_n >$ of real numbers **converges to a real number** A if and only if for every $\epsilon > 0$ there exists an $N \in \mathbf{N}$ such that for every $n \in \mathbf{N}$ if $n > N$, then $|a_n - A| < \epsilon$.

The number A is called the **limit** of the sequence and usually we write $\lim_{n\to\infty} a_n = A$ or $a_n \to A$ to denote that the sequence $< a_n >$ converges to the limit A.

If a sequence $< a_n >$ does not converge, we say the sequence is **divergent** or "the sequence **diverges**."

In the definition of convergence of a sequence, the real number $\epsilon > 0$ is the error tolerance and the natural number N, which usually depends on the choice of ϵ, tells us how far out in the sequence we must go in order to guarantee that for all $n > N$ the distance between a_n and A is less than ϵ.

Let c be a real constant. The constant sequence $c, c, c, \cdots$ converges to the limit c. To prove this, let $\epsilon > 0$ be given. Choose $N = 1$. Then for all natural numbers $n > N$, we have $|c - c| = 0 < \epsilon$. It follows from the definition of convergence that $\lim_{n\to\infty} c = c$.

Now let us use the definition of convergence to prove that the sequence $\left\langle \frac{1}{n} \right\rangle$ converges to 0. Our reasoning and "scratch work" for the required proof proceeds as follows. Given any real number $\epsilon > 0$, we must determine how to choose a natural number N such that if $n > N$, then $\left|\frac{1}{n} - 0\right| < \epsilon$. First, we notice that if $n > N > 0$, then $\frac{1}{N} > \frac{1}{n} > 0$. Consequently, if $n > N$, then $\left|\frac{1}{n} - 0\right| = \frac{1}{n} < \frac{1}{N}$ and $\frac{1}{N} < \epsilon$, provided N is any natural number greater than $\frac{1}{\epsilon}$. By the Archimedean property of the real numbers, there always exists such an N. Thus, our theorem and proof that $\lim_{n\to\infty} \frac{1}{n} = 0$ reads as follows.

Theorem 8.1 The sequence $\left\langle \frac{1}{n} \right\rangle$ converges to 0.

Proof: Let $\epsilon > 0$ be given. By the Archimedean property of the real numbers, there exists a natural number $N > \frac{1}{\epsilon}$. For all $n > N$,

$$\left| \frac{1}{n} - 0 \right| = \left| \frac{1}{n} \right| = \frac{1}{n} < \frac{1}{N} < \epsilon.$$

Hence, $\lim_{n \to \infty} \frac{1}{n} = 0$.

EXAMPLE 1 Proving the Limit of a Sequence

a. Estimate the limit of the sequence whose general term is $a_n = \frac{3n^2+1}{2n^2+5}$.

b. Prove the estimate is correct.

SOLUTION

a. Calculating the value of a_n for $n = 10, 100$, and 1000, we find

$$a_{10} = \frac{301}{205} \approx 1.468292, \qquad a_{100} = \frac{30001}{20005} \approx 1.499675$$

$$a_{1000} = \frac{3000001}{2000005} \approx 1.499997.$$

(The symbol $\approx$ is read "is approximately.") Based on these calculations, our estimate for the limit of the sequence is $1.5 = 3/2$.

b. **Scratch work**: Given $\epsilon > 0$, we must determine a natural number N such that if $n > N$, then

$$\left| \frac{3n^2+1}{2n^2+5} - \frac{3}{2} \right| < \epsilon.$$

Calculating, we find

$$\left| \frac{3n^2+1}{2n^2+5} - \frac{3}{2} \right| = \left| \frac{6n^2+2-6n^2-15}{2(2n^2+5)} \right| = \frac{13}{2(2n^2+5)} < \frac{13}{4n^2}.$$

If $n > N > 0$, then $\frac{1}{N} > \frac{1}{n} > 0$ and $\frac{1}{N^2} > \frac{1}{n^2}$. Hence, for $n > N$,

$$\left| \frac{3n^2+1}{2n^2+5} - \frac{3}{2} \right| < \frac{13}{4n^2} < \frac{13}{4N^2} < \epsilon, \tag{1}$$

provided N is chosen so that $\frac{13}{4N^2} < \epsilon$. Solving this last inequality for N, we see that (1) will be satisfied by any natural number $N > \sqrt{\frac{13}{4\epsilon}}$.

Our proof that $\lim_{n\to\infty} \frac{3n^2+1}{2n^2+5} = \frac{3}{2}$ reads as follows.

Proof: Let $\epsilon > 0$ be given. By the Archimedean property of real numbers, there exists a natural number $N > \sqrt{\frac{13}{4\epsilon}}$. For all $n > N$,

$$\left|\frac{3n^2+1}{2n^2+5} - \frac{3}{2}\right| < \left|\frac{-13}{2(2n^2+5)}\right| < \frac{13}{2(2n^2+5)} < \frac{13}{4n^2} < \frac{13}{4N^2} < \epsilon.$$

Hence, $\lim_{n\to\infty} \frac{3n^2+1}{2n^2+5} = \frac{3}{2}$.

How do we prove a sequence diverges? First, we write the definition of convergence more symbolically, so it will be easier to negate and write correctly in English. The symbolic definition of convergence is

$$(2) \quad \lim_{n\to\infty} a_n = A \Leftrightarrow (\forall\epsilon > 0)(\exists N \in \mathbf{N})(\forall n \in \mathbf{N})[(n > N) \Rightarrow (|a_n - A| < \epsilon)]$$

Recalling that the negation of the statement $P \Rightarrow Q$ is $\neg(\neg P \vee Q) \equiv P \wedge (\neg Q)$, we negate (2) and obtain the following definition

$$(3) \quad \lim_{n\to\infty} a_n \neq A \Leftrightarrow (\exists\epsilon > 0)(\forall N \in \mathbf{N})(\exists n \in \mathbf{N})[(n > N) \wedge (|a_n - A| \geq \epsilon)]$$

Statement (3) says: "The sequence $< a_n >$ does not converge to the limit A if and only if there exists a positive real number, ϵ, such that for every natural number, N, there exists a natural number $n > N$ and $|a_n - A| \geq \epsilon$." In the next example, we use this statement to prove the sequence $\left\langle \frac{1+(-1)^n}{2} \right\rangle$ diverges.

EXAMPLE 2 Proving a Sequence Diverges

Prove that the sequence whose general term is $a_n = \frac{1+(-1)^n}{2}$ diverges.

SOLUTION

Scratch work: Let A be any real number and suppose $\lim_{n\to\infty} a_n = A$. The terms of the sequence are $0, 1, 0, 1, \cdots$. Since the distance between the two values 0 and 1 is 1, we choose $\epsilon = 1/2$. (Any value of ϵ less than 1/2 will work as well.) We must show for any A and any $N \in \mathbf{N}$ there is an $n > N$ for which

$|a_n - A| \geq 1/2$. We will break the proof into two parts. In the first part, we will assume $A \geq 1/2$; while in the second part, we will assume $A < 1/2$. When $A \geq 1/2$ and N is any natural number, we choose n to be any odd natural number greater than N. Then $a_n = 0$ and $|a_n - A| = |0 - A| = |A| = A \geq 1/2$. When $A < 1/2$ and N is any natural number, we choose n to be any even natural number greater than N. Then $a_n = 1$. Since $A < 1/2$, we have $0 < \frac{1}{2} - A$ and, therefore, $\frac{1}{2} < 1 - A$. Hence, $|1 - A| > \frac{1}{2}$.

The following is a proof that the sequence $\left\langle \frac{1 + (-1)^n}{2} \right\rangle$ diverges.

Proof: Let A be any real number and let $a_n = \frac{1 + (-1)^n}{2}$. Choose $\epsilon = 1/2$ and let N be any natural number. Suppose that $\lim_{n \to \infty} a_n = A$. We divide the proof into two cases.

Case 1 If $A \geq 1/2$, we choose n to be any odd natural number greater than N. Then $a_n = 0$ and $|a_n - A| = A \geq 1/2$. Hence, $\lim_{n \to \infty} a_n \neq A$, if $A \geq 1/2$.

Case 2 If $A < 1/2$, we choose n to be any even natural number greater than N. Then $a_n = 1$. Since $A < 1/2$, we have $0 < \frac{1}{2} - A$ and, therefore, $\frac{1}{2} < 1 - A$. Hence, $|1 - A| > 1/2$. Thus, $\lim_{n \to \infty} a_n \neq A$, if $A < 1/2$.

Therefore, the sequence $\left\langle \frac{1 + (-1)^n}{2} \right\rangle$ diverges.

The following theorem states that the limit of a convergent sequence is unique.

Theorem 8.2 If a sequence converges, its limit is unique.

Proof: << As with most uniqueness theorems, we assume there are two distinct objects which satisfy the specified condition and then reach a contradiction.>> We prove this theorem by contradiction. Assume there is a sequence $< a_n >$ and two real numbers A and B such that $A \neq B$, $a_n \to A$, and $a_n \to B$. Since $A \neq B$, we have $|A - B| > 0$. Let $\epsilon = |A - B|/3$. Since $a_n \to A$, there exists an $N_1 \in \mathbf{N}$ such that $n > N_1$ implies $|a_n - A| < \epsilon$. And since $a_n \to B$, there exists an $N_2 \in \mathbf{N}$ such that $n > N_2$ implies $|a_n - B| < \epsilon$. Let $N = \max\{N_1, N_2\}$. Then by the triangle inequality, for $n > N$ we have

$$|A - B| = |A - a_n + a_n - B| \leq |A - a_n| + |a_n - B| < \epsilon + \epsilon = 2|A - B|/3$$

Since $|A - B| > 0$, we have proven $1 < 2/3$. Contradiction. Consequently, if a sequence converses, its limit is unique.

For obvious reasons, the next theorem is called the "squeeze" or "sandwich" theorem for sequences. This theorem permits us to determine and prove the limit of some sequences without appealing to the definition.

Theorem 8.3 If $< a_n >$, $< b_n >$, and $< c_n >$ are sequences such that $a_n \le b_n \le c_n$ for all $n \in \mathbf{N}$ and if $\lim_{n\to\infty} a_n = \lim_{n\to\infty} c_n = L$, then $\lim_{n\to\infty} b_n = L$.

Proof: Let $\epsilon > 0$ be given. By the definition of convergent sequence, there exists an $N \in \mathbf{N}$ such that for all $n > N$, we have $|a_n - L| < \epsilon$ and $|c_n - L| < \epsilon$ or

$$(4) \quad L - \epsilon < a_n < L + \epsilon \quad \text{and} \quad (5) \quad L - \epsilon < c_n < L + \epsilon.$$

From the left inequality of (4), the hypothesis $a_n \le b_n \le c_n$ for all $n \in \mathbf{N}$, and the right inequality in (5), we have for all $n > N$

$$L - \epsilon < a_n \le b_n \le c_n < L + \epsilon$$

Thus, $n > N$ implies $|b_n - L| < \epsilon$ and by definition $\lim_{n\to\infty} b_n = L$.

We can use theorem 8.3 to determine the limit of the sequence

$$\left\langle \frac{\cos(n^2 + 2n)}{n} \right\rangle$$

without having to use the definition of convergence to prove the result. Since the range of the cosine function is $[-1, 1]$, we have $-1 \le \cos(n^2 + 2n) \le 1$. Dividing by $n > 0$, we find

$$\frac{-1}{n} \le \frac{\cos(n^2 + 2n)}{n} \le \frac{1}{n}.$$

By the "squeeze" theorem with $a_n = \dfrac{-1}{n} \to 0$, with $b_n = \dfrac{\cos(n^2 + 2n)}{n}$, and with $c_n = \dfrac{1}{n} \to 0$, we conclude

$$b_n = \frac{\cos(n^2 + 2n)}{n} \to 0.$$

We need the following boundedness definitions for sequences. Notice the similarities in these definitions and the definitions for bounded sets.

DEFINITIONS Bounded Above, Bounded Below, and Bounded Sequences

Let $< a_n >$ be a sequence of real numbers.

The sequence $< a_n >$ is **bounded above** if and only if there exists a real number U such that $a_n \leq U$ for all $n \in \mathbf{N}$.

The sequence $< a_n >$ is **bounded below** if and only if there exists a real number L such that $L \leq a_n$ for all $n \in \mathbf{N}$.

If a sequence $< a_n >$ **bounded** if and only if there exists a positive real number M such that $|a_n| \leq M$ for all $n \in \mathbf{N}$.

For instance, the sequence $0, 1, 0, 1, \cdots$ is bounded above by 1 (and any larger real number), it is bounded below by 0 (and any other smaller real number), and it is bounded by 1 (and any larger real number). The sequence $2, 1, 4, \frac{1}{2}, 8, \frac{1}{4}, 16, \frac{1}{8}, \cdots$ is not bounded above and, therefore, it is not bounded; however, it is bounded below by 0. The sequence $\frac{1}{2}, -1, \frac{2}{3}, -2, \frac{3}{4}, -3, \frac{4}{5}, \cdots$ is bounded above by 1; it is not bounded below; and, hence, it is not bounded. And the sequence $1, -1, 2, -2, 3, -3, \cdots$ is not bounded above or below.

The next theorem states that every convergent sequence is bounded. The number 1 used for ϵ in the proof is arbitrary—we could use any other positive real number.

Theorem 8.4 Every convergent sequence is bounded.

Proof: Let $< a_n >$ be a convergent sequence with limit A. Letting $\epsilon = 1$ in the definition of convergence, there exists a natural number N such that $|a_n - A| < 1$ for all $n > N$. Because $||a_n| - |A|| < |a_n - A|$, for all $n > N$ we have $||a_n| - |A|| < 1$ which implies for all $n > N$ that $|a_n| - |A| < 1$ and, hence, $|a_n| < 1 + |A|$. Let $B = \max\{|a_1|, |a_2|, \ldots, |a_N|, 1 + |A|\}$. Then $|a_n| \leq B$ for all $n \in \mathbf{N}$. That is, the sequence $< a_n >$ is bounded.

Stated in the more usual "if ..., then ..." form, theorem 8.4 says: "If a sequence converges, then it is bounded." The contrapositive of theorem 8.4 is the theorem which says: "If a sequence is not bounded, then the sequence is not convergent." Consequently, we have a valuable method, which does not involve the definition of convergence, for showing that a sequence does not converge. For example, we know from the contrapositive of theorem 8.4

that the sequences $< \log n >$, $< n >$, $< n^2 >$, and $< 2^n >$ do not converge, because they are not bounded.

EXERCISES 8.1

1. Write an expression for the general term of each of the given sequences.

a. $\frac{4}{3}, \frac{6}{5}, \frac{8}{7}, \frac{10}{9}, \frac{12}{11}, \cdots$ b. $\frac{3}{2}, \frac{5}{4}, \frac{7}{6}, \frac{9}{8}, \frac{11}{10}, \cdots$

c. $-1, \frac{1}{2}, -\frac{1}{6}, \frac{1}{24}, -\frac{1}{120}, \cdots$ d. $1, \frac{1}{3}, \frac{3}{5}, \frac{3}{7}, \frac{5}{9}, \frac{5}{11}, \cdots$

e. $\frac{1}{2}, -\frac{2}{5}, \frac{3}{8}, -\frac{4}{11}, \frac{5}{14}, \cdots$ f. $-\frac{\sin 3}{4}, \frac{\sin 5}{9}, -\frac{\sin 7}{16}, \frac{\sin 9}{25}, -\frac{\sin 11}{36}, \cdots$

2. For $n \in \mathbf{N}$, what is $\left\{3 \sin\left(\frac{n\pi}{4}\right)\right\}$? For $n \in \mathbf{N}$, what is $\left\langle 3 \sin\left(\frac{n\pi}{4}\right)\right\rangle$?

3. Estimate the limit of the sequence whose general term is given and then prove the estimate is correct using the definition of convergence.

a. $a_n = -7 + \frac{1}{n}$ b. $a_n = \frac{3 - 3n}{n}$

c. $a_n = \frac{6n - 3}{2n + 4}$ d. $a_n = \frac{5n - 7}{7n + 5}$

e. $a_n = \frac{3n^2 + 1}{6n^2 + 1}$ f. $a_n = 4^{-n}$

g. $a_n = \frac{\sqrt{n} + 2}{4\sqrt{n} + 7}$ h. $a_n = n\left(\sqrt{1 + \frac{1}{n}} - 1\right)$

4. For the following sequences, which are defined recursively, (i) give an explicit formula for the nth term, (ii) estimate the limit of the sequence, and (iii) use the definition of convergence to prove your estimate is correct.

a. $a_1 = 1, \quad a_n = a_{n-1} + \frac{1}{2^{n-1}} \quad$ for $n \geq 2$

b. $a_1 = 1, \quad a_n = \frac{a_{n-1}}{n} \quad$ for $n \geq 2$

c. $a_1 = 2, \quad a_n = \frac{(n-1)a_{n-1}}{n} \quad$ for $n \geq 2$

5. Give an example of a sequence of negative numbers which converges to a nonnegative number.

6. Prove the sequence $< (-1)^n >$ diverges.

7. a. Give an example of two sequences $< a_n >$ and $< b_n >$ which diverge, but the sequence $< a_n + b_n >$ converges.

   b. Give an example of two sequences $< a_n >$ and $< b_n >$ which diverge, but the sequence $< a_n b_n >$ converges.

8. Give an example of two sequences $< a_n >$ and $< b_n >$ such that $a_n \rightarrow A$, $b_n \rightarrow B$, and $a_n < b_n$ for all $n \in N$, but A is not less than B.

9. Prove that if $a_n < b_n$ for all $n \in \mathbf{N}$ and if $a_n \rightarrow A$ and $b_n \rightarrow B$, then $A \leq B$.

10. For $n \in \mathbf{N}$,

$$\sqrt{3 - \frac{2}{n^2}} \leq b_n \leq \sqrt{3 - \frac{1}{n^2}}$$

   a. What is $\lim_{n\rightarrow\infty} b_n$? b. Why?

11. For $n \in \mathbf{N}$,

$$1 - \frac{1}{6n^2} < \frac{\sin\frac{1}{n}}{2n\left(1 - \cos\frac{1}{n}\right)} < 1$$

   a. What is $\lim_{n\rightarrow\infty} \dfrac{\sin\frac{1}{n}}{2n\left(1 - \cos\frac{1}{n}\right)}$? b. Give reasons why?

12. For $n \in \mathbf{N}$,

$$\frac{1}{2} - \frac{1}{24n^2} < n^2\left(1 - \cos\frac{1}{n}\right) < \frac{1}{2}$$

   a. What is $\lim_{n\rightarrow\infty} n^2\left(1 - \cos\frac{1}{n}\right)$? b. Give reasons why?

13. Prove that $a_n \rightarrow 0$ if and only if $|a_n| \rightarrow 0$.

14. a. Prove that if $a_n \rightarrow A$, then $|a_n| \rightarrow |A|$.

   b. Give a counterexample to the following statement: "If $|a_n| \rightarrow |A|$, then $a_n \rightarrow A$."

15. For each given sequence, determine if the sequence is bounded above, bounded below, or bounded.

a. $\left\langle \frac{5}{n} \right\rangle$ b. $\left\langle \frac{3\cos n}{n} \right\rangle$

c. $\left\langle \left(\frac{-3}{4}\right)^n \right\rangle$ d. $\left\langle \left(\frac{-4}{3}\right)^n \right\rangle$

e. $\left\langle \frac{3n+2}{5n} \right\rangle$ f. $\left\langle \frac{3^n+4^n}{7^n} \right\rangle$

g. $< \sqrt{n} >$ h. $< \sqrt{n+1} - \sqrt{n} >$

16. Prove that if the sequence $< a_n >$ is bounded and if the sequence $< b_n >$ converges to zero, then the sequence $< a_n b_n >$ converges to zero.

17. Given that the sequences $< a_n >$ and $< b_n >$ are both bounded. Prove that the sequences $< a_n + b_n >$, $< a_n - b_n >$, and $< a_n b_n >$ are bounded.

8.2 Limit Theorems for Sequences

As the expression for the general term of a sequence becomes more complicated, it becomes more difficult to prove what the limit of the sequence is from the definition. Consequently, it is important for us to prove some theorems about limits which will allow us to calculate limits more easily. The next four theorems contain some basic results regarding the relationship between limits and algebraic combinations of sequences. Each proof of a theorem will require using the triangle inequality, as do many proofs in real analysis. Recall the triangle inequality says: "For real numbers x and y, the inequality $|x+y| \le |x| + |y|$ is valid."

The first theorem we prove states: "The sum of two convergent sequences is a convergent sequence, and the limit of the sum of the sequences is the sum of the limits of the sequences."

Theorem 8.5 If the sequence $< a_n >$ converges to A and the sequence $< b_n >$ converges to B, then the sequence $< a_n + b_n >$ converges to $A + B$.

Scratch Work: To prove $a_n + b_n \to A + B$, we must prove that for every $\epsilon > 0$ there exists a natural number N such that if $n > N$, then $|(a_n + b_n) - (A + B)| < \epsilon$. From the hypotheses of the theorem, we know

(i) for every $\epsilon_1 > 0$ there exists a natural number N_1 such that $n > N_1$ implies $|a_n - A| < \epsilon_1$
and

(ii) for every $\epsilon_2 > 0$ there exists a natural number N_2 such that $n > N_2$ implies $|b_n - B| < \epsilon_2$
It follows from (i) and (ii) that by making N_1 sufficiently large, we can make $|a_n - A|$ as small as we like and by making N_2 sufficiently large, we can make $|b_n - B|$ as small as we like. We must use this information to make the quantity $|(a_n + b_n) - (A + B)| < \epsilon$. Thus, we need to be able to rewrite $|(a_n + b_n) - (A + B)|$ in terms of $|a_n - A|$ and $|b_n - B|$. Rearranging terms and using the triangle inequality, we find

$$|(a_n + b_n) - (A + B)| = |(a_n - A) + (b_n - B)| \leq |(a_n - A)| + |(b_n - B)|.$$

Given $\epsilon > 0$, if we choose $\epsilon_1 = \epsilon/2$ and $\epsilon_2 = \epsilon/2$, then there exists an N_1 such that $n > N_1$ implies $|a_n - A| < \epsilon/2$ and there exists an N_2 such that $n > N_2$ implies $|b_n - B| < \epsilon/2$. Choosing $N = \max\{N_1, N_2\}$, we have for $n > N$ both $|a_n - A| < \epsilon/2$ and $|b_n - B| < \epsilon/2$; consequently, $|(a_n + b_n) - (A + B)| < \epsilon$. << NOTE: We could have selected $\epsilon_1 < \epsilon/4$ and $\epsilon_2 < 3\epsilon/4$ or any other combination for which $\epsilon_1 + \epsilon_2 \leq \epsilon$ and proved the result.>> The following is our proof of theorem 8.5.

Proof: Let $\epsilon > 0$ be given. Since $a_n \to A$ there exists a natural number N_1 such that $n > N_1$ implies $|a_n - A| < \epsilon/2$. Since $b_n \to B$ there exists a natural number N_2 such that $n > N_2$ implies $|b_n - B| < \epsilon/2$. Let $N = \max\{N_1, N_2\}$. For $n > N$, we have

$$|(a_n+b_n)-(A+B)| = |(a_n-A)+(b_n-B)| \leq |(a_n-A)|+|(b_n-B)| < \frac{\epsilon}{2}+\frac{\epsilon}{2} = \epsilon.$$

By mathematical induction, theorem 8.5 can be extended to the sum of any finite number of sequences.

The proof of the next theorem is very similar to the proof of theorem 8.5. You are asked to prove this theorem in exercise 1. The theorem says: "The difference of two convergent sequences is a convergent sequence, and the limit of the difference is the difference of the limits."

Theorem 8.6 If the sequence $< a_n >$ converges to A and the sequence $< b_n >$ converges to B, then the sequence $< a_n - b_n >$ converges to $A - B$.

As we might anticipate, the product of two convergent sequences is a convergent sequence, and the limit of the product is the product of the limits. We now state and prove this theorem.

Theorem 8.7 If the sequence $< a_n >$ converges to A and the sequence $< b_n >$ converges to B, then the sequence $< a_n b_n >$ converges to AB.

Scratch Work: As in the proof of theorem 8.5, by choosing N sufficiently large we can make both $|a_n - A|$ and $|b_n - B|$ as small as we want.

Given any $\epsilon > 0$, we must determine how to choose N large enough so that $|a_n b_n - AB| < \epsilon$. Thus, we need to discover how to rewrite $|a_n b_n - AB|$ in terms of $|a_n - A|$ and $|b_n - B|$. A technique that is sometimes used in analysis is to subtract and then add the same quantity to an expression. In this case, we can subtract and add either Ab_n or $a_n B$ to $a_n b_n - AB$. We subtract and add Ab_n and use the triangle inequality to obtain

$$
\begin{aligned}
(1) \quad |a_n b_n - AB| &= |a_n b_n - Ab_n + Ab_n - AB| = |(a_n - A)b_n + A(b_n - B)| \\
&\leq |a_n - A||b_n| + |A||b_n - B|
\end{aligned}
$$

Since $b_n \to B$, the sequence $< b_n >$ is bounded by theorem 8.4. Thus, there exists a positive real number M such that $|b_n| \leq M$ for all $n \in \mathbf{N}$. Given $\epsilon > 0$, we choose N_1 sufficiently large so that $|a_n - A| < \dfrac{\epsilon}{2M}$ for $n > N_1$. Then the first term on the right hand side of (1) satisfies

$$|a_n - A||b_n| < \frac{\epsilon}{2M} M = \epsilon/2$$

Given $\epsilon > 0$ we would like to choose N_2 large enough so that $|b_n - B| < \dfrac{\epsilon}{2|A|}$; however, A could be zero and we are not allowed to divide by zero. So, instead, we choose N_2 sufficiently large so that $|b_n - B| < \dfrac{\epsilon}{2(|A|+1)}$ for $n > N_2$. This choice will ensure that the denominator is nonzero and that the second term on the right hand side of (1) satisfies

$$|A||b_n - B| < |A| \frac{\epsilon}{2(|A|+1)} < \epsilon/2$$

For N sufficiently large, both terms on the right hand side of (1) are less than $\epsilon/2$ and $|a_n b_n - AB| < \epsilon$. Hence, our proof of theorem 8.7 reads:

Proof: Let $\epsilon > 0$ be given. Since $b_n \to B$, the sequence $< b_n >$ is bounded by theorem 8.4. Thus, there exists a positive integer M such that $|b_n| \leq M$ for all $n \in \mathbf{N}$. Since $a_n \to A$, there exists an $N_1 \in \mathbf{N}$ such that $|a_n - A| < \dfrac{\epsilon}{2M}$ for $n > N_1$. Since $b_n \to B$, there exists an $N_2 \in \mathbf{N}$, such that $|b_n - B| < \dfrac{\epsilon}{2(|A|+1)}$ for $n > N_2$. Choose $N = \max\{N_1, N_2\}$. For $n > N$,

$$
\begin{aligned}
|a_n b_n - AB| &= |a_n b_n - Ab_n + Ab_n - AB| = |(a_n - A)b_n + A(b_n - B)| \\
&\leq |a_n - A||b_n| + |A||b_n - B| < \frac{\epsilon}{2M} M + |A| \frac{\epsilon}{2(|A|+1)} \\
&< \frac{\epsilon}{2} + \frac{\epsilon}{2} = \epsilon
\end{aligned}
$$

By mathematical induction, theorem 8.7 can be extended to the product of any finite number of sequences.

We would like to prove a theorem about the quotient of two convergent sequences which essentially says: "If the sequence $< a_n >$ converges to A and the sequence $< b_n >$ converges to B, then the sequence $\left\langle \frac{a_n}{b_n} \right\rangle$ converges to $\frac{A}{B}$." However, this statement is false, because the sequence $\left\langle \frac{a_n}{b_n} \right\rangle$ is not defined unless $b_n \neq 0$ for all natural numbers and because $\frac{A}{B}$ is not defined unless $B \neq 0$. So we must include the conditions $b_n \neq 0$ for all $n \in \mathbf{N}$ and $B \neq 0$ in the hypotheses of our theorem. Because $\frac{a_n}{b_n} = a_n \left(\frac{1}{b_n} \right)$, we can use theorem 8.7 about the product of two sequences to prove our quotient theorem, provided we can prove the following lemma.

Lemma 8.1 If $b_n \neq 0$ for all $n \in \mathbf{N}$ and if the sequence $< b_n >$ converges to $B \neq 0$, then the sequence $\left\langle \frac{1}{b_n} \right\rangle$ converges to $\frac{1}{B}$.

Scratch Work: For $\epsilon > 0$ we must determine how to choose a natural number N such that for $n > N$,

$$\left| \frac{1}{b_n} - \frac{1}{B} \right| < \epsilon \qquad \text{or} \qquad \left| \frac{B - b_n}{b_n B} \right| = \frac{|b_n - B|}{|b_n||B|} < \epsilon.$$

Since $b_n \to B$, by choosing N sufficiently large, we can make the quantity $|b_n - B|$ as small as we like. The quantity $|B|$ which appears in the denominator is not zero, but we must be certain the quantity $|b_n|$ which also appears in the denominator is bounded away from zero. That is, we must show there is some $M > 0$ and some $N_1 \in \mathbf{N}$ such that $|b_n| \geq M > 0$ for all $n > N_1$. Since $B \neq 0$ and $b_n \to B$, there exists a natural number N_1 such that $|b_n - B| < |B|/2$ for $n > N_1$. Since by the triangle inequality

$$|B| = |B - b_n + b_n| = |(B - b_n) + b_n| \leq |(B - b_n)| + |b_n| < \frac{|B|}{2} + |b_n|,$$

we have for $n > N_1$ that $0 < \frac{|B|}{2} < |b_n|$. Given any $\epsilon > 0$, since $b_n \to B$, we can also choose a natural number N_2 such that for $n > N_2$,

$$|b_n - B| < \frac{\epsilon |B|^2}{2}.$$

The proof of lemma 8.1 reads as follows.

Proof: Let $\epsilon > 0$ be given. Since $B \neq 0$ and since the sequence $< b_n >$ converges to B, there exists an $N_1 \in \mathbf{N}$ such that $|b_n - B| < |B|/2$ for $n > N_1$. Because

$$|B| = |B - b_n + b_n| \leq |(B - b_n)| + |b_n| < \frac{|B|}{2} + |b_n|,$$

we have for all $n > N_1$ that $0 < \frac{|B|}{2} < |b_n|$. Also since $< b_n >$ converges to B, there exists an $N_2 \in \mathbf{N}$ such that for $n > N_2$

$$|b_n - B| < \frac{\epsilon |B|^2}{2}.$$

Choose $N = \max\{N_1, N_2\}$. Then for $n > N$,

$$\left| \frac{1}{b_n} - \frac{1}{B} \right| = \left| \frac{B - b_n}{b_n B} \right| = \frac{|b_n - B|}{|b_n||B|} < \frac{|b_n - B|}{\frac{|B|}{2}|B|} < \frac{\frac{\epsilon |B|^2}{2}}{\frac{|B|^2}{2}} = \epsilon.$$

Hence, $\frac{1}{b_n} \to \frac{1}{B}$.

The next theorem follows easily from lemma 8.1 and theorem 8.7.

Theorem 8.8 If $b_n \neq 0$ for all $n \in \mathbf{N}$, if the sequence $< b_n >$ converges to $B \neq 0$, and if the sequence $< a_n >$ converges to A, then the sequence $\left\langle \frac{a_n}{b_n} \right\rangle$ converges to $\frac{A}{B}$.

Proof: By lemma 8.1, the sequence $\left\langle \frac{1}{b_n} \right\rangle$ converges to $\frac{1}{B}$. Since $< a_n >$ converges to A and since $\frac{a_n}{b_n} = a_n \left(\frac{1}{b_n} \right)$, the sequence $\left\langle \frac{a_n}{b_n} \right\rangle = \left\langle a_n \left(\frac{1}{b_n} \right) \right\rangle$ converges to $A \left(\frac{1}{B} \right) = \frac{A}{B}$ by theorem 8.7.

The next example illustrates how to use theorems 8.5 through 8.8 to calculate a limit.

EXAMPLE 1 Calculating the Limit of a Sequence

Calculate

$$\lim_{n \to \infty} \frac{2 + 3n - 4n^3}{5n^3 + 6n^2 - 7}$$

SOLUTION

The general term of this sequence is a rational function of n. The highest power to which n appears in the numerator and denominator is $p = 3$. So, we multiply both numerator and denominator by $\frac{1}{n^p} = \frac{1}{n^3}$ and obtain

$$\frac{(2+3n-4n^3)\left(\frac{1}{n^3}\right)}{(5n^3+6n^2-7)\left(\frac{1}{n^3}\right)} = \frac{\frac{2}{n^3}+\frac{3}{n^2}-4}{5+\frac{6}{n}-\frac{7}{n^3}}$$

Applying theorems 8.5 through 8.8 and taking limits, we find

$$\begin{aligned}
\lim_{n\to\infty} \frac{2+3n-4n^3}{5n^3+6n^2-7} &= \frac{\lim_{n\to\infty}\left[(2+3n-4n^3)\left(\frac{1}{n^3}\right)\right]}{\lim_{n\to\infty}\left[(5n^3+6n^2-7)\left(\frac{1}{n^3}\right)\right]} \\
&= \frac{\lim_{n\to\infty}(\frac{2}{n^3}+\frac{3}{n^2}-4)}{\lim_{n\to\infty}(5+\frac{6}{n}-\frac{7}{n^3})} \\
&= \frac{\lim_{n\to\infty}\frac{2}{n^3}+\lim_{n\to\infty}\frac{3}{n^2}-\lim_{n\to\infty}4}{\lim_{n\to\infty}5+\lim_{n\to\infty}\frac{6}{n}-\lim_{n\to\infty}\frac{7}{n^3}} \\
&= \frac{0+0-4}{5+0+0} = \frac{-4}{5}
\end{aligned}$$

EXERCISES 8.2

1. Prove theorem 8.6.

In exercises 2-7, use theorems 8.5 through 8.8 to prove the given statement.

2. If the sequence $< a_n >$ converges to A and k is any real constant, then the sequence $< a_n + k >$ converges to $A + k$.

3. If the sequence $< a_n >$ converges to A and k is any real constant, then the sequence $< ka_n >$ converges to kA.

4. Let $< a_n >$ and $< b_n >$ be sequences. If $< a_n >$ converges and $< a_n + b_n >$ converges, then $< b_n >$ converges.

5. The sum of a convergent sequence and a divergent sequence is a divergent sequence.

6. If the sum and the difference of two sequences converges, then both sequences converge.

7. Let $< a_n >$ and $< b_n >$ be sequences. If $< a_n >$ converges to $A \neq 0$ and $< a_n b_n >$ converges, then $< b_n >$ converges.

8. If the sequence $< a_n >$ converges to A, then the sequence $< b_n >$ where $b_n = (a_{n+1} + a_{n+2} + \cdots + a_{2n})/n$ converges to A.

9. If $a_n \geq 0$ for all $n \in \mathbf{N}$ and if the sequence $< a_n >$ converges to $A > 0$, then $< \sqrt{a_n} >$ converges to $\sqrt{A}$.
(HINT: If $A > 0$, then $\sqrt{a_n} - \sqrt{A} = \dfrac{a_n - A}{\sqrt{a_n} + \sqrt{A}}$.)

10. Calculate the limit of each of the following sequences.

a. $\left\langle \dfrac{4n^2 + 2n + 1}{3n^2 - 1} \right\rangle$ b. $\left\langle \dfrac{4n^2 + 2n + 1}{3n^3 - 1} \right\rangle$

c. $\left\langle 1 - \left| \dfrac{\cos n}{n} \right| \right\rangle$ d. $\left\langle \dfrac{(-1)^n}{3n + 1} \right\rangle$

e. $< \sqrt{n + 5} - \sqrt{n} >$ f. $< \sqrt{n^2 + n} - \sqrt{n} >$

g. $\left\langle \dfrac{n}{1 + \dfrac{1}{n}} - n \right\rangle$ h. $\left\langle \dfrac{1}{\sqrt{1 + \dfrac{1}{n}}} - n \right\rangle$

11. Given that the sequence $< a_n >$ converges to A and the sequence $< b_n >$ converges to B, find the limit of the sequence

$$\left\langle \frac{3a_n + 2nb_n}{5n + 1} \right\rangle$$

8.3 Monotone Sequences and Subsequences

In this section, we define monotone sequences and subsequences. We prove that every bounded monotone sequence converges—our first theorem which allows us to prove a sequence converges without knowing the limit in advance. We prove that if a sequence converges, then every subsequence converges to the limit of the sequence. And we prove that if two subsequences converge to different limits, then the sequence diverges. Next, we show that every sequence of real numbers has a monotone subsequence. Finally, we state and prove a classical and fundamental theorem of real analysis—the Bolzano-Weierstrass Theorem.

DEFINITIONS Monotone Increasing and Monotone Decreasing Sequences

A sequence $< a_n >_{n=1}^{\infty}$ of real numbers is **monotone increasing** if and only if $a_n \leq a_{n+1}$ for all $n \in \mathbf{N}$.

A sequence $< a_n >_{n=1}^{\infty}$ of real numbers is **monotone decreasing** if and only if $a_n \geq a_{n+1}$ for all $n \in \mathbf{N}$.

A sequence $< a_n >$ is **monotone** if it is either monotone increasing or monotone decreasing.

The sequences $< n >$, $< 1, 2, 2, 3, 3, 3, 4, 4, 4, 4, \cdots >$, $< 4, 4, 4, \cdots >$, and $< 10 - \frac{1}{n} >$ are all monotone increasing sequences, and the sequences $< -n >$, $< -1, -1, -2, -2, -3, -3, \cdots >$, $< 4, 4, 4, \cdots >$, and $< 10 + \frac{1}{n} >$ are all monotone decreasing sequences. The sequences $< n >$, $< 1, 2, 2, 3, 3, 3, \cdots >$, $< -n >$, and $< -1, -1, -2, -2, -3, -3, \cdots >$ are unbounded and diverge; while the sequences $< 4, 4, 4, \cdots >$, $< 10 - \frac{1}{n} >$, and $< 10 + \frac{1}{n} >$ are all bounded and converge. These results are not coincidental as we shall see when we prove the Monotone Convergence Theorem.

Earlier, we mentioned that one property which distinguishes the ordered field of rational numbers from the ordered field of real numbers is the **property of completeness**. This property is used to prove many fundamental theorems of real analysis. Sometimes the property of completeness is called the Axiom of Completeness or the Completeness Axiom, although it is actually a theorem which was proved independently in 1872 by both Dedekind and Cantor. Dedekind used cuts to prove the theorem; while Cantor used Cauchy sequences.

Completeness Property of the Real Numbers Every nonempty subset of real numbers which is bounded above has a least upper bound.

The completeness property is an existence theorem which says that given a nonempty set S of real numbers which is bounded above there exists a real number which is the least upper bound for S. Geometrically, the completeness property says there are no "holes" in the real number line. Furthermore, the completeness property could be stated in terms of sets which are bounded below as follows: "Every nonempty subset of real numbers which is bounded below has a greatest lower bound." These two statements are equivalent because if S is the nonempty set which is bounded below, then $-S$ is a nonempty set which is bounded above and by the completeness property has least upper bound m. Hence, $-m$ is the greatest lower bound for S. So either statement may serve as the completeness property for real numbers. We will need the completeness property in order to prove the monotone convergence theorem.

Theorem 8.9 The Monotone Convergence Theorem A monotone sequence is convergent if and only if it is bounded.

Proof: Suppose that $< a_n >$ is a convergent monotone sequence. Since the sequence is convergent, it is bounded by theorem 8.4.

Now suppose that $< a_n >$ is a monotone increasing sequence which is bounded. Let $S = \{a_n \mid n \in \mathbf{N}\}$. Since S is a nonempty, by the completeness property there exists a greatest lower bound, call it A. We will prove the sequence $< a_n >$ converges to A. Let $\epsilon > 0$ be given. Since $A - \epsilon$ is not an upper bound for S, there exists an $N \in \mathbf{N}$ such that $A - \epsilon < a_N$. Because the sequence $< a_n >$ is a monotone increasing and A is an upper bound for S, we have for all $n > N$

$$A - \epsilon < a_N \leq a_n \leq A < A + \epsilon.$$

Hence, $|a_n - A| < \epsilon$ for all $n > N$ and, therefore, $\lim_{n\to\infty} a_n = A$.

The proof for the case in which the sequence $< a_n >$ is a monotone decreasing and bounded is similar.

In order to use the definition of convergence to prove that a sequence converges, it is necessary to know the value of the limit in advance. However, it is often difficult or impossible to determine the value of the limit of a sequence. The monotone convergence theorem allows us to prove that a sequence converges without finding its limit, but we must be able to prove that the sequence is monotone and bounded.

EXAMPLE 1 Proving a Sequence Converges Using the Monotone Convergence Theorem

Define the sequence $< a_n >$ recursively by $a_1 = 1$ and $a_{n+1} = \sqrt{2 + a_n}$ for $n \geq 1$.

a. Show that the sequence $< a_n >$ is bounded.

b. Show that the sequence $< a_n >$ is monotone.

c. Find $\lim_{n\to\infty} a_n$.

SOLUTION

a. Calculating the first few terms of the sequence, we find

$$a_2 = \sqrt{2+1} = \sqrt{3} \approx 1.7320508$$

$$a_3 = \sqrt{2+\sqrt{3}} \approx 1.93185165$$

$$a_4 = \sqrt{2+\sqrt{2+\sqrt{3}}} \approx 1.98288972$$

$$a_5 = \sqrt{2+\sqrt{2+\sqrt{2+\sqrt{3}}}} \approx 1.99571785$$

Based on our calculations, it appears the sequence is monotone increasing and is bounded above by 2.

We prove by induction that $a_n < 2$ for all $n \in \mathbf{N}$. Let $P(n)$ be the statement "$a_n < 2$." The statement $P(1)$ is $1 < 2$ which is true. Now, we assume that $P(n)$ is true for some natural number n. That is, we assume that $a_n < 2$. Then from the recursive definition $a_{n+1} = \sqrt{2+a_n} < \sqrt{2+2} = \sqrt{4} = 2$. Hence, $P(n+1)$ is true and by induction $a_n < 2$ for all $n \in \mathbf{N}$.

b. Next, we prove by induction that $a_n \leq a_{n+1}$ for all $n \in \mathbf{N}$. Let $Q(n)$ be the statement "$a_n \leq a_{n+1}$." The statement $Q(1)$ is $a_1 \leq a_2$ or $1 \leq \sqrt{3}$ which is true. Now, we assume that $Q(n)$ is true for some natural number n. Thus, we assume that $a_n \leq a_{n+1}$. Adding 2 to this inequality, we get $0 < 2 + a_n \leq 2 + a_{n+1}$, because $a_n \geq 0$. Taking square roots and using the recursive definition, we find

$$a_{n+1} = \sqrt{2+a_n} \leq \sqrt{2+a_{n+1}} = a_{n+2}.$$

Thus, the statement $Q(n+1)$ is true and by induction $a_n \leq a_{n+1}$ for all $n \in \mathbf{N}$. That is, the sequence $< a_n >$ is monotone increasing.

c. By the monotone convergence theorem, $\lim_{n\to\infty} a_n = A$. Squaring the recursive formula, yields $a_{n+1}^2 = 2 + a_n$. Taking the limit as n approaches infinity, we find that A must satisfy the equation $A^2 = 2 + A$. Solving the quadratic equation $A^2 - A - 2 = 0$, yields $A = -1$ or $A = 2$. Since $< a_n >$ is an increasing sequence and $a_1 = 1$, the limit of the given sequence is $A = 2$.

Let $< a_n >$ be a sequence of real numbers. By deleting a finite number of terms from $< a_n >$ and leaving the remaining terms in the same relative order as the original sequence, we obtain a new sequence which is called a subsequence of $< a_n >$. Notice we could delete no terms, in which case, we would have the original sequence. That is, the sequence $< a_n >$ is a subsequence of itself. By deleting an infinite number of terms from the sequence $< a_n >$ so long as an infinite number of terms still remain, we also obtain a subsequence of $< a_n >$. The following is a formal definition of a subsequence.

DEFINITION Subsequence

Let $< a_n >_{n=1}^{\infty}$ be a sequence of real numbers and let $< n_k >_{k=1}^{\infty}$ be any sequence of natural numbers such that $n_1 < n_2 < n_3 < \cdots$. The sequence $< a_{n_k} >_{k=1}^{\infty}$ is a **subsequence** of the sequence $< a_n >_{n=1}^{\infty}$.

By induction on k, it follows from the definition of a subsequence, that $n_k \geq k$ for all $k \in \mathbf{N}$. When $n_k = k$ for all $k \in \mathbf{N}$, we obtain the original sequence.

Example 2. The sequences

$$\left\langle \frac{1}{2k} \right\rangle_{k=1}^{\infty} = \frac{1}{2}, \frac{1}{4}, \frac{1}{6}, \frac{1}{8}, \cdots, \qquad \left\langle \frac{1}{k^2} \right\rangle_{k=1}^{\infty} = 1, \frac{1}{4}, \frac{1}{9}, \frac{1}{16}, \cdots, \quad \text{and}$$

$$\left\langle \frac{1}{2^k} \right\rangle_{k=1}^{\infty} = \frac{1}{2}, \frac{1}{4}, \frac{1}{8}, \frac{1}{16}, \cdots$$

are all subsequences of the sequence

$$\left\langle \frac{1}{n} \right\rangle_{n=1}^{\infty} = 1, \frac{1}{2}, \frac{1}{3}, \frac{1}{4}, \cdots$$

Observe that the sequence $\left\langle \frac{1}{n} \right\rangle$ converges to 0 as do all of the subsequences $\left\langle \frac{1}{2k} \right\rangle$, $\left\langle \frac{1}{k^2} \right\rangle$, and $\left\langle \frac{1}{2^k} \right\rangle$.

Example 3. The sequence

$$\left\langle \frac{1 + (-1)^n}{2} \right\rangle_{n=1}^{\infty} = 0, 1, 0, 1, \cdots$$

diverges, but the subsequence

$$\left\langle \frac{1 + (-1)^{2k-1}}{2} \right\rangle_{k=1}^{\infty} = 0, 0, 0, 0, \cdots$$

converges to 0 and the subsequence

$$\left\langle \frac{1 + (-1)^{2k}}{2} \right\rangle_{k=1}^{\infty} = 1, 1, 1, 1, \cdots$$

converges to 1.

In example 2, the original sequence converged to 0 as did all three subsequences we examined. In example 3, the sequence diverged, but we found two convergent subsequences—one converged to the limit 0 while the other converged to the value 1. The next theorem pertains to these results.

Theorem 8.10 Let $< a_n >$ be a sequence of real numbers.

(i) If the sequence $< a_n >$ converges to A, then every subsequence converges to A.

(ii) If the sequence $< a_n >$ has two subsequences which converge to different limits, then the sequence $< a_n >$ diverges.

Proof: (i) Suppose $< a_n >$ converges to A and let $< a_{n_k} >$ be any subsequence of $< a_n >$. Since $a_n \to A$, given any $\epsilon > 0$ there exists an $N \in \mathbf{N}$ such that $n > N$ implies $|a_n - A| < \epsilon$. Since for all $k \in \mathbf{N}$, we have $n_k \geq k$. Therefore, $|a_{n_k} - A| < \epsilon$ for $k > N$. Hence, by definition of convergence, the subsequence $< a_{n_k} >$ converges to A.

(ii) Our proof is by contradiction. Let $< a_{n_k} >$ and $< a_{n_\ell} >$ be subsequences of the sequence $< a_n >$ and suppose that $a_{n_k} \to L$ and $a_{n_\ell} \to M$ where $L \neq M$. Assume the sequence $< a_n >$ converges to A. Either L or M is not A because $L \neq M$. Therefore, $< a_n >$ has a subsequence that does not converge to A which contradicts (i).

The sequence $< n >$ is monotone increasing and so is every subsequence. The sequence $\left\langle \frac{1}{n} \right\rangle$ is monotone decreasing and so is every subsequence. The sequence

$$< a_n > = \frac{2}{3}, \frac{1}{3}, \frac{3}{4}, \frac{1}{4}, \frac{4}{5}, \frac{1}{5}, \frac{5}{6}, \frac{1}{6}, \cdots$$

is divergent, but it has a monotone increasing subsequence

$$< a_{2k-1} > = \frac{2}{3}, \frac{3}{4}, \frac{4}{5}, \frac{5}{6}, \cdots$$

which converges to 1 and it has a monotone decreasing subsequence

$$< a_{2k} > = \frac{1}{3}, \frac{1}{4}, \frac{1}{5}, \frac{1}{6}, \cdots$$

which converges to 0.

The sequence

$$< b_n > = 1, 9, 2, 9\frac{1}{2}, 3, 9\frac{2}{3}, 4, 9\frac{3}{4}, \cdots$$

diverges and has a monotone increasing subsequence

$$< b_{2k-1} > = 1, 2, 3, 4, \cdots$$

which diverges and a monotone increasing subsequence

$$< b_{2k} > = 9, 9\frac{1}{2}, 9\frac{2}{3}, 9\frac{3}{4}, \cdots$$

which converges to 10.

The following theorem is a constructive theorem. The proof of this theorem contains an algorithm for constructing a monotone subsequence for any given sequence of real numbers.

Theorem 8.11 Every sequence of real numbers has a monotone subsequence.

Proof: Let $< a_n >$ be an arbitrary sequence of real numbers. Define S to be the set of natural numbers

$$S = \{n \in \mathbf{N} \mid a_n \text{ is a lower bound for the set } \{a_{n+1}, a_{n+2}, a_{n+3}, \cdots\}\}$$

Either S is an infinite set or S is a finite set.

Case 1 Suppose S is an infinite set. Testing each natural number n in order, we generate the sequence of indices contained in S and the associated terms of the sequence such that $n_1 < n_2 < n_3 < \cdots$ and $a_{n_1} < a_{n_2} < a_{n_3} < \cdots$. The subsequence $< a_{n_k} >_{k=1}^{\infty}$ constructed in this manner is a monotone increasing subsequence of $< a_n >$.

Case 2 Suppose S is a finite set. Let N be any natural number that is an upper bound for the set S. For any $n_1 \in \mathbf{N}$ such that $n_1 > N$, we have $n_1 \notin S$, so a_{n_1} is not a lower bound for the set $\{a_{n_1+1}, a_{n_1+2}, a_{n_1+3}, \cdots\}$. Hence, there is a natural number $n_2 > n_1$ such that $a_{n_2} < a_{n_1}$. Continuing in this manner, we generate a sequence of indices and associated terms of the sequence such that $n_1 < n_2 < n_3 < \cdots$ and $a_{n_1} > a_{n_2} > a_{n_3} > \cdots$. Therefore, the subsequence $< a_{n_k} >_{k=1}^{\infty}$ is a monotone decreasing subsequence of $< a_n >$.

We now state and prove the Bolzano-Weierstrass theorem. This theorem was first stated and proved by Bernard Bolzano (1781-1848) in 1817. However, his proof of the theorem was lost. This theorem also seems to have been known by Cauchy. About fifty years after Bolzano proved the theorem, it was reproved by Karl Weierstrass (1815-1897), who is known as the father of modern analysis. The Bolzano-Weierstrass Theorem is a fundamental result in real analysis.

Theorem 8.12 The Bolzano-Weierstrass Theorem Every bounded sequence of real numbers has a convergent subsequence.

Proof: Let $< a_n >$ be a bounded sequence of real numbers. By theorem 8.11 there exists a monotone subsequence $< a_{n_k} >$. The sequence $< a_{n_k} >$ is bounded, because it is a subsequence of the bounded sequence $< a_n >$. By the Monotone Convergence Theorem (theorem 8.9) the subsequence $< a_{n_k} >$ converges.

EXERCISES 8.3

In exercises 1-6, give an example of a sequence with the specified properties.

1. The sequence is bounded, but it is not monotone.

2. The sequence is monotone, but it is not bounded.

3. The sequence is neither bounded nor monotone.

4. The sequence is convergent, but it is not monotone.

5. The sequence is monotone, but it is not convergent.

6. The sequence is neither convergent nor monotone.

In exercises 7-12, prove that each sequence is monotone and bounded. Then find the limit of the sequence.

7. $a_1 = 1$ and $a_{n+1} = (a_n + 5)/4$ for $n \geq 1$.

8. $a_1 = 2$ and $a_{n+1} = 2 - \dfrac{1}{a_n}$ for $n \geq 1$.

9. $a_1 = 2$ and $a_{n+1} = 2 - \dfrac{1}{2a_n}$ for $n \geq 1$.

10. $a_1 = 2$ and $a_{n+1} = \dfrac{a_n}{2} + \dfrac{1}{a_n}$ for $n \geq 1$.

    (**HINT**: From $0 \leq (x - y)^2 = x^2 - 2xy + y^2$, it follows that $xy \leq (x^2 + y^2)/2$.)

11. $a_1 = 4$ and $a_{n+1} = \sqrt{6 + a_n}$ for $n \geq 1$.

12. $a_1 = 1$ and $a_{n+1} = \sqrt{3 + 2a_n}$ for $n \geq 1$.

13. Let $0 < a_1 < b_1$ and define

$$a_{n+1} = \sqrt{a_n b_n} \quad \text{and} \quad b_{n+1} = \frac{a_n + b_n}{2} \quad \text{for} \quad n \geq 1.$$

    Prove that the sequences $< a_n >$ and $< b_n >$ converge and their limits are the same.

14. For $n \in \mathbf{N}$, let $a_n = (-1)^n(3 - \dfrac{5}{n})$.

    a. Write a formula for the subsequence $< a_{2n} >$.

    b. What is $\lim_{n\to\infty} a_{2n}$?

c. Write a formula for the subsequence $< a_{2n-1} >$.

d. What is $\lim_{n\to\infty} a_{2n-1}$?

e. What can you say about the sequence $< a_n >$?

15. Let $< a_n >$ be the sequence whose general term is $a_n = 2\cos\dfrac{n\pi}{2}$. Find three subsequences of the sequence $< a_n >$ that have different limits. What are the limits? Does the sequence $< a_n >$ converge or diverge?

16. Give an example of a sequence with the specified properties.

a. The sequence has subsequences with limits $\sqrt{2}$, e, π, and $\sqrt{11}$.

b. The sequence is unbounded and it has a convergent subsequence.

c. The sequence has no convergent subsequence.

17. For each of the following sequences, find the monotone subsequence generated by the algorithm given in the proof of theorem 8.11.

a. $2, 1, 0, 1, 2, 1, 2, 3, 2, 3, 4, 3, 4, 5, \cdots$

b. $-1, 1, 0, \frac{1}{2}, \frac{1}{2}, \frac{1}{3}, \frac{2}{3}, \frac{1}{4}, \frac{3}{4}, \frac{1}{5}, \frac{4}{5}, \cdots$

8.4 Cauchy Sequences

In order to use the definition of convergence of a sequence to prove that a particular sequence converges, we must know the limit of the sequence. However, we can prove that a monotone sequence converges without knowing its limit by simply proving the sequence is bounded. A large number of sequences converge which are not monotone. Consequently, we would like a condition for convergence which does not require the sequence to be monotone and which does not require knowing the limit of the sequence in advance. Such a condition does exist. It is called the **Cauchy criterion** in honor of the French mathematician Augustin-Louis Cauchy (1789-1857). Cauchy is the second most prolific mathematician in history—second only to Euler.

By definition of convergence, a sequence $< a_n >$ converges to the limit A, provided for all n sufficiently large ($n > N$) all of the terms a_n are close to A (that is, $|a_n - A|$ is small). If for n sufficiently large all of the terms a_n are near A, then for $n, m > N$ all of the terms a_n and a_m are near one another (that is, $|a_n - a_m|$ is small). Hence, we have the following definition of a Cauchy sequence.

DEFINITION Cauchy Sequence

A sequence $< a_n >_{n=1}^{\infty}$ of real numbers is a **Cauchy sequence** if and only if for every $\epsilon > 0$ there exists a natural number N such that for all $n, m > N$, if $n, m > N$, then $|a_n - a_m| < \epsilon$.

Our first theorem states that all convergent sequences are Cauchy sequences.

Theorem 8.13 Every convergent sequence is a Cauchy sequence.

Proof: Let $< a_n >$ be a sequence of real numbers that converges to the limit A. Let $\epsilon > 0$ be given. Since $a_n \to A$, there exists an $N \in \mathbf{N}$ such that if $n > N$, then $|a_n - A| < \epsilon/2$. For $n, m > N$, we have by subtracting and adding A and then using the triangle inequality

$$\begin{aligned}|a_n - a_m| &= |a_n - A + A - a_m| = |(a_n - A) + (A - a_m)| \\ &\leq |a_n - A| + |a_m - A| < \frac{\epsilon}{2} + \frac{\epsilon}{2} = \epsilon.\end{aligned}$$

Hence, the convergent sequence $< a_n >$ is a Cauchy sequence.

The next theorem states that Cauchy sequences are bounded—as are convergent sequences.

Theorem 8.14 Every Cauchy sequence is bounded.

Proof: Let $< a_n >$ be a Cauchy sequence. Choose $\epsilon = 1$. Since the sequence $< a_n >$ is Cauchy, there exists an $N \in \mathbf{N}$ such that for $n, m > N$ we have $|a_n - a_m| < 1$. Since $N+1 > N$, it follows that $|a_n - a_{N+1}| < 1$ for all $n > N$. Because $||a_n| - |a_{N+1}|| \leq |a_n - a_{N+1}|$, for all $n > N$ we have $||a_n| - |a_{N+1}|| < 1$ which implies that $|a_n| - |a_{N+1}| < 1$ and, hence, $|a_n| < 1 + |a_{N+1}|$. Let $B = \max\{|a_1|, |a_2|, \ldots, |a_N|, 1 + |a_{N+1}|\}$. Then $|a_n| \leq B$ for all $n \in \mathbf{N}$ and the Cauchy sequence $< a_n >$ is bounded.

The following theorem says if a Cauchy sequence has a convergent subsequence, then the Cauchy sequence converges to the limit of the subsequence.

Theorem 8.15 If $< a_n >$ is a Cauchy sequence of real numbers and if the subsequence $< a_{n_k} >$ converges to A, then the sequence $< a_n >$ converges to A.

Proof: Suppose that $< a_{n_k} >$ is a convergent subsequence of the Cauchy sequence $< a_n >$ and suppose $a_{n_k} \to A$. Since $< a_n >$ is a Cauchy sequence, there exists a natural number N such that $|a_n - a_m| < \epsilon/2$ for all $n, m > N$. Since $a_{n_k} \to A$, there exists a natural number K such that $|a_{n_k} - A| < \epsilon/2$ for

$k > K$. Let $M = \max\{N, K\}$. Choose n_k such that $k > M$. Then $n_k > N$. Hence, if $n > M$, then by the triangle inequality

$$|a_n - A| = |a_n - a_{n_k} + a_{n_k} - A| \leq |a_n - a_{n_k}| + |a_{n_k} - A| < \frac{\epsilon}{2} + \frac{\epsilon}{2} = \epsilon.$$

Consequently, the Cauchy sequence $< a_n >$ converges to A.

In the next theorem, we use the Bolzano-Weierstrass theorem and theorems 8.14 and 8.15 to prove that every Cauchy sequence converges.

Theorem 8.16 Every Cauchy sequence of real numbers is a convergent sequence.

Proof: Let be a Cauchy sequence of real numbers. By theorem 8.14, the sequence is bounded. By the Bolzano-Weierstrass theorem (theorem 8.12), the Cauchy sequence has a convergent subsequence. By theorem 8.15, the Cauchy sequence is a convergent sequence.

Theorems 8.13 and 8.16 can be combined into the following single theorem.

Theorem 8.17 A sequence of real numbers converges if and only if it is a Cauchy sequence.

Provided we can prove a sequence is a Cauchy, we can prove the sequence converges without knowing the limit of the sequence. In example 1, we prove a sequence converges by proving the sequence is Cauchy.

EXAMPLE 1 Proving a Sequence Is a Cauchy Sequence

Prove the sequence $< a_n >= \left\langle \frac{3n+4}{2n} \right\rangle$ is a Cauchy sequence and, therefore, it is convergent.

SOLUTION

Scratch Work Let $\epsilon > 0$ be given. We must show how to choose a natural number N such that $|a_n - a_m| < \epsilon$ for all $n, m > N$. Assume for definiteness that $n \leq m$. We want to select N so that for $m \geq n > N$,

$$\left| \frac{3n+4}{2n} - \frac{3m+4}{2m} \right| = \left| \frac{6mn + 8m - 6mn - 8n}{4nm} \right|$$
$$= \frac{8|m-n|}{4nm} \leq \frac{2m}{nm} = \frac{2}{n} < \frac{2}{N} < \epsilon.$$

Hence, we choose N to be any natural number greater than $2/\epsilon$.

So our proof that the sequence $\left\langle \frac{3n+4}{2n} \right\rangle$ is a Cauchy sequence is as follows.

Proof: Let $\epsilon > 0$ be given. Choose N to be any natural number such that $N > 2/\epsilon$. Then for all $m \geq n > N$,

$$|a_n - a_m| = \left|\frac{3n+4}{2n} - \frac{3m+4}{2m}\right| = \frac{2|m-n|}{nm} \leq \frac{2m}{nm} = \frac{2}{n} < \frac{2}{N} < \epsilon.$$

Thus, the sequence $\left\langle \frac{3n+4}{2n} \right\rangle$ is a Cauchy sequence and it is convergent.

EXERCISES 8.4

1. Give an example of a sequence that is monotone but not Cauchy.
2. Give an example of a sequence that is Cauchy but not monotone.
3. Give an example of a sequence that is bounded but not Cauchy.
4. Use the definition of Cauchy to prove that the sequence

$$< a_n >= \left\langle \frac{n}{n+4} \right\rangle$$

is a Cauchy sequence.

Chapter 9

Proofs from Group Theory

The branch of mathematics which examines, compares, and contrasts algebraic structures is called abstract algebra. Topics from abstract algebra appear in many other branches of mathematics, because much of contemporary mathematics is algebraic in nature. In this chapter, we will introduce and briefly study one of the more fundamental algebraic structures—the group.

9.1 Binary Operations and Algebraic Structures

When we add two real numbers a and b, we perform the operation of addition and denote the resulting real number by $a + b$. Likewise, when we multiply two real numbers, we perform the operation of multiplication and denote the resulting real number by $a \cdot b$, or simply, ab. The operations of addition and multiplication both assign to an ordered pair (a, b) of real numbers a unique real number—namely, $a + b$ and $a \cdot b$, respectively. These operations are examples of **binary operations**, which we now define formally.

DEFINITION Binary Operation

Let S be a nonempty set. A **binary operation on** S is a function from $S \times S$ to S.

We usually denote a binary operation on the set S by $\circ$, $+$, $\cdot$, or $*$ and call the operation multiplication or addition even when the operation has nothing whatsoever to do with multiplication or addition. If the image of the ordered pair $(a, b) \in S \times S$ under the operation $\circ$ is c, instead of writing $\circ((a, b)) = c$ we write $a \circ b = c$. Also, we often omit the operation symbol $\circ$ and just write $ab = c$ when it is clearly understood what operation we are using.

An **algebraic structure** is a nonempty set S, a nonempty collection of operations on S, and a collection (which may be empty) of relations on S. For example, the set of integers $\mathbf{Z}$ under the usual operation of addition is an algebraic structure which we denote by $(\mathbf{Z}, +)$. The set of real numbers, $\mathbf{R}$, under the operations of addition, $+$, and multiplication, $\cdot$, and the relation less than, $<$, is an algebraic structure called an ordered field and is denoted by $(\mathbf{R}, +, \cdot, <)$. The set of complex numbers, $\mathbf{C}$, under $+$ and $\cdot$ is an unordered field and denoted by $(\mathbf{C}, +, \cdot)$. Let A be a nonempty set and let $\mathcal{P}(A)$ be the power set of A. The power set $\mathcal{P}(A)$ under the operation of union, $\cup$, under the operation of intersection, $\cap$, and under the operation of difference, $-$, are all algebraic structures. That is, $(\mathcal{P}(A), \cup)$, $(\mathcal{P}(A), \cap)$, and $(\mathcal{P}(A), -)$ are all algebraic structures. However, $(\mathbf{N}, -)$ is not an algebraic structure since subtraction, $-$, is not a binary operation on the set of natural numbers (because, for example, $4, 7 \in \mathbf{N}$ but $4 - 7 = -3 \notin \mathbf{N}$). Let $\div$ represent the operation of division. The structure $(\mathbf{R}, \div)$ is not an algebraic structure, but $(\mathbf{R} - \{0\}, \div)$ is. Why?

An algebraic structure $(S, *)$ may satisfy all, some, or none of the following special properties.

DEFINITIONS Associativity, Existence of an Identity, Existence of Inverses, and Commutativity

Let $(S, *)$ be an algebraic system. Then

G1. **Associativity:** $(S, *)$ is **associative** if and only if $(x*y)*z = x*(y*z)$ for all $x, y, z \in S$.

G2. **Existence of an Identity:** $(S, *)$ has an **identity element** (or simply an **identity**) if and only if there exists an $e \in S$ such that $x * e = e * x = x$ for all $x \in S$.

G3. **Existence of Inverses:** $(S, *)$ has inverses if and only if S has an identity element e and for each element $x \in S$ there is an element $y \in S$, called the **inverse** of x, such that $x * y = y * x = e$.

G4. **Commutativity:** $(S, *)$ is **commutative** if and only if $x * y = y * x$ for all $x, y \in S$.

Several comments are in order. When $(S, *)$ satisfies the associativity property, G1, we may write $x * y * z$ without any parentheses because $(x * y) * z = x * (y * z) = x * y * z$. That is, as long as the factors x, y, z appear in the same order and $(S, *)$ is associative, we may write $(x * y) * z$, $x * (y * z)$, or $x * y * z$ interchangeably. When $(S, *)$ is not associative, there exist some elements $x, y, z \in S$ for which $(x * y) * z \neq x * (y * z)$. Two elements $x, y \in S$ are said to **commute** if $x*y = y*x$. Property G2 says if $(S, *)$ has an identity element e, then e commutes with every element $x \in S$ and, furthermore, $x*e = e*x = x$. A structure $(S, *)$ cannot satisfy property G3 unless it satisfies property G2.

When $(S, *)$ satisfies properties G2 and G3, then every element $x \in S$ has an inverse y with which it commutes and $x * y = y * x = e$. Observe if y is an inverse of x, then x is an inverse of y—that is, inverses occur in pairs. When $(S, *)$ satisfies property G4, every element in S commutes with all elements in S.

Example 1. The set of integers with the usual operation of addition, $(\mathbf{Z}, +)$, satisfies properties G1 through G4. Property G1 is the associative law of addition for the integers—namely, $(a + b) + c = a + (b + c)$. The identity element is 0, since $a + 0 = 0 + a = a$ for all $a \in \mathbf{Z}$. For $a \in \mathbf{Z}$, the additive inverse is $-a \in \mathbf{Z}$ and $a + (-a) = (-a) + a = 0$. Property G4 is the commutative law of addition for the integers—that is, $a + b = b + a$ for all $a, b \in \mathbf{Z}$.

Example 2. The structure $(\mathbf{N}, +)$ where $+$ is addition for the natural numbers satisfies properties G1 and G4, but it does not satisfy properties G2 and G3. Property G2 is not satisfied because $0 \notin \mathbf{N}$. And property G3 is not satisfied, because $(\mathbf{N}, +)$ has no identity element—that is, property G2 is not satisfied.

Example 3. The set of integers under the usual operation of subtraction, $(\mathbf{Z}, -)$, does not satisfy any of the properties G1 through G4.

Example 4. The set of rational numbers with the usual operation of multiplication, $(\mathbf{Q}, \cdot)$, satisfies properties G1, G2, and G4. However, $(\mathbf{Q}, \cdot)$ does not satisfy property G3, because $0 \in \mathbf{Q}$ has no inverse. That is, there is no element $b \in \mathbf{Q}$ such that $0 \cdot b = b \cdot 0 = 1$.

Example 5. The structure $(\mathbf{Q} - \{0\}, \cdot)$ satisfies properties G1 through G4. The identity is the element 1, and for every $a \in \mathbf{Q} - \{0\}$ the inverse is the rational number $1/a$.

Example 6. Let $S = \{p, q, r\}$ and let $*$ be the binary operation defined on S by $x * y = y$ for all $x, y \in S$. For arbitrary $x, y, z \in S$, $(x * y) * z = z$ and $x * (y * z) = y * z = z$. Hence, $(S, *)$ is associative. In order for an element $e \in S$ to be the identity of $(S, *)$, we must have $x * e = e * x = x$ for all $x \in S$. However, by definition of $*$, for every element $e \in S$ and arbitrary $x \in S$ such that $x \neq e$, we have $x * e = e \neq x$. Hence, $(S, *)$ has no identity; and, consequently, no element of S has an inverse. Also $(S, *)$ is not commutative, since for $x \neq y$, we have $x * y = y$ and $y * x = x$—that is, $x * y \neq y * x$ for $x \neq y$.

Example 7. Let $S = \mathbf{N} \cup \{0\}$ and let $\circ$ be the binary operation defined on S by $a \circ b = |a - b|$ for all $a, b \in S$. The structure $(S, \circ)$ is not associative, because

$$(2 \circ 3) \circ 4 = |2 - 3| \circ 4 = |-1| \circ 4 = 1 \circ 4 = |1 - 4| = |-3| = 3$$

but

$$2 \circ (3 \circ 4) = 2 \circ |3 - 4| = 2 \circ |-1| = 2 \circ 1 = |2 - 1| = |1| = 1.$$

The element $0 \in S$ is the identity, because $a \circ 0 = |a - 0| = |a| = a$ and $0 \circ a = |0 - a| = |-a| = a$ for all $a \in S$. Each element in S is its own inverse since $a \circ a = |a - a| = |0| = 0$ for all $a \in S$. Let $a, b \in S$, since $|a - b| = |b - a|$, we have $a \circ b = |a - b| = |b - a| = b \circ a$. Hence, $(S, \circ)$ is commutative.

If S is a finite set, then the **order** of the algebraic structure $(S, \circ)$ is the number of elements in S, which is denoted by $|S|$. When S is infinite, we say $(S, \circ)$ has infinite order. For structures with small finite order, it is often convenient to represent the binary operation with a table called a **Cayley table** or **operation table**. The table is named in honor of the English mathematician Arthur Cayley (1821-1895). In 1849, Cayley published two important articles on abstract groups in which he defined an abstract group and used a table to display group multiplication. In 1854, Cayley proved every group can be represented as a group of permutations. A Cayley table for the structure $(S, *)$ of order n is an $n \times n$ array of products such that the product $x * y$ appears in row x and column y. For example, let $S = \{a, b, c\}$ and let $*$ be the operation defined by table 9.1. We see from this table that $a * a = c$, $a * b = a$, $a * c = b$, and so forth. Notice that the row labels in the left-hand column and the column labels across the top of the table are arranged in the same order—namely, a, b, c. By constructing Cayley tables in this manner, we can check easily for commutativity. When the Cayley table of a structure is symmetric with respect to the **main diagonal**, which extends from the upper left corner to the lower right corner of the body of the table, the structure is commutative. That is, when the entry in the (x, y) position is identical to the entry in the (y, x) position for all $x, y \in S$, then the structure is commutative. Due to the symmetry of table 9.1, $(S, *)$ is commutative.

$*$	**a**	**b**	**c**
a	c	a	b
b	a	b	c
c	b	c	a

Table 9.1

To find the identity element of a Cayley table, if there is one, we look for a row label in which the entries of the row match the column labels. Then, we check the column with the same label as the row label to see if the column entries match the row labels, or not. From table 9.1, we see that the element b is the identity, because the entries of row b are a, b, c and the entries of column b are a, b, c also. The **left-inverse of an element** y, if there is one, is the element x which satisfies $x * y = e$ where e is the identity. To find the left-inverse of the element a in table 9.1, we search through the column labeled a for the entry b—the identity. From table 9.1, we see $c * a = b$, so c is the left-inverse of a. The **right-inverse of an element** y, if there is one, is the element z which satisfies $y * z = e$. To find the right-inverse of the element a in table 9.1, we search through the row labeled a for the entry b. Since b appears in the column labeled c, we have $a * c = b$; and, therefore, c is the right-inverse of a. Because the left and right-inverse of a are both c, the element c is the inverse of a. We see from table 9.1 that the element b has inverse b and the element c has inverse a. Thus, $(S, *)$ satisfies property G3. To prove $(S, *)$ defined by table 9.1 is associative, we must verify for all possible 27 triples (x, y, z) that $(x * y) * z = x * (y * z)$. This is tedious to do by hand, but easy to do by computer. In order to show some structure is not associative, it is sufficient to find a single triple (x, y, z) for which $(x * y) * z \neq x * (y * z)$.

The operation $\circ$ of table 9.2 is not associative, because $(a \circ a) \circ b = b \circ b = c$ but $a \circ (a \circ b) = a \circ b = b$. The identity is the element c. The element b is the left-inverse of a and b, the element c is the left-inverse of c, the element a is the right-inverse of b, the element b is the right-inverse of b, and the element c is the right-inverse of c. Consequently, a has no inverse, b is the inverse of b, and c is the inverse of c. Thus, $(S, \circ)$ does not satisfy property G3. The operation $\circ$ is not commutative, because $a \circ b = b$ but $b \circ a = c$.

$\circ$	**a**	**b**	**c**
a	b	b	a
b	c	c	b
c	a	b	c

Table 9.2

$+$	**a**	**b**	**c**
a	a	c	b
b	c	b	a
c	b	a	c

Table 9.3

The operation $+$ of table 9.3 is not associative since $(a + b) + c = c + c = c$ but $a + (b + c) = a + a = a$. There is no identity element; and, consequently, there are no inverses. However, the operation $+$ is commutative, because the Cayley table 9.3 is symmetric with respect to the main diagonal.

The following theorem states that an algebraic structure can have at most one identity.

Theorem 9.1 Let $(S, *)$ be an algebraic structure. If $(S, *)$ has an identity, it is unique.

Proof: Assume to the contrary that $(S, *)$ has two distinct identity elements e and e'. By the definition of an identity, for every $a \in S$, (1) $a * e = e * a = a$ and (2) $a * e' = e' * a = a$. In equation (1), we let $a = e'$. The right-hand side of the resulting equation is (3) $e * e' = e'$. Letting $a = e$ in equation (2), we obtain (4) $e * e' = e$. It follows from (3) and(4) that $e' = e$, which is a contradiction. Thus, if an identity of an algebraic structure exists, it is unique.

An algebraic structure which is associative and has an identity is called a **monoid**. The following theorem states that if an element of a monoid has an inverse, then the inverse is unique.

Theorem 9.2 Let $(S, *)$ be an associative structure with identity e. If $a \in S$ has an inverse, then the inverse is unique.

Proof: Suppose that $(S, *)$ is associative with identity e and suppose that there exists an $a \in S$ and elements $x, y \in S$ such that x and y are inverses of a and $x \neq y$. Since x and y are inverses of a, (5) $a * x = e$ and (6) $y * a = e$. Multiplying equation (5) on the left by y, yields (7) $y * (a * x) = y * e = y$. By associativity of $*$ and equation (6), we have (8) $y*(a*x) = (y*a)*x = e*x = x$. It follows from (7) and (8) that $y = x$, which is a contradiction.

EXERCISES 9.1

1. In each part of this question, a set is specified and an $*$ is defined on that set. Determine whether $*$ is a binary operation on the specified set or not. For those $*$ which are not binary operations, explain why they are not.

   a. Set $\mathbf{N}$ where $*$ is defined by $x * y = x/y$.

   b. Set $\mathbf{N}$ where $*$ is defined by $x * y = x^y$.

   c. Set $\mathbf{N}$ where $*$ is defined by $x * y = xy$.

   d. Set $\mathbf{N}$ where $*$ is defined by $x * y = x + y$.

   e. Set $\mathbf{N}$ where $*$ is defined by $x * y = |x - y|$.

   f. Set $\mathbf{Z}$ where $*$ is defined by $x * y = 2xy - x + 1$.

   g. Set $\mathbf{Q}$ where $*$ is defined by $x * y = x/y$.

   h. Set $\mathbf{Q} - \{0\}$ where $*$ is defined by $x * y = x/y$.

   i. Set $\mathbf{Q}^+$, the set of positive rational numbers, where $*$ is defined by $x * y = \sqrt{xy}$.

   j. Set $\mathbf{R}^+$, the set of positive real numbers, where $*$ is defined by $x * y = \sqrt{xy}$.

2. Which binary operations found in exercise 1 are associative? Which are commutative? Which have an identity? What is the identity?

3. Answer the following questions for the Cayley tables 9.4 through 9.7.

   a. Is the operation associative?

   b. Is the operation commutative?

   c. Is there an identity element? If so, what is it?

   d. List elements which have inverses and their inverses.

∘	**0**	**1**
0	0	0
1	0	0

Table 9.4

+	**0**	**1**
0	0	1
1	1	0

Table 9.5

*	**0**	**1**
0	0	0
1	0	1

Table 9.6

#	**0**	**1**
0	0	0
1	1	1

Table 9.7

4. Construct the Cayley table for the set $S = \{-1, 0, 1\}$ under the usual operation of multiplication $\cdot$.

   a. Is the operation $\cdot$ associative?

   b. Is the operation $\cdot$ commutative?

   c. What is the identity element?

   d. Which elements have inverses and what are they?

   e. Does this structure satisfy property G3?

9.2 Groups

Recall that a polynomial of degree n in one variable x is a function of the form $p(x) = a_n x^n + a_{n-1} x^{n-1} + \cdots + a_1 x + a_0$ where n is a positive integer, $a_n, a_{n-1}, \ldots, a_1, a_0$ are real numbers, and $a_n \neq 0$. The linear equation $ax = b$ where a and b are both positive could be solved algebraically and geometrically by civilizations which existed prior to recorded history. Approximately 2000 B. C. the ancient Babylonians knew how to solve the quadratic equation $ax^2 + bx + c = 0$ algebraically by both the method of completing the square and by substitution into the quadratic formula. They could solve certain special cubic equations as well. About 1515, the Italian Scipio del Ferro (ca. 1465-1526) discovered the algebraic solution of the cubic equation of the form $x^3 + bx = c$. He did not publish his result, but revealed its solution to his pupil Antonio Maria Fior. In 1535, Niccolo Fontana (1499-1557) claimed to have solved algebraically the cubic equation of the form $x^3 + ax^2 = c$. (Fontana was also known as Tartaglia (the stammerer), because of a saber wound he received when he was only thirteen years old at the hands of the French in the 1512 massacre at Brescia.) Fior believed Fontana was simply bragging and challenged Fontana to a public contest. Fontana accepted the challenge, and a few days before the contest he discovered how to solve cubic equations of the type Fior could solve. Knowing how to solve both kinds of cubic equations, Fontana triumphed completely. However, Fontana did not publish his methods of solution. In 1539, Girolamo Cardano (Cardan) (1501-1576), an unscrupulous man who practiced medicine in Milan, met with Fontana who revealed his methods of solution to Cardan under a pledge of secrecy. In 1540, Zuanne de Tonini da Coi proposed a problem to Cardan which required the solution of a quartic equation. Cardan was unable to solve the equation, but gave it to his student Ludovico Ferrari (1522-1565). Ferrari solved the problem and in the process showed how to reduce the solution of all quartic equations of the form $x^4 + px^2 + qx + r = 0$ to the solution of a cubic equation. In effect, Ferrari solved the general quartic equation $x^4 + ax^3 + bx^2 + cx + d = 0$, since it reduces to $x^4 + px^2 + qx + r = 0$ by means of a simple linear transformation.

Prior to 1545, mathematicians accepted only positive real numbers as roots of polynomial equations. In his *Ars Magna*, which was published in 1545, Cardan accepted negative real numbers as roots of polynomials and used the square root of negative numbers in computations. He demonstrated that both $x = 5 - \sqrt{-15}$ and $x = 5 + \sqrt{-15}$ were solutions of the equation $x^2 + 40x = 10x$. Cardan included Fontana's solution of the general cubic equation and Ferrari's solution of the general quartic equation in *Ars Magna*. Thus, *Ars Magna* contains the first published solution of both the general cubic equation and the general quartic equation, although neither was the work of the author.

In his *Arithmetica philosophica* of 1608, Peter Roth (1580-1617) appears to be the first writer to explicitly state the Fundamental Theorem of Algebra which says that an nth degree polynomial has n roots. The first rigorous proof was given by Gauss in 1799. Later, in 1849, a simpler proof was given.

The Italian physician and mathematician Paola Ruffini (1765-1822) was the first person to claim that a polynomial of fifth degree could not be solved algebraically. In 1799, he published unaccepted proofs that the roots of fifth and higher degree polynomials cannot be written in terms of its coefficients using radicals and the standard operations of arithmetic. Furthermore, in this article Ruffini introduces groups of permutations. Disappointed by the reaction to his proofs, Ruffini published additional proofs in 1803, 1805, and again in 1813. In 1824, the Norwegian mathematician Niels Henrik Abel successfully proved the result. In 1832, the French mathematician Évariste Galois (1811-1832) gave the first definition of a group; however, it did not appear in print until Joseph Liouville published Galois' papers fourteen years later in 1846. The first abstract definition of a group was given by Cayley in 1854. The term "group" became the accepted standard terminology in 1863 when Camille Jordan wrote a commentary on Galois' work.

DEFINITION Group

A **group** is a nonempty set G with a binary operation $\circ$ that satisfies the following three properties:

G1. **Associativity:** $(G, \circ)$ is associative. That is, $(x \circ y) \circ z = x \circ (y \circ z)$ for all $x, y, z \in G$.

G2. **Existence of an Identity:** There exists an identity element $e \in G$. That is, there exists an $e \in G$ such that $x \circ e = e \circ x = x$ for all $x \in G$.

> G3. **Existence of Inverses:** Every element $x \in G$ has an inverse $x^{-1} \in G$. That is, for all $x \in G$ there exists an $x^{-1} \in G$ such that $x \circ x^{-1} = x^{-1} \circ x = e$.

To prove that a structure $(G, \circ)$ is a group, we must verify (i) the set G is nonempty, (ii) the operation $\circ$ is a binary operation from $G \times G$ to G, and (iii) properties G1, G2, and G3 are all satisfied. When it is clearly understood what operation is associated with the group, we usually denote the group $(G, \circ)$ by just G.

If a group G has property G4—that is, if $ab = ba$ for all $a, b \in G$, then we say the group is an **abelian group** in honor of Niels Abel. A group is **nonabelian**, if there is some pair $a, b \in G$ such that $ab \neq ba$. The structures $(\mathbf{Z}, +)$, $(\mathbf{Q}, +)$, $(\mathbf{R}, +)$, and $(\mathbf{C}, +)$ are all abelian groups. In each group, 0 is the identity and the inverse of a is $-a$. The structure $(\mathbf{N}, +)$ is not a group, because there is no identity element. The structure $(\mathbf{Z}, -)$ is not a group, because the operation $-$ is not associative. The structures $(\mathbf{Q}, \cdot)$ and $(\mathbf{R}, \cdot)$ are not groups, because the element 0 has no inverse. However, $(\mathbf{Q} - \{0\}, \cdot)$, $(\mathbf{Q}^+, \cdot)$, $(\mathbf{R} - \{0\}, \cdot)$, and $(\mathbf{R}^+, \cdot)$ are abelian groups.

Let $Z_n = \{0, 1, \ldots, n-1\}$. For $n \geq 1$, $(Z_n, +)$ where $+$ denotes addition modulo n is an abelian group. The identity element is 0, and for $m \in Z_n$ and $m \neq 0$ the inverse of m is $n - m$. The group Z_n is called the **group of integers modulo** n. The structure $(Z_n, \cdot)$ where $\cdot$ is multiplication modulo n is not a group for any n, because the element 0 has no inverse.

It follows from theorems 9.1 and 9.2 that a group has a unique identity and each element of the group has a unique inverse. The following theorem says the left and right cancellation laws hold in any group.

Theorem 9.3 Let G be a group. Then the left and right cancellation laws hold in G. That is,
(1) **Left Cancellation Law:** For every $a, b \in G$, if $ab = ac$, then $b = c$.
and
(2) **Right Cancellation Law:** For every $a, b \in G$, if $ba = ca$, then $b = c$.

Proof: << We will prove (1) and leave (2) for you to prove as an exercise.>> Suppose that G is a group; that $a, b, c \in G$; and that (3) $ab = ac$. Since $a^{-1} \in G$, we multiply both the left side and right side of equation (3) on the left by a^{-1} and use the properties of associativity, inverse, and the identity e, to obtain

$$(4)\quad a^{-1}(ab) = (a^{-1}a)b = eb = b \quad \text{and} \quad (5)\quad a^{-1}(ac) = (a^{-1}a)c = ec = c.$$

It follows from equations (3), (4), and (5) that $b = a^{-1}(ab) = a^{-1}(ac) = c$.

For a finite group G, it follows from the left cancellation law that no element can occur more than once in any row a of the Cayley table for the group. Also, since $a(a^{-1}b) = b$ for every $a, b \in G$, every element b occurs exactly once in every row a. In a like manner, it follows from the right cancellation law that every element occurs exactly once in every column a of the Cayley table. Hence, we have the theorem: If G is a finite group, then every row and every column of the Cayley table for the group is a permutation of the elements of the group. The converse of this theorem is false. In the exercises, we will ask you to provide an example to prove that the converse is false.

The next theorem says the inverse of the inverse of an element in a group is the element itself and the inverse of the product of two elements of a group is the product of the inverses of the elements written in reverse order.

Theorem 9.4 If G is a group and $a, b \in G$, then (6) $(a^{-1})^{-1} = a$ and (7) $(ab)^{-1} = b^{-1}a^{-1}$.

Proof: Let G be a group and let $a \in G$. Since $a^{-1} \in G$ is the inverse of a, we have (8) $aa^{-1} = a^{-1}a = e$ where e is the identity of G. Because a satisfies equation (8), the property of the unique inverse of a^{-1}, it follows that $(a^{-1})^{-1} = a$.

Let G be a group and let $a, b \in G$. Since $a, b \in G$, we have $ab \in G$ and $(ab)^{-1} \in G$. The element $(ab)^{-1}$ is the unique element $x \in G$ such that (9) $(ab)x = x(ab) = e$. We claim $x = b^{-1}a^{-1}$. Substituting $x = b^{-1}a^{-1}$ in the left hand side of equation (9), we find by associativity, definition of the inverse, and definition of the identity that

$$(ab)x = (ab)(b^{-1}a^{-1}) = ((ab)b^{-1})a^{-1} = (a(bb^{-1}))a^{-1} = (ae)a^{-1} = aa^{-1} = e$$

Likewise, substituting for $x = b^{-1}a^{-1}$ in the center expression of equation (9), we see

$$x(ab) = (b^{-1}a^{-1})(ab) = ((b^{-1}a^{-1})a)b = (b^{-1}(a^{-1}a))b = (b^{-1}e)b = b^{-1}b = e$$

Hence, $x = b^{-1}a^{-1}$ satisfies equation (9) and, by definition, $b^{-1}a^{-1}$ is the inverse of ab.

The following theorem states that linear equations in one variable with coefficients from a group G have a unique solution which is an element of the group.

Theorem 9.5 Let G be a group. If $a, b \in G$, then there is a unique $x \in G$ such that (10) $ax = b$ and there exists a unique $y \in G$ such that (11) $ya = b$.

Proof: << We will prove equation (10) has a unique solution and let you prove equation (11) has a unique solution in the exercises.>> Let e be the identity of the group and let a^{-1} be the inverse of a. Consider $x = a^{-1}b \in G$. We have (12) $ax = a(a^{-1}b) = (aa^{-1})b = eb = b$. From equation (12), we see that $x = a^{-1}b$ is one solution of the equation $ax = b$.

Suppose x_1 and $x_2 \neq x_1$ are both solutions of (10) $ax = b$. Thus, $ax_1 = b$ and $ax_2 = b$. Consequently, $ax_1 = ax_2$ and by the left cancellation law $x_1 = x_2$, which contradicts $x_2 \neq x_1$. So, $x = a^{-1}b$ is the unique solution of (10) $ax = b$.

EXERCISES 9.2

1. Which of the following algebraic structures is a group?

   a. $(S, \cdot)$ where S is the set of rational numbers in the interval $(0, 1]$ and $\cdot$ is the usual operation of multiplication.

   b. $(T, \cdot)$ where T is the set of irrational numbers in the interval $(0, 1]$ and $\cdot$ is the usual operation of multiplication.

   c. $(U, \circ)$ where $U = \{1, 2, 4\}$ and $\circ$ is multiplication modulo 5.

   d. $(U, *)$ where $U = \{1, 2, 4\}$ and $*$ is multiplication modulo 7.

   e. $(V, +)$ where $V = \{0, 2, 4\}$ and $+$ is addition modulo 6.

   f. $(W, \circ)$ where $W = \{1, 3, 9\}$ and $\circ$ is the operation of multiplication modulo 10.

   g. $(E, +)$ where E is the set of even integers under the operation of addition.

   h. $(O, \cdot)$ where O is the set of odd integers under the operation of multiplication.

   i. $(X, *)$ where $X = \mathbf{Q} - \{1\}$ and $*$ is the operation defined by $a * b = a + b - ab$ for all $a, b \in X$.

   j. $(\mathbf{Z}, \circ)$ where $\circ$ is the operation defined by $a \circ b = a + b + 1$ for all $a, b \in \mathbf{Z}$.

2. For those structures of exercises 1 which are not groups, explain why they are not groups.

3. For those structures of exercise 1 which are groups, answer the following questions:

   a. What is the identity of the group?

   b. What is the inverse of each element of the group?

   c. Is the group abelian?

4. Prove the right cancellation law. That is, prove that if G is a group, if $a, b, c \in G$, and if $ba = ca$, then $b = c$.

5. Give an example of an algebraic structure $(G, *)$ in which G has three elements, each row and each column of the Cayley table for G under the operation $*$ contains each element of G exactly once, but G is not a group.

6. Complete the partial Cayley tables given in tables 9.8 and 9.9.

$\circ$	**a**	**b**	**c**	**d**
a				b
b		b	c	
c	d		b	
d			a	

Table 9.8

$*$	**1**	**2**	**3**	**4**	**5**	**6**
1				4		6
2	2	3			4	
3		4	2			
4		5		1		3
5	5		4	3	1	
6			5			

Table 9.9

a. What is the identity element of each group?

b. Is the group abelian?

7. Let $G = \{e, a, b, c\}$ be a group such that $a^2 = b^2 = c^2 = e$ where e is the identity. Construct a Cayley table for G. Is G abelian?

8. Let G be a group with identity e. Prove that if $a^2 = e$ for all $a \in G$, then G is abelian. (**HINT:** See exercise 7.)

9. Let G be a group and let $a_1, a_2, \ldots, a_n \in G$. Prove by mathematical induction that $(a_1 a_2 \cdots a_n)^{-1} = a_n^{-1} \cdots a_2^{-1} a_1^{-1}$ for all $n \in \mathbf{N}$.

10. Let G be an abelian group and let $a, b \in G$. Prove by mathematical induction that $(ab)^n = a^n b^n$ for all $n \in \mathbf{N}$.

11. Prove that if G is a group and if $a, b \in G$, then there exists a unique $y \in G$ such that $ya = b$.

12. Let G be a group and let $a, b, c \in G$. Prove that each of the following equations has a unique solution x:

    (i) $abx = c$ (ii) $axb = c$ (iii) $a^{-1}xa = c$

13. Prove that G is an abelian group if and only if $(ab)^{-1} = a^{-1}b^{-1}$ for all $a, b \in G$.

14. Prove that G is an abelian group if and only if $(ab)^2 = a^2b^2$ for all $a, b \in G$.

15. A square array of real numbers $\begin{pmatrix} a & b \\ c & d \end{pmatrix}$ is called a 2×2 **matrix**. Addition of 2×2 matrices is defined by

$$\begin{pmatrix} a_1 & b_1 \\ c_1 & d_1 \end{pmatrix} + \begin{pmatrix} a_2 & b_2 \\ c_2 & d_2 \end{pmatrix} = \begin{pmatrix} a_1 + a_2 & b_1 + b_2 \\ c_1 + c_2 & d_1 + d_2 \end{pmatrix}$$

and multiplication is defined by

$$\begin{pmatrix} a_1 & b_1 \\ c_1 & d_1 \end{pmatrix} \cdot \begin{pmatrix} a_2 & b_2 \\ c_2 & d_2 \end{pmatrix} = \begin{pmatrix} a_1a_2 + b_1c_2 & a_1b_2 + b_1d_2 \\ c_1a_2 + d_1c_2 & c_1b_2 + d_1d_2 \end{pmatrix}$$

Let $M_{2,2}$ denote the set of all 2×2 matrices with real entries. The structure $(M_{2,2}, +)$ is an abelian group.

a. What is the identity of this group?

b. What is the inverse of $\begin{pmatrix} a & b \\ c & d \end{pmatrix}$?

c. Prove that $(M_{2,2}, +)$ is abelian.

The structure $(M_{2,2}, \cdot)$ where $\cdot$ denotes matrix multiplication is not a group, because the matrix $\begin{pmatrix} 0 & 0 \\ 0 & 0 \end{pmatrix}$ has no inverse. The **determinant** of the matrix $A = \begin{pmatrix} a & b \\ c & d \end{pmatrix}$ is the real number $\det A = ad - bc$. The set of all 2×2 matrices with real entries and nonzero determinant

$$GL(2, \mathbf{R}) = \left\{ \begin{pmatrix} a & b \\ c & d \end{pmatrix} \;\middle|\; a, b, c, d \in \mathbf{R} \text{ and } ad - bc \neq 0 \right\}$$

is a nonabelian group under the operation of matrix multiplication.

d. Verify that the product of two matrices in the set $GL(2, \mathbf{R})$ is a 2×2 matrix and is in $GL(2, \mathbf{R})$.

e. What is the identity element of $GL(2, \mathbf{R})$?

f. Prove that the inverse of the matrix $A = \begin{pmatrix} a & b \\ c & d \end{pmatrix}$ in $GL(2, \mathbf{R})$ is the matrix

$$A^{-1} = \begin{pmatrix} \dfrac{d}{\det A} & \dfrac{-b}{\det A} \\ \dfrac{-c}{\det A} & \dfrac{a}{\det A} \end{pmatrix}$$

g. Prove that $GL(2, \mathbf{R})$ is nonabelian by displaying two matrices A and B in $GL(2, \mathbf{R})$ such that $AB \neq BA$.

16. The set $G = \left\{ \begin{pmatrix} a & a \\ a & a \end{pmatrix} \;\middle|\; a \in \mathbf{R} \text{ and } a \neq 0 \right\}$ is a group under matrix multiplication.

a. What is the identity of this group?

b. What is the inverse of each element in the group?

c. Is the group abelian?

9.3 Subgroups and Cyclic Groups

An algebraic structure $(S, *)$ often has embedded within it one or more algebraic substructures $(T, *)$ with the same operations and relations as the original structure. These substructures can be useful in describing the original structure. For $(T, *)$ to be a substructure, it is necessary that the operation $*$ be a function from $T * T$ to T. That is, the restriction of the binary operation $*$ to the set T must also be a binary operation. This property is called **closure** and is defined as follows.

DEFINITION Closed Under the Operation $*$
Let $(S, *)$ be an algebraic structure with binary operation $*$ and let T be a subset of S. The set T is **closed under the operation** $*$ if and only if $x * y \in T$ for all $x, y \in T$.

The two statements "T is closed under $*$." and "$*$ is an operation on T." will be used interchangeably. For example, the algebraic structure $(\mathbf{R}, +)$, the set of real numbers under the operation of addition, is closed under the operation $+$ as are all of the substructures $(\mathbf{Q}, +)$, $(\mathbf{Z}, +)$, $(\mathbf{N}, +)$, and $(\mathbf{A}, +)$ where $A = \{-1, 0, 1\}$.

Many times a group has embedded within it one or more groups. These groups are called subgroups.

DEFINITION Subgroup
Let $(G, \circ)$ be a group and let H be a nonempty subset of G. The structure $(H, \circ)$ is a **subgroup of** G if and only if $(H, \circ)$ is a group.

To verify a nonempty subset H of a group $(G, \circ)$ is a subgroup of G, we must prove $(H, \circ)$ is closed under the operation $\circ$ and $(H, \circ)$ satisfies properties G1 through G3.

Let $(G, \circ)$ be a group with identity e. By definition, the group $(G, \circ)$ is a subgroup of itself and it is easy to verify that $(\{e\}, \circ)$ is a subgroup of G. The subgroup $(\{e\}, \circ)$ is called the **identity subgroup** or **trivial subgroup** of G.

We can prove a subset H of the group $(G, \circ)$ is not a subgroup of G, by showing that one of the following three things is true.

1. Show that H is not closed under the operation $\circ$. That is, we find two elements $x, y \in H$ such that $x \circ y \notin H$.

2. Show that H has no identity element.

3. Show that some element in H does not have an inverse which is in H. That is, we find an element $x \in H$ such that $x^{-1} \notin H$.

Consider the group $(\mathbf{Z}, +)$. The set of odd integers, $\mathbf{O}$, is not a subgroup of $(\mathbf{Z}, +)$, because (i) the sum of any two odd integers is an even integer, because (ii) $(\mathbf{O}, +)$ has no identity element, and because (iii) no element of $\mathbf{O}$ has an inverse since $\mathbf{O}$ has no identity. That is, $(\mathbf{O}, +)$ fails to be a subgroup of $(\mathbf{Z}, +)$ not for just one of the reasons listed above, but for all three reasons.

The following theorem states two basic results common to all subgroups—namely, the identity of a subgroup is the identity of the group and the inverse of an element in a subgroup is the inverse of the element in the group.

Theorem 9.6 Let $(G, \circ)$ be a group with identity e and let $(H, \circ)$ be a subgroup of G. Then (i) $e \in H$ and (ii) if $x \in H$, then the inverse of x in H is the inverse of x in G.

Proof: Suppose that the identity of the subgroup H is the element i. Then (1) $i \circ i = i$. Since e is the identity of G, (2) $e \circ i = i$. From (1) and (2), it follows that $i \circ i = i = e \circ i$ and by the right cancellation law $i = e$.

Let $x \in H$ and suppose that the inverse of x in H is z. That is, suppose that (3) $x \circ z = z \circ x = e$. Since $x \in H$, the element $x \in G$ and its inverse in G is x^{-1} which has the property (4) $x \circ x^{-1} = x^{-1} \circ x = e$. It follows from (3) and (4) that $x \circ z = x \circ x^{-1}$ and by the left cancellation law that $z = x^{-1}$.

The next theorem is an excellent example of the purpose of some theorems—to reduce the effort necessary to verify a fact. Theorem 9.7 greatly reduces the amount of work required to prove that a particular set is a subgroup of a given group. Using the definition of a subgroup to prove that a set is a subgroup is often very tedious and laborious. Theorem 9.7 states two necessary and sufficient conditions for proving that a subset H of the group $(G, \circ)$ with identity e is a subgroup of G. The first condition is that H is nonempty. This condition is usually easy to prove and is often verified by proving $e \in H$. The second condition is that if $a, b \in H$, then $a \circ b^{-1} \in H$. It is normally easier to prove that $a \circ b^{-1} \in H$ than it is to prove both that H is closed under the operation $\circ$ and that if $x \in H$, then $x^{-1} \in H$.

Theorem 9.7 Let $(G, \circ)$ be a group with identity e and let H be a subset of G. The algebraic structure $(H, \circ)$ is a group if and only if (i) H is nonempty and (ii) if $a, b \in H$, then $a \circ b^{-1} \in H$.

Proof: First, suppose that H is a subgroup of G. Because H is a group, $e \in H$ and $H \neq \emptyset$. Let $a, b \in H$. By the existence of inverses property $b^{-1} \in H$ and by the closure property for H, we have $a \circ b^{-1} \in H$.

Now suppose that $H \neq \emptyset$ and that (5) $a, b \in H \Rightarrow a \circ b^{-1} \in H$. We must prove that the properties G1 through G3 are true for H and that H is closed under the operation $\circ$.

G1: Let $x, y, z \in H$. Since $H \subseteq G$, $x, y, z \in G$ and since G is associative, $(x \circ y) \circ z = x \circ (y \circ z)$. Thus, $(H, \circ)$ is associative.

G2: Since $H \neq \emptyset$, there exists some $x \in H$. Letting $a = x$ and $b = x$ in equation (5), we see that $x \circ x^{-1} = e \in H$.

G3: Suppose that $y \in H$. Choosing $a = e$ and $b = y$ in equation (5), we find that $e \circ y^{-1} = y^{-1} \in H$.

Closure: Let $x, y \in H$. We just showed that $y^{-1} \in H$. Letting $a = x$ and $b = y^{-1}$ in equation (5), we see that $x \circ (y^{-1})^{-1} = x \circ y \in H$. Therefore, H is closed under the operation $\circ$.

Since H is a subset of G; since H is closed under $\circ$; and since $(H, \circ)$ satisfies properties G1, G2, and G3; $(H, \circ)$ is a subgroup of G.

EXAMPLE 1 Proving a Subset of a Group Is a Subgroup

Let $(G, *)$ be an abelian group with identity e. Prove that the set $H = \{x \in G \mid x * x = e\}$ is a subgroup of G.

SOLUTION

<< We will use theorem 9.7 to establish H is a subgroup of G.>>

First, we observe that H is nonempty, because $e * e = e$. Therefore, $e \in H$. Next, we suppose that $a, b \in H$. From the definition of H, (6) $a * a = e$ and (7) $b * b = e$. To prove $a * b^{-1} \in H$, we must prove $(a * b^{-1}) * (a * b^{-1}) = e$. Since G is associative,

$$(8) \qquad (a * b^{-1}) * (a * b^{-1}) = ((a * b^{-1}) * a) * b^{-1} = (a * (b^{-1} * a)) * b^{-1}$$

Since G is abelian, $b^{-1} * a = a * b^{-1}$. Substituting for $b^{-1} * a$ in equation (8), we see by associativity and equations (6) and (7) that

$$\begin{aligned}(a * b^{-1}) * (a * b^{-1}) &= (a * (a * b^{-1})) * b^{-1} = ((a * a) * b^{-1}) * b^{-1} \\ &= (e * b^{-1}) * b^{-1} = b^{-1} * b^{-1} = (b * b)^{-1} = e^{-1} = e\end{aligned}$$

Hence, by theorem 9.7, H is a subgroup of G.

The next theorem is also very useful in proving a subset of a group is a subgroup.

Theorem 9.8 Let G be a group with identity e and let H be a nonempty subset of G. The set H is a subgroup of G if and only if H satisfies the two properties:

(i) Closure: If $a, b \in H$, then $ab \in H$.

(ii) Existence of Inverses: If $a \in H$, then $a^{-1} \in H$.

Proof: Suppose that H is a subgroup of G. By definition of a group, H is closed under the operation on the group. Thus, property (i) is satisfied. Also by theorem 9.6 part (ii) since H is a group, every element in H has an inverse which is an element of H. Therefore, property (ii) is satisfied.

Now, let H be a nonempty subset of the group G which satisfies properties (i) and (ii). By property (i), the set H is closed under the operation on the group. The associativity property, G1, is proved as in the proof of theorem 9.7. Since $H \neq \emptyset$, there exists some $x \in H$. By property (ii), $x^{-1} \in H$ and by property (i), $xx^{-1} = e \in H$. Thus, H contains the identity element of G and $xe = ex = x$ for all $x \in H$. Consequently, property G2, the existence of an identity property, is satisfied. Property (ii) is property G3, the existence of inverses, which by hypothesis is assumed to be true. Therefore, H is a subgroup of G.

EXAMPLE 2 Proving a Subset of a Group Is a Subgroup

Let $(G, *)$ be an abelian group with identity e. Prove that the set $H = \{x * x \mid x \in G\}$ is a subgroup of G.

SOLUTION

$<<$ We will use theorem 9.8 to establish H is a subgroup of G.$>>$

To prove property (i) of theorem 9.8, we assume that $a * a$, $b * b \in H$. Then by definition of H, we have $a \in G$ and $b \in G$. Since G is a group, $a * b \in G$. Since the operation $*$ is associative and commutative (because G is abelian),

$$\begin{aligned}(a * a) * (b * b) &= ((a * a) * b) * b = (a * (a * b)) * b \\ &= (a * (b * a)) * b = ((a * b) * a) * b = (a * b) * (a * b)\end{aligned}$$

Hence, by definition of H, we have $(a*b)*(a*b) \in H$. Therefore, property (i) of theorem 9.8 is satisfied.

To prove property (ii) of theorem 9.8 is true, we assume that $a * a \in H$. By definition of H, we have $a \in G$. Since G is a group, the element $a^{-1} \in G$ and by closure $a^{-1} * a^{-1} \in G$. By theorem 9.4, the inverse of $a * a$ is $(a * a)^{-1} = a^{-1} * a^{-1}$ which is an element of H since $a^{-1} \in G$.

Because H under the operation $*$ satisfies properties (i) and (ii) of theorem 9.8, the algebraic structure $(H, *)$ is a subgroup of G.

We now introduce exponential notation for elements in a group. Let G be a group with identity e. For $a \in G$ define $a^0 = e$, $a^1 = a$, and for $n \in \mathbf{N}$ recursively define $a^{n+1} = a^n a$. Since $a \in G$, we have $a^{-1} \in G$ and for $n \in \mathbf{Z}$ and $n < 0$, we define $a^n = (a^{-1})^{-n}$. Then for $m, n \in \mathbf{Z}$, we have the familiar rules of exponents:

$$a^m a^n = a^{m+n} \quad \text{and} \quad (a^m)^n = a^{mn}$$

We can prove the inverse of a^n is $(a^n)^{-1} = a^{-n}$. Furthermore, it is true that $(ab)^n = a^n b^n$ if and only if G is abelian.

The following theorem is useful in proving that a nonempty, finite subset H of a group G is a subgroup. Using theorem 9.9 to prove a nonempty, finite subset is a subgroup is usually much easier than using either theorem 9.7 or theorem 9.8 to prove a nonempty, infinite subset is a subgroup.

Theorem 9.9 Let H be a nonempty, finite subset of a group G. If H is closed under the operation on G, then H is a subgroup of G.

Proof: The hypothesis of this theorem is property (i) of theorem 9.8. Thus, in order to prove this theorem, we need to prove only property (ii) of theorem 9.8—namely, if $a \in H$, then $a^{-1} \in H$. So let $a \in H$. If $a = e$, the identity of G, then $a^{-1} = e = a \in H$. Now, suppose $a \in H$ and $a \neq e$. Consider the set $P = \{a, a^2, a^3, \ldots\}$ of positive powers of a. Since H is assumed to be closed, each element of P is in H. The set P is finite because the set H is finite. Therefore, there are positive integers i and j such that $i > j$ and $a^i = a^j$. Multiplying the last equation on the right by a^{-j}, we see that $a^{i-j} = e$. Because $a \neq e$, we have $i - j > 1$ and $i - j - 1 > 0$. Therefore, $a^{i-j-1} \in P$. Furthermore, $a^{i-j-1}a = aa^{i-j-1} = a^{i-j} = e$. Hence, $a^{i-j-1} = a^{-1} \in P$.

For example, the nonempty, finite subset $H = \{-1, 1\}$ of the infinite abelian group $(\mathbf{Z} - \{0\}, \cdot)$, where $\cdot$ is the usual operation of multiplication, is a subgroup of $\mathbf{Z} - \{0\}$ by theorem 9.9 because $(-1) \cdot (-1) = 1 \in H$, $(-1) \cdot 1 = -1 \in H$, and $(1) \cdot (1) = 1 \in H$.

The proof of theorem 9.9 provides us with a method for producing subgroups of a group by generating subsets which consist of powers of the elements of the group. We formalize this result in theorem 9.10.

Theorem 9.10 Let G be a group and let $a \in G$. Then $\langle a \rangle = \{a^n \mid n \in \mathbf{Z}\}$ is a subgroup of G.

Proof: << We use theorem 9.7 to prove that $\langle a \rangle$ is a subgroup of G.>>

(i) The set $\langle a \rangle$ is nonempty, because $a \in \langle a \rangle$.

(ii) Let $a^m, a^n \in \langle a \rangle$. Then $a^m(a^n)^{-1} = a^m a^{-n} = a^{m-n} \in \langle a \rangle$.

So by theorem 9.7, $\langle a \rangle$ is a subgroup of G.

We now define a cyclic subgroup generated by an element a, a cyclic group, and a generator of a group.

DEFINITION **Cyclic Subgroup Generated by** a
Let a be an element of a group G. The **cyclic subgroup generated by** a is $\langle a \rangle = \{a^n \mid n \in \mathbf{Z}\}$.

DEFINITION Cyclic Group and Generator

A group G is called a **cyclic group** and a is called a **generator for** G if there exists an $a \in G$ such that $\langle a \rangle = G$.

The next theorem states that "Every cyclic group is abelian." The contrapositive, which is logically equivalent, says that "If a group is not abelian, then the group is not cyclic." The contrapositive of theorem 9.11 is often used to show that a particular group is not cyclic.

Theorem 9.11 If G is a cyclic group, then G is abelian.

Proof: Let a be a generator for the cyclic group G. If $x, y \in G = \langle a \rangle$, then there exist integers m and n such that $x = a^m$ and $y = a^n$. Multiplying and using the properties of exponents, we find

$$xy = a^m a^n = a^{m+n} = a^{n+m} = a^n a^m = yx$$

Hence, the group G is abelian.

DEFINITION Order of an Element of a Group

Let a be an element of a group G. The **order** of a, denoted by $|a|$, is the order of the cyclic subgroup $\langle a \rangle$ generated by a. That is, $|a| = |\langle a \rangle|$.

Let us examine the group $(U, *)$ where $U = \{1, 3, 7, 9\}$ and $*$ is multiplication modulo 10. The order of the group is $|U| = 4$. The element 1 is the identity of the group. The subgroup generated by the identity is always the set containing only the identity. In this example, $\langle 1 \rangle = \{1\}$ and $|1| = 1$. Since $3^0 = 1$, $3^1 = 3$, $3^2 = 3 * 3 = 9$, and $3^3 = 3^2 * 3 = 9 * 3 \equiv 7 \pmod{10}$, the cyclic subgroup generated by 3 is $\langle 3 \rangle = \{1, 3, 9, 7\} = U$. Therefore, U is a cyclic group and 3 is a generator for the group. The element 7 is also a generator for the group, because $7^0 = 1$, $7^1 = 7$, $7^2 = 7 * 7 \equiv 9 \pmod{10}$, and $7^3 = 7^2 * 7 = 9 * 7 \equiv 3 \pmod{10}$. Calculating powers of 9, we find $9^0 = 1$, $9^1 = 9$, $9^2 = 9 * 9 \equiv 1 \pmod{10}$, and $9^3 = 9^2 * 9 = 1 * 9 = 9$. So the cyclic subgroup generated by 9 is $\langle 9 \rangle = \{1, 9\}$ and the order of 9 is $|9| = 2$. Summarizing, we have $|1| = 1$, $|3| = 4$, $|7| = 4$, $|9| = 2$, and $|U| = 4$. This example illustrates that the generator of a group is not unique. What relationship, if any, do you notice between the order of the elements of the group and the order of the group?

Now consider the group $(\mathbf{Z}_4, +)$ where $+$ is addition modulo 4. The Cayley table for $(\mathbf{Z}_4, +)$ is displayed in table 9.10. The identity for this group is the element 0, so the subgroup generated by 0 is $\langle 0\rangle = \{0\}$ and the order of the element 0 is $|0| = 1$. Because the Cayley table 9.10 is symmetric, the group $(\mathbf{Z}_4, +)$ is abelian. Next, we calculate the subgroups generated by the nonidentity elements using table 9.10.

For the element 1, we have $1 = 1$, $1 + 1 = 2$, $(1 + 1) + 1 = 2 + 1 = 3$, and $((1 + 1) + 1) + 1 = 3 + 1 = 0$. So $\langle 1\rangle = \{1, 2, 3, 0\} = \mathbf{Z}_4$ and $|1| = 4$. Hence, $(\mathbf{Z}_4, +)$ is a cyclic group with generator 1.

For 2, we have $2 = 2$, $2+2 = 0$, $(2+2)+2 = 0+2 = 2$, and $((2+2)+2)+2 = 2 + 2 = 0$. Therefore, $\langle 2\rangle = \{2, 0\}$ and $|2| = 2$.

For 3, we have $3 = 3$, $3+3 = 2$, $(3+3)+3 = 2+3 = 1$, and $((3+3)+3)+3 = 1 + 3 = 0$. Thus, $\langle 3\rangle = \{3, 2, 1, 0\} = \mathbf{Z}_4$ and $|3| = 4$.

Summarizing, $(\mathbf{Z}_4, +)$ is a cyclic, abelian group of order 4 with generators 1 and 3 and $|2| = 2$. What relationship, if any, do you notice between the order of the elements of the group and the order of the group?

+	**0**	**1**	**2**	**3**
0	0	1	2	3
1	1	2	3	0
2	2	3	0	1
3	3	0	1	2

Table 9.10

$\oplus$	**0**	**1**	**2**	**3**
0	0	1	2	3
1	1	0	3	2
2	2	3	0	1
3	3	2	1	0

Table 9.11

The Cayley table for the Klein 4-group is shown in table 9.11. The group is abelian, because the table is symmetric. The identity is the element 0, $\langle 0\rangle = \{0\}$, and $|0| = 1$. Calculating, from table 9.11:

For 1, we have $1 \oplus 1 = 0$, $(1 \oplus 1) \oplus 1 = 0 \oplus 1 = 1$. Hence, $\langle 1\rangle = \{0, 1\}$ and $|1| = 2$.

For 2, we have $2 \oplus 2 = 0$, $(2 \oplus 2) \oplus 2 = 0 \oplus 2 = 2$. Therefore, $\langle 2\rangle = \{0, 2\}$ and $|2| = 2$.

For 3, we have $3 \oplus 3 = 0$, $(3 \oplus 3) \oplus 3 = 0 \oplus 3 = 3$. Thus, $\langle 3\rangle = \{0, 3\}$ and $|3| = 2$.

Since no element of the group is of order 4, the Klein 4-group is not cyclic. Consequently, the Klein 4-group is an example of an abelian group which is not cyclic.

Theorem 9.12 tells us how to calculate the finite order of an element of a group and how to write the cyclic subgroup generated by the element.

Theorem 9.12 Let G be a group with identity e and let a be an element of G of order n. Then n is the smallest positive integer such that $a^n = e$ and $\langle a\rangle = \{e, a, a^2, \ldots, a^{n-1}\}$.

Proof: Because the element a has finite order n, the cyclic subgroup $\langle a\rangle$ is finite. Therefore, the powers of a cannot all be distinct. Hence, there exist some integers i and j such that $i > j$ and $a^i = a^j$. Multiplying on the right by a^{-j}, we have $a^i a^{-j} = a^j a^{-j}$ or $a^{i-j} = e$ for $i - j > 0$. Thus, $a^p = e$ for the positive integer $p = i - j$. Let $P = \{q \in \mathbf{N} \mid a^q = e\}$. Since P is nonempty (because $p \in P$), by the Well-Ordering Principle there exists a least positive integer $m \in P$.

The elements e, a, a^2, ..., a^{m-1} are all distinct. For if $a^r = a^s$ and $0 \leq s < r < m$, then $a^{r-s} = e$ and $0 < r - s < m$, which contradicts m is the least element of P.

By definition $\langle a \rangle = \{a^t \mid t \in \mathbf{Z}\}$. Let $t \in \mathbf{Z}$. By the division algorithm, there exist integers k and ℓ such that $t = km + l$ where $0 \leq l < m$. Hence, $a^t = a^{km+l} = (a^m)^k a^l = e^k a^l = ea^l = a^l$ where $0 \leq l < m$. That is, every element in $\langle a \rangle$ has the form $a^0 = e$, a^1, a^2, ..., a^{m-1}. Since we have already proven these elements are distinct, $\langle a \rangle = \{e, a, a^2, \ldots, a^{m-1}\}$ and $n = m$ by definition of the order of a.

EXAMPLE 3 Finding the Order and Subgroup Generated by an Element of a Group

Let $M = \{1, 3, 5, 9, 11, 13\}$ and let $*$ denote multiplication modulo 14. Given $(M, *)$ is a group, find the order of the element 9 and the cyclic subgroup $\langle 9 \rangle$.

SOLUTION

The identity of the group is 1. Computing successive powers of 9, we find $9^0 = 1$, $9^1 = 9$, $9^2 \equiv 11 \pmod{14}$, $9^3 \equiv 1 \pmod{14}$. Hence, by theorem 9.12 the order of 9 is $|9| = 3$ and the cyclic subgroup generated by 9 is $< 9 > = \{9^0, 9^1, 9^2\} = \{1, 9, 11\}$.

EXERCISES 9.3

1. Is $(\mathbf{N}, +)$ a subgroup of $(\mathbf{Z}, +)$?

2. Is $\{0, 2, 4\}$ a subgroup of $(\mathbf{Z}_7, +)$?

3. Is $\{1, 2, 4\}$ a subgroup of $(\mathbf{Z}_7 - \{0\}, \cdot)$?

4. Give an example of a group G and two subgroups H and K of G such that $H \cup K$ is not a subgroup of G.

5. Prove that if H and K are subgroups of a group G, then $H \cap K$ is a subgroup of G.

6. Prove that if $\{H_\lambda \mid \lambda \in \Lambda\}$ is a family of subgroups of a group G, then $\bigcap_{\lambda \in \Lambda} H_\lambda$ is a subgroup of G.

7. Prove that if H is a subgroup of the group K and if K is a subgroup of the group G, then H is a subgroup of G.

8. The **center**, $Z(G)$, of a group G is the subset of G whose elements commute with every element of G. Symbolically,

$$Z(G) = \{g \in G \mid gx = xg \text{ for all } x \in G\}.$$

Prove that the center of a group G is a subgroup of G.

9. Find the center of the group whose Cayley table appears table 9.12.

*	**0**	**1**	**2**	**3**	**4**	**5**
0	0	1	2	3	4	5
1	1	2	0	5	3	4
2	2	0	1	4	5	3
3	3	4	5	0	1	2
4	4	5	3	2	0	1
5	5	3	4	1	2	0

Table 9.12

10. Find the center of the group whose Cayley table appears in table 9.13.

*	**0**	**1**	**2**	**3**	**4**	**5**	**6**	**7**
0	0	1	2	3	4	5	6	7
1	1	2	3	0	6	7	5	4
2	2	3	0	1	5	4	7	6
3	3	0	1	2	7	6	4	5
4	4	6	5	7	0	2	3	1
5	5	7	4	6	2	0	1	3
6	6	5	7	4	1	3	0	2
7	7	4	6	5	3	1	2	0

Table 9.13

11. Let a be a fixed element of a group G. The **centralizer of a in G**, $C(a)$, is the subset of G whose elements commute with a. Symbolically, $C(a) = \{g \in G \mid ga = ag\}$. Prove that the centralizer of a in G is a subgroup of G.

12. Let a be any element of a group G. Prove that the center of G is a subgroup of the centralizer of a in G. That is, prove that $Z(G)$ is a subset of $C(a)$ for every $a \in G$.

13. Let H be a subgroup of a group G and let a be a fixed element in G. Prove that $K = \{a^{-1}ha \mid h \in H\}$ is a subgroup of G.

14. Find the order of each element in the following groups.

    a. $(\mathbf{Z}, +)$ b. $(\mathbf{Z}_6, +)$ where + denotes addition modulo 6

    c. Table 9.12 d. Table 9.13.

15. Find one cyclic subgroup and one noncyclic subgroup of the noncyclic group represented by table 9.13.

Appendix A

Reading and Writing Mathematical Proofs

During semesters I teach a course that includes mathematical proofs, several students usually ask the following questions:

1. "How do I learn to read mathematical proofs?"

and

2. "How do I learn to write mathematical proofs?"

The short answer to the first question is "You learn to read proofs by reading many well-written proofs by several different authors." And, the short answer to the second question is "You learn to write proofs by reading well-written proofs, by emulating them when you write your own proofs, and by having other people read your proofs and criticize them constructively." The remainder of this appendix is devoted to the long answers to these two questions.

Reading Mathematical Proofs

When you begin to read a proof, you need to have paper and pencil available in order to make notes and verify computations. First, you need to identify the hypotheses and conclusions of the theorem. Then, you may wish to make a quick first reading of the entire proof to see what methods the author employed in proving the theorem and to determine what definitions and theorems are used in the proof. You should write down complete statements for the definitions and theorems that appear in the proof. During the first reading of the proof, you should not attempt to verify the logic used and you should not carefully check the computations made. After you have looked up and learned the definitions and theorems of the proof, you should read and verify the proof line-by-line. You must provide justifications and perform computations that the author has omitted either on purpose or by mistake. Even when a mathematical proof is well-written, reading the proof may still be a slow process. Furthermore, reading a proof that is poorly written is usually very difficult and exceedingly frustrating. Reading a proof may be difficult (1) because the order in which the steps of the written proof are presented does not correspond to the order in which the original proof was developed,

(2) because the author does not tell you explicitly what proof techniques he is using, (3) because all justifications are not provided, or (4) because what should have been presented in several different steps in the proof has been combined into one single step.

Writing Mathematical Proofs

A mathematical proof is a convincing argument which combines elements of the English language, logic, and mathematics. It is not possible to give a recipe or prescription for writing a clear, informative, and interesting proof. Every written proof is a communication between the author and the reader; and, therefore, its effectiveness depends upon the skills, temperament, taste, and knowledge of both the author and reader. Since most mathematical writing is factual, the simple declarative sentence is the best means of communication in a proof.

English Usage. Writing is more difficult than speaking, because the tone of your voice helps you make your meaning clear when you speak. When writing a proof use correct diction, grammar, punctuation, and spelling. Write correctly, carefully, and clearly. Write simple prose. Start every sentence with a capital letter and end it with a period, question mark, or exclamation point. When possible, write two short sentences instead of one long, convoluted sentence. Use the active voice instead of the passive voice. That is, write "We have observed that ..." instead of "It has been observed that .... " The word "we" is often used in order to avoid using the passive voice. Also, when writing mathematical proofs, articles, books, and so forth, use the word "we" instead of the word "I." The word "we" refers to the author(s) and the reader working together as a team to prove a theorem or learn a fact.

Mathematical Usage. Mathematics is symbolic-oriented and has a language of its own. Therefore, mathematical writing is a combination of English words and mathematical symbols. Always strive to use the words of logic and mathematics correctly. The purpose of using good mathematical language is to make it easy for the reader to understand a mathematical document. The author's aim should be to anticipate difficulties the reader may encounter and to alleviate them. A good approach when preparing a written mathematical document is to pretend that the document is to be read aloud. The basic aim in writing a mathematical proof is to communicate a newly discovered fact.

Symbol usage. Starting a sentence with a mathematical symbol is discouraged. For instance, writing: "$(S, \circ)$ is a cyclic group." is discouraged. Instead, write: "The group $(S, \circ)$ is cyclic." Rewrite the sentence: "$x^2 - 3x + 2 = 0$ has two distinct real roots." as "The equation $x^2 - 3x + 2 = 0$ has two distinct real roots."

A mathematical symbol represents a word or phrase. For example, it is possible for "=" to mean (1) *equals*, (2) *is equal to*, (3) *be equal to*, or (4) *which is equal to*. Thus, we could write:

"Hence, $x = 0$." which may be read "Hence, x equals zero." or "Hence, x is equal to zero."

"Let $x = 0$." which may be read "Let x be equal to zero."

"Then $x^2 + y^2 = y^2 = 1$." which may be read "Then x squared plus y squared equals y squared which is equal to one."

Do not combine English words and mathematical symbols improperly. Instead of writing: "Every even natural number > 2 may be written as a sum of two prime numbers." you should write: "Every even natural number greater than 2 may be written as the sum of two prime numbers." The sentence: "The equation $(x-1)(x-2) = 0$ implies $x = 1$ or 2." is not written correctly. It should be written as "The equation $(x-1)(x-2) = 0$ implies $x = 1$ or $x = 2$."

When possible, symbols that are not a part of a list should be separated by words. For example, instead of writing: "The sequence a_n, $n < m$ is ...." write: "The sequence a_n where $n < m$ is ...." And, instead of writing: "In addition to 1, 2 is a root of the polynomial $x^2 - 3x + 2 = 0$." write: "In addition to 1, the number 2 is a root of the polynomial $x^2 - 3x + 2 = 0$."

Display important mathematical expressions, long expressions, and multi-level expressions on a line by themselves. The following two expressions are displayed.

$$(2x-3)^3 = (2x)^3 - 3(2x)^2(3) + 3(2x)(3^2) - 3^3 = 8x^3 - 36x^2 + 54x - 27$$

$$\frac{x - \dfrac{1}{x}}{x + \dfrac{2}{x}}$$

When a displayed expression will be referred to later, give a reference number to the expression. The reference number may appear at the left margin or right margin of the line on which the expression appears. For example,

(1) $(2x-3)^3 = (2x)^3 - 3(2x)^2(3) + 3(2x)(3^2) - 3^3 = 8x^3 - 36x^2 + 54x - 27$

When mathematical expressions are joined by equal signs or inequalities, it is good idea to display them on more than one line and align the equal signs and inequality symbols as illustrated below:

(2)
$$\begin{aligned}(2x-3)^3 &= (2x)^3 - 3(2x)^2(3) + 3(2x)(3^2) - 3^3\\ &= 8x^3 - 36x^2 + 54x - 27\end{aligned}$$

When a long inequality or equation such as (2) is written within the text and not displayed and when it is necessary to break the expression at the end of a line, the expression should be broken after one of the relation symbols $=$, $\leq$, $<$, $>$, or $\geq$ or after one of the operation symbols $+$ or $-$. That is, equation (2) should be written in line as $(2x-3)^3 = (2x)^3 - 3(2x)^2(3) + 3(2x)(3^2) - 3^3 =$ $8x^3 - 36x^2 + 54x - 27$ or it should be written as $(2x-3)^3 = (2x)^3 - 3(2x)^2(3) +$ $3(2x)(3^2) - 3^3 = 8x^3 - 36x^2 + 54x - 27$.

Developing a Proof and Writing It. The goal of mathematics is to seek truth. Usually, this is done by formulating a conjecture or having one presented to you. Your task is to decide if the conjecture is true or false. If the conjecture is false, then your task becomes one of constructing a counterexample. If you are successful in constructing a counterexample, then you may want to return to the original conjecture to determine if you can modify it and make it a true statement. You can often do this by adding one or more hypotheses or by making the conclusion weaker. Once you have constructed a statement that you believe to be true, your task is to prove it is true. In chapter 2, several methods for proving theorems were presented and discussed. Suppose you want to prove the theorem: "If A, then B." First, look up and then understand all of the definitions which appear in A and B. Locate and copy any theorems that have A as hypothesis and any theorems that have B as conclusion. It is possible, but not too likely, that you will find two theorems: "If A, then C." and "If C, then B." In this case, your proof is simply: A implies C and C implies B; therefore, A implies B by transitive inference. If this does not happen, your time is not wasted, because you will have seen proofs of A implies X for different statements X and proofs of Y implies B for different statements Y. Thus, you will have seen some techniques that may help you start your proof and some techniques that may help you end your proof. To begin the development of your proof, write the hypotheses on separate lines at the top of a piece of paper. Then write the definitions of the terms that appear in the hypotheses. After each hypothesis and definition write formulas and equations that are associated with the hypothesis and definition. These formulas and definitions will give you some idea of how to start your proof. The theorems and proofs you have looked up earlier may give you some different ideas of how to start your proof also. At the bottom of the page, write the conclusion of the theorem. Immediately above, write the definitions that appear in the conclusion. Above that, write the formulas or equations that are associated with the conclusion and definitions. Next, you try to develop the logical thread of the proof by moving forward from the hypotheses as much as possible and backwards from the conclusion as much as possible. The object is to alternate moving forward and backward logically until you arrive at the same statement located somewhere in the middle of the proof. If you do not succeed, it may be necessary for you to apply this proof-development technique to other kind of proof—a proof by contraposition, a proof by contradiction, a proof by cases, etc. Once you have developed a sequence of statements, each of which is deducible from its predecessor or from some known theorem, that extends from the hypotheses to the conclusion, you have you a first rough draft of your proof. You should put your proof away for awhile. Then later reread the proof very carefully and make changes that will improve it.

Now that you have a proof, you must write it for others to read. When writing a proof, you must keep in mind the knowledge and mathematical maturity of your readers. You cannot communicate effectively with your readers until you know who they are. In your case, your proofs should be written for your instructor; but, they should be written at a level that your classmates can understand. The first proofs you write should contain more detail and justification than the ones your write later. It requires practice and experience to know what you may omit and what you should not omit. Your rough draft is an outline of your proof and is probably arranged in the order in which you want to present the proof. However, you should never be satisfied with your first draft as your final proof. The next step is to add more English to the proof, being sure to follow the rules cited above. Next, type your proof and reread it. As you read, be sure to challenge every word, phrase, sentence, and paragraph. Did you choose the best possible words? Is there some verbiage that can be removed? Are the sentences presented in the best order? Are the paragraphs organized properly? Be certain that the statements of the proof follow logically from one another. Read your proof aloud to yourself. Does it sound right? Continue to rewrite the proof again and again until you are totally satisfied with the result. Then have someone else read the proof and constructively criticize it. Then rewrite the proof one final time.

The following is an example of developing a proof and then writing it. The theorem and proof are relatively simple; but, they should provide you with an idea of how the proof writing process goes. As in the text, comments will appear between << and >>. Suppose that a homework assignment is to prove the theorem: "The product of an even integer and an odd integer is an even integer." First, we rewrite the theorem as a logically equivalent conditional statement, so that we may more easily identify the hypotheses and conclusion. A logically equivalent form of the theorem is "If m is an even integer and n is an odd integer, then the product mn is an even integer." We write the hypotheses at the top of a piece of paper and the conclusion at the bottom. Then we add definitions under the hypotheses and above the conclusion as shown on the next page.

Hypothesis: m is an even integer.
<< In a proof, you should not start a sentence with a symbol; however, this is "scratch work," so no one will see it except you. >>
Definition: An integer is a member of the set $\mathbf{Z} = \{\ldots, -2, -1, 0, 1, 2, \ldots\}$.
Definition: m is an even integer if and only if there exists an integer j such that $m = 2j$.
Hypothesis: n is an odd integer.
Definition: n is an odd integer if and only if there exists an integer k such that $n = 2k + 1$.
<< It is very important that the symbol, k, used in defining n is different from the symbol, j, used in defining m. When the original definitions of even and odd integers were given, it was permissible to use the same symbol in both definitions. However, the simultaneous definition $m = 2j$ and $n = 2j - 1$ implies $n = m - 1$. That is, n is one less than m instead of an arbitrary odd integer.>>
<< An alternate definition someone might use for an odd integer is $n = 2k - 1$. When you are reading a proof, be sure you know and understand the author's definition of an entity.>>.

*

Definition: The product mn is an even integer provided mn is an integer and mn can be written as $mn = 2\ell$ where ℓ is an integer.
Conclusion: The product mn is an even integer.

Next, at location * we compute the product mn and use the associative law of multiplication for integers to obtain

$$mn = (2j)(2k + 1) = 2[j(2k + 1)]$$

Letting $\ell = j(2k + 1)$, we see that $mn = 2\ell$.

We note that mn is an integer, because by hypotheses m and n are integers and because the integers are closed under multiplication.

Also, we observe that $\ell = j(2k + 1)$ is an integer because j and $2k + 1 = n$ are integers and the integers are closed under multiplication.

We now have a rough draft of our proof, and we need to write it for others to read. We start by typing the word "Theorem" at the left margin followed by the original statement of the theorem. Then, we skip a line and type the word "Proof:" followed by our proof. Here is our completed theorem and proof.

Theorem The product of an even integer and an odd integer is an even integer.

Proof: We prove this statement by proving the logically equivalent statement: If m is an even integer and n is an odd integer, then the product mn is an even integer. Assume that m is an even integer. Then by definition, there exists an integer j such that $m = 2j$. Assume that n is an odd integer. Then by definition, there exists an integer k such that $n = 2k + 1$. Computing mn and using the associative law of multiplication for integers, we obtain

$$mn = (2j)(2k + 1) = 2[j(2k + 1)] = 2\ell \quad \text{where} \quad \ell = j(2k + 1).$$

Because m and n are integers and because the integers are closed under multiplication, mn is an integer. Because j and $2k+1 = n$ are integer and because the integers are closed under multiplication, $\ell = j(2k+1)$ is an integer. Since $mn = 2\ell$ and ℓ is an integer, the integer mn is even by definition.

As you write a proof, you need to keep your reader informed. At the beginning of a proof, you should tell the reader what proof technique you will use unless the proof is a direct proof. If the proof is a direct proof, you may just assume the hypothesis. For various types of proofs, here are a few examples of some sentences you might use to start a proof of the conditional statement: "If A, then B."

Direct Proof

Proof: We assume A.

Proof by Contraposition

Proof: We prove this theorem by proving its contrapositive. That is, we prove that if $\neg B$, then $\neg A$. Assume $\neg B$.

Proof by Contradiction

Proof: We prove this statement by contradiction. Assume A and $\neg B$.

or

Proof: We assume to the contrary that A and $\neg B$.

Proof by Cases

Proof: We prove this theorem by considering three cases.

Proof by Induction

Proof: We prove this result by mathematical induction. Let $P(n)$ be the statement ....

Answers to Selected Exercises

Chapter 1

Exercises 1.1

The following sentences are statements: 1, 7, 8, 10.

11. The number $\sqrt{2}$ is not rational. Or, the number $\sqrt{2}$ is irrational.

13. $7 > 5$

15. Not every even integer greater than two can be written as the sum of two prime numbers.

17. a. (1) $\neg I$ (2) $T \wedge (\neg I)$ (3) $T \wedge I$ (4) $I \wedge T$ (5) $T \wedge (\neg I)$

17. b. (1) ABC is not a triangle.
(2) ABC is a triangle, or ABC is isosceles.
(3) ABC is not a triangle, and ABC is not isosceles.
(4) It is not the case that ABC is a triangle and ABC is isosceles.
(5) ABC is not a triangle, or ABC is not isosceles.

Exercises 1.2

1. F 3. T 5. T 7. F 9. T 11. T 13. T

15. a. T b. The truth value is the same as the truth value of R. c. F

17.

P	P $\wedge$ P
T	T
F	F

Therefore, P $\wedge$ P $\equiv$ P

19.

P	Q	$P \wedge Q$	$P \wedge (P \vee Q)$
T	T	T	T
T	F	T	T
F	T	T	F
F	F	F	F

Therefore, $P \wedge (P \vee Q) \equiv P$

21.

P	Q	R	$P \wedge Q$	$(P \wedge Q) \wedge R$
T	T	T	T	T
T	T	F	T	F
T	F	T	F	F
T	F	F	F	F
F	T	T	F	F
F	T	F	F	F
F	F	T	F	F
F	F	F	F	F

P	Q	R	$Q \wedge R$	$P \wedge (Q \wedge R)$
T	T	T	T	T
T	T	F	F	F
T	F	T	F	F
T	F	F	F	F
F	T	T	T	F
F	T	F	F	F
F	F	T	F	F
F	F	F	F	F

Therefore, $(P \wedge Q) \wedge R \equiv P \wedge (Q \wedge R)$

23.

P	Q	$\neg Q$	$P \wedge (\neg Q)$	$\neg (P \wedge (\neg Q))$
T	T	F	F	T
T	F	T	T	F
F	T	F	F	T
F	F	T	F	T

There are no tautologies and no contradictions.

25. Let R be the statement $Q \vee (\neg Q)$ and let S be $P \wedge (\neg(Q \vee (\neg Q)))$.

P	Q	¬ Q	Q ∨ (¬ Q)	R	S
T	T	F	T	F	F
T	F	T	T	F	F
F	T	F	T	F	F
F	F	T	T	F	F

The statement $Q \vee (\neg Q)$ is a tautology and the statements $\neg(Q \vee (\neg Q))$ and $P \wedge (\neg(Q \vee (\neg Q)))$ are both contradictions.

27. The statement $(P \wedge (Q \vee (\neg R))) \vee ((\neg P) \vee R)$ is a tautology.

29. I do not sweeten my tea with sugar, and I do not sweeten my tea with honey.

31. I drink my coffee with sugar, or I drink my coffee with cream.

33. I go to the opera, or I do not go to the theater.

35. $(\neg P) \wedge Q$ 37. $P \wedge Q$ 39. P 41. exclusive 43. exclusive

45. Let R be the statement (P ∨ Q) ∧ (¬ (P ∧ Q)).

P	Q	P ∨ Q	¬ (P ∧ Q)	R
T	T	T	F	F
T	F	T	T	T
F	T	T	T	T
F	F	F	T	F

$(P \vee Q) \wedge (\neg(P \wedge Q)) \equiv P \nabla Q$

Exercises 1.3

In exercises 1 through 8, let H denote the hypothesis, C denote the conclusion, CS denote the conditional statement, and N denote the negation. As usual, T denotes true and F denotes false.

1. (1) H: New York is on the East Coast. C: Los Angeles is on the West Coast.

   (2) H is T, C is T, and CS is T

   (3) N: New York is on the East Coast and Los Angeles is not on the West Coast.

3. (1) H: The number $\sqrt{2}$ is irrational. C: The number π is rational.

   (2) H is T, C is F, and CS is F

   (3) N: The number $\sqrt{2}$ is irrational and the number π is not rational.

5. (1) H: $2 < 3$. C: $2^3 > 3^2$.

   (2) H is T, C is F, and CS is F

   (3) N: $2 < 3$ and not $2^3 > 3^2$.

7. (1) H: $2^3 > 3^2$. C: $2 < 3$.

   (2) H is F, C is T, and CS is T

   (3) N: $2^3 > 3^2$ and not $2 < 3$.

9. Converse: If Los Angeles is on the West Coast, then New York is on the East Coast.

   Inverse: If New York is not on the East Coast, then Los Angeles is not on the West Coast.

   Contrapositive: If Los Angeles is not on the West Coast, then New York is not on the East Coast.

11. Converse: If the number π is rational, then the number $\sqrt{2}$ is irrational.
    Inverse: If the number $\sqrt{2}$ is not irrational, then the number π is not rational.

    Contrapositive: If the number π is not rational, then the number $\sqrt{2}$ is not irrational.

13. Converse: If $2^3 > 3^2$, then $2 < 3$.
    Inverse: If not $2 < 3$, then not $2^3 > 3^2$.
    Contrapositive: If not $2^3 > 3^2$, then not $2 < 3$.

15. Converse: If $2 < 3$, then $2^3 > 3^2$.
    Inverse: If not $2^3 > 3^2$, then not $2 < 3$.
    Contrapositive: If not $2 < 3$, then not $2^3 > 3^2$.

17. T 19. F 21. Q is T and R is T.

23.

P	Q	$P \wedge Q$	$P \Rightarrow (P \wedge Q)$
T	T	T	T
T	F	F	F
F	T	F	T
F	F	F	T

25.

P	Q	$P \wedge Q$	$(P \wedge Q) \Rightarrow P$
T	T	T	T
T	F	F	T
F	T	F	T
F	F	F	T

$(P \wedge Q) \Rightarrow P$ is a tautology.

27.

P	Q	$P \vee Q$	$P \wedge Q$	$(P \vee Q) \Rightarrow (P \wedge Q)$
T	T	T	T	T
T	F	T	F	F
F	T	T	F	F
F	F	F	F	T

29. Let R be the statement $(Q \wedge (P \Rightarrow Q)) \Rightarrow P$.

P	Q	$P \Rightarrow Q$	$Q \wedge (P \Rightarrow Q)$	R
T	T	T	T	T
T	F	F	F	T
F	T	T	T	F
F	F	T	F	T

31.

P	Q	$P \vee Q$	$(P \vee Q) \Leftrightarrow P$
T	T	T	T
T	F	T	T
F	T	T	F
F	F	F	T

33.

P	Q	$P \Rightarrow Q$	$Q \Rightarrow P$	$(P \Rightarrow Q) \Leftrightarrow (Q \Rightarrow P)$
T	T	T	T	T
T	F	F	T	F
F	T	T	F	F
F	F	T	T	T

35. $(\neg P) \vee Q \vee R$

37. $P \wedge ((\neg Q) \vee P)$

39. $(\neg P \vee Q) \wedge (\neg Q \vee P)$

Exercises 1.4

1. Let R be the statement $P \Rightarrow (\neg Q)$,

   let S be the statement $(P \wedge Q) \wedge (P \Rightarrow \neg Q)$,

   let U be the statement $P \wedge (\neg Q)$

   and let V be the statement $[(P \wedge Q) \wedge (P \Rightarrow \neg Q)] \Rightarrow (P \wedge (\neg Q))$.

P	Q	$P \wedge Q$	$\neg Q$	R	S	U	V
T	T	T	F	F	F	F	T
T	F	F	T	T	F	T	T
F	T	F	F	T	F	F	T
F	F	F	T	T	F	F	T

$P \wedge Q,\ P \Rightarrow \neg Q\ \therefore\ P \wedge (\neg Q)$ is a valid argument.

3. Let S be the statement $(P \wedge Q) \wedge (R \Rightarrow P)$,

   and let U be the statement $[(P \wedge Q) \wedge (R \Rightarrow P)] \Rightarrow (Q \vee R)$.

P	Q	R	$P \wedge Q$	$R \Rightarrow P$	S	$Q \vee R$	U
T	T	T	T	T	T	T	T
T	T	F	T	T	T	T	T
T	F	T	F	T	F	T	T
T	F	F	F	T	F	F	T
F	T	T	F	F	F	T	T
F	T	F	F	T	F	T	T
F	F	T	F	F	F	T	T
F	F	F	F	T	F	F	T

$P \wedge Q,\ R \Rightarrow P\ \therefore\ Q \vee R$ is a valid argument.

7. P is T and Q is T. 9. P is F, Q is F, and R is T.

11. valid 13. valid 15. valid 17. valid

5. Let S be the statement $P \Rightarrow Q$, let U be the statement $Q \Rightarrow R$,

   let V be the statement $(P \Rightarrow Q) \wedge (Q \Rightarrow R) \wedge (\neg Q)$, and let W be the statement $[(P \Rightarrow Q) \land (Q \Rightarrow R) \land (\neg Q)] \Rightarrow (\neg R)$.

P	Q	R	S	U	¬ Q	V	¬ R	W
T	T	T	T	T	F	F	F	T
T	T	F	T	F	F	F	T	T
T	F	T	F	T	T	F	F	T
T	F	F	F	T	T	F	T	T
F	T	T	T	T	F	F	F	T
F	T	F	T	F	F	F	T	T
F	F	T	T	T	T	T	F	F
F	F	F	T	T	T	T	T	T

$P \Rightarrow Q,\ \ Q \Rightarrow R,\ \ \neg Q\ \ \therefore \neg R$ is an invalid argument.

19. 2. rule of conjunctive simplification

    3. rule of conjunctive simplification
    4. rule of disjunction

21. 4. rule of transitive inference

    5. rule of conjunctive simplification
    6. rule of detachment
    7. rule of conjunctive simplification
    8. rule of conjunction

23. 2. definition of implication

    3. definition of implication
    4. rule of substitution
    5. associative law for disjunction
    6. commutative law for disjunction
    7. rule of substitution
    8. associative law for disjunction
    9. definition of implication
    10. rule of substitution
    11. definition of implication

25. $A\ \therefore\ B \Rightarrow A$

1.	A	premise
2.	$\neg B \lor A$	1, rule of disjunction
3.	$B \Rightarrow A$	2, definition of implication

27. $E \wedge F \therefore E \vee F$

1. $E \wedge F$	premise
2. E	1, rule of conjunctive simplification
3. $E \vee F$	2, rule of disjunction

29. $J \Rightarrow (K \wedge L) \therefore J \Rightarrow K$

1. $J \Rightarrow (K \wedge L)$	premise
2. $\neg J \vee (K \wedge L)$	1, definition of implication
3. $(\neg J \vee K) \wedge (\neg J \vee L)$	2, distributive law for disjunction
4. $\neg J \vee K$	3, rule of conjunctive simplification
5. $J \Rightarrow K$	4, definition of implication

31. $(P \vee Q) \Rightarrow R \therefore P \Rightarrow R$

1. $(P \vee Q) \Rightarrow R$	premise
2. $\neg(P \vee Q) \vee R$	1, definition of implication
3. $\neg(P \vee Q) \equiv (\neg P) \wedge (\neg Q)$	a De Morgan law
4. $((\neg P) \wedge (\neg Q)) \vee R$	2,3, rule of substitution
5. $((\neg P) \vee R) \wedge ((\neg Q) \vee R)$	distributive law for disjunction
6. $\neg P \vee R$	5, rule of conjunctive simplification
7. $P \Rightarrow R$	6, definition of implication

33. $V \Rightarrow W,\ V \vee W \therefore W$

1. $V \Rightarrow W$	premise
2. $\neg V \vee W$	1, definition of implication
3. $V \vee W$	premise
4. $(\neg V \vee W) \wedge (V \vee W)$	2,3, rule of conjunction
5. $((\neg V) \wedge V) \vee W$	distribution law for disjunction
6. $(\neg V) \wedge V \equiv f$	law of the excluded middle
7. $f \vee W$	5,6, rule of substitution
8. W	7, a contradiction law

35. $(A \vee B) \Rightarrow (\neg C),\ B \vee C,\ (B \vee (\neg A)) \Rightarrow D,\ A \therefore \neg B \vee D$

1. A	premise
2. $A \vee B$	1, rule of disjunction
3. $(A \vee B) \Rightarrow (\neg C)$	premise
4. $\neg C$	2,3, rule of detachment
5. $B \vee C$	premise
6. B	4,5, rule of disjunctive syllogism
7. $B \vee (\neg A)$	6, rule of disjunction
8. $(B \vee (\neg A)) \Rightarrow D$	premise
9. D	7,8, rule of detachment
10. $(\neg B) \vee D$	9, rule of disjunction

37. $(\mathrm{T} \wedge \mathrm{M}) \Rightarrow \mathrm{V},\ \mathrm{T} \vee \mathrm{A}\ \therefore\ \neg \mathrm{A} \Rightarrow ((\neg \mathrm{M}) \vee \mathrm{V})$

1.	$\mathrm{T} \vee \mathrm{A}$	premise
2.	$\mathrm{A} \vee \mathrm{T}$	1, commutative law for disjunction
3.	$\mathrm{A} \equiv \neg(\neg \mathrm{A})$	double negation law
4.	$\neg(\neg \mathrm{A}) \vee \mathrm{T}$	2,3, rule of substitution
5.	$\neg \mathrm{A} \Rightarrow \mathrm{T}$	4, definition of implication
6.	$(\mathrm{T} \wedge \mathrm{M}) \Rightarrow \mathrm{V}$	premise
7.	$\neg(\mathrm{T} \wedge \mathrm{M}) \vee \mathrm{V}$	6, definition of implication
8.	$\neg(\mathrm{T} \wedge \mathrm{M}) \equiv ((\neg \mathrm{T}) \vee (\neg \mathrm{M}))$	a De Morgan law
9.	$((\neg \mathrm{T}) \vee (\neg \mathrm{M})) \vee \mathrm{V}$	7,8, rule of substitution
10.	$(\neg \mathrm{T}) \vee ((\neg \mathrm{M}) \vee \mathrm{V})$	9, associative law for disjunction
11.	$\mathrm{T} \Rightarrow ((\neg \mathrm{M}) \vee \mathrm{V})$	10, definition of implication
12.	$\neg \mathrm{A} \Rightarrow ((\neg \mathrm{M}) \vee \mathrm{V})$	5,11, rule of transitive inference

39. $\mathrm{I} \Rightarrow (\mathrm{E} \Rightarrow \mathrm{A}),\ \neg \mathrm{A} \wedge \mathrm{I}\ \therefore\ \neg \mathrm{E}$

41. $\mathrm{A} \Rightarrow \mathrm{B},\ \mathrm{B} \Rightarrow \neg \mathrm{C},\ \mathrm{C} \Rightarrow \neg \mathrm{D},\ \mathrm{B} \Rightarrow \neg \mathrm{E},\ \neg \mathrm{D} \Rightarrow \mathrm{F},\ (\neg \mathrm{E}) \vee (\neg \mathrm{F})$

$\therefore\ \mathrm{E} \Rightarrow \neg(\mathrm{A} \vee \mathrm{C})$

43. $\mathrm{I} \Leftrightarrow (\mathrm{A} \vee \mathrm{C}),\ (\mathrm{M} \vee \mathrm{S}) \Rightarrow \mathrm{C},\ \mathrm{S} \wedge \neg \mathrm{A}\ \therefore\ \mathrm{C},\ \mathrm{I}$

Exercises 1.5

---

1. $\{ 2 \}$ 3. $\{\ldots, -8, -4, 0, 4, 8, \ldots\}$ 5. $\{ \ldots, 1, 2, 3, 4 \}$

7. $\{ 1, 2, 3, 4, \ldots \}$ 9. $\{ \ldots, -3, -2, -1 \}$

11. $(\forall x \in \mathrm{U})(\mathrm{I}(x) \Rightarrow \mathrm{E}(x))$ False 13. $(\forall x \in \mathrm{U})(\mathrm{I}(x) \Rightarrow \neg \mathrm{R}(x))$ False

15. $(\exists x \in \mathrm{U})(\mathrm{I}(x) \wedge \neg \mathrm{R}(x))$ True 17. $(\exists x \in \mathrm{U})(\mathrm{R}(x) \wedge \mathrm{I}(x))$ True

19. $(\forall x \in \mathrm{U})(\mathrm{R}(x) \Rightarrow \neg \mathrm{I}(x))$ False

21. Let N$*$ denote the negation of statement $*$.

N11. Some isosceles triangles are not equilateral triangles.

N12. Some equilateral triangles are not isosceles triangles.

N13. Some isosceles triangles are right triangles.

N14. No isosceles triangles are equilateral triangles.

N15. All isosceles triangles are right triangles.

N16. No right triangles are equilateral triangles.

N17. No right triangles are isosceles triangles.

N18. Some equilateral triangles are right triangles.

N19. Some right triangles are isosceles triangles.

N20. Some equilateral triangles are not isosceles triangles.

23. There exists a natural number x such that $2x > 0$. True

25. There exists a natural number x such that $2x \leq 0$. False

27. There exists a unique integer x such that $x^2 = 0$. True

29. There exists a unique integer x such that $x^2 = x$. False

31. There exists a unique real number x such that $x = \sqrt{7}$. True

33. $(\forall x \in \mathbf{N})((x \text{ is a prime}) \Rightarrow (x \text{ is odd}))$ False

35. $(\exists! x \in \mathbf{N})((x \text{ is even}) \wedge (x \text{ is a prime}))$ True

37. Let N$*$ denote the negation of statement $*$.

N22. $(\exists x \in \mathbf{R})(2x \leq 0)$

N23. $(\forall x \in \mathbf{N})(2x \leq 0)$

N24. $(\exists x \in \mathbf{N})((x \text{ is a prime}) \wedge (x \text{ is even}))$

N25. $(\forall x \in \mathbf{N})(2x > 0)$

N26. $(\exists x \in \mathbf{N})((x \text{ is a prime}) \wedge (x \neq 2) \wedge (x \text{ is even}))$

N27. $(\forall x \in \mathbf{Z})((x^2 \neq 0) \vee (\exists y \in \mathbf{Z})((y^2 = 0) \wedge (y \neq x)))$

N28. $(\forall x \in \mathbf{N})((x \text{ is not a prime}) \vee (x \text{ is even}) \vee (\exists y \in \mathbf{N})((y \text{ is a prime}) \wedge$ $(y \text{ is not odd}) \wedge (y \neq x)))$

N29. $(\forall x \in \mathbf{Z})((x^2 \neq x) \vee (\exists y \in \mathbf{Z})((y^2 = y) \wedge (y \neq x)))$

N30. $(\forall x \in \mathbf{R})((e^x \neq 1) \vee (\exists y \in \mathbf{Z})((e^y = 1) \wedge (y \neq x)))$

N31. $(\forall x \in \mathbf{R})((x \neq \sqrt{7}) \vee (\exists y \in \mathbf{Z})((y = \sqrt{7}) \wedge (y \neq x)))$

N33. Some primes are not odd.

N34. No primes are even.

N35. For all natural numbers either x is odd or x is not a prime or there exists a natural number $y \neq x$ which is an even prime.

N36. There does not exist a unique smallest natural number.

39. $(\exists m \in \mathbf{N})(\forall n \in \mathbf{N})(m > n)$ False

41. $(\forall m \in \mathbf{N})(\forall n \in \mathbf{N})(m > n)$ False

43. $(\forall x \in \mathbf{Z})(\exists y \in \mathbf{Z})(x = 2y)$ False

45. $(\exists! x \in \mathbf{Q})(\forall y \in \mathbf{Q})(x + y = 0)$ False

47. $(\forall x \in \mathbf{Q})(\exists! y \in \mathbf{Q})(x + y = 1)$ True

49. $(\forall x \in \mathbf{R})((x > 0) \Rightarrow (\exists n \in \mathbf{N})(\frac{1}{n} < x))$ True

51. $(\forall x \in \mathbf{R})(\forall y \in \mathbf{R})(\forall z \in \mathbf{R})((x < y) \Rightarrow (x + z < y + z))$ True

Review Exercises

Sentences 3 and 4 are statements.

7. The number π is not rational.

9. It is not cloudy and the sun is not shining.

11. $2 + 3 = 4$ and $5 + 6 \neq 7$

13. I finish my homework and it does not rain but I do not play tennis.

15. a. If you are old enough, then you may vote.
    b. If the gasoline engine is running, then there is fuel in the tank.
    c. If you may run for the United States Senate, then you are at least thirty-five years of age.
    d. If there is a garden, then there is rain.
    e. If a triangle is equilateral, then the triangle is isosceles.
    f. If yesterday is Thursday, then today is Friday.

17. a. Converse: If you may vote, then you are old enough.
    Inverse: If you are not old enough, then you may not vote.
    Contrapositive: If you may not vote, then you are not old enough.

    b. Converse: If there is fuel in the tank, then the gasoline engine is running.
    Inverse: If the gasoline engine is not running, then there is no fuel in the tank.
    Contrapositive: If there is no fuel in the tank, then the gasoline engine is not running.

    c. Converse: If you are at least thirty-five years of age, then you may run for the United States Senate.
    Inverse: If you may not run for the United States Senate, then you are not at least thirty-five years of age.
    Contrapositive: If you are not at least thirty-five years of age, then you may not run for the United States Senate.

    d. Converse: If there is rain, then there is a garden.
    Inverse: If there is no garden, then there is no rain.
    Contrapositive: If there is no rain, then there is no garden.

e. Converse: If the triangle is isosceles, then the triangle is equilateral.
Inverse: If a triangle is not equilateral, then the triangle is not isosceles.
Contrapositive: If the triangle is not isosceles, then the triangle is not equilateral.

f. Converse: If today is Friday, then yesterday is Thursday.
Inverse: If yesterday is not Thursday, then today is not Friday.
Contrapositive: If today is not Friday, then yesterday is not Thursday.

19. Arguments b and c are valid.

21. (a) $\neg P \vee \neg Q$ (b) $(P \vee Q) \wedge (P \vee R)$ (c) $\neg(P \vee Q)$
(d) $P \wedge (Q \vee R)$ (e) $P \wedge \neg Q$ (f) $P \wedge (P \vee Q)$

23. $(\forall x)(M(x) \Rightarrow \neg P(x))$

25. $(\forall x)(M(x) \Rightarrow \neg P(x))$

27. $(\exists x)(M(x) \wedge P(x))$

29. $(\forall x \in \mathbf{Z})(\exists! y \in \mathbf{Z})(x + y = 0)$

31. $(\exists! x \in \mathbf{R})(\forall y \in \mathbf{R})(x + y = x)$

Chapter 2

Exercises 2.1

---

1. No No No No

3. No Yes No No

5. No No $m = 1, n = 2, p = 3$

7. 1. Theorem 2.1; 2. Z11; 3. 1, 2, substitution; 4. Z7; 5. 3, 4, substitution; 6. Z11

9. 1. Z5; 2. Z3; 3. 1, 2, substitution; 4. Z3; 5. Z11; 6. Z3; 7. 5, 6, substitution; 8. 3, 7, substitution; 9. Z10; 10. 8, 9, substitution; 11. definition of subtraction

11. 1. Z7; 2. Z11; 3. 1, 2, substitution 4. Theorem 2.1; 5. 3, 4, substitution; 6. Z4 and Z9; 7. 5, 6, substitution; 8. Z4; 9. 7, 8, substitution; 10. Z11; 11. 9, 10, substitution; 12. Theorem 2.6.

Exercises 2.2.1

1. Assume m is an even integer. By definition of even, there exists an integer k such that $m = 2k$. Adding 1, we find $m + 1 = 2k + 1$. Hence, by definition $m + 1$ is odd.

3. Assume m is an odd integer. By the contrapositive of theorem 2.9, m^2 is an odd integer. By exercise 2, $m^2 + 1$ is an even integer.

5. Let m and n be even integers. Then there exist integers r and s such that $m = 2r$ and $n = 2s$. The integer $m + n = 2r + 2s = 2(r + s)$. Since $r + s$ is an integer, $m + n$ is even.

7. Assume m is an even integer and n is an odd integer. Then there exist integers k and ℓ such that $m = 2k$ and $n = 2\ell + 1$. Hence, $m + n = 2k + 2\ell + 1 = 2(k + \ell) + 1$. Since $k + \ell$ is an integer, $m + n$ is an odd integer.

9. We prove this theorem by proving its contrapositive—namely, if $m + 1$ is an odd integer, then m is an even integer. Since $m + 1$ is assumed to be odd, there exists an integer k such that $m + 1 = 2k + 1$. By a cancellation property of addition (Theorem 2.6), $m = 2k$ and m is even.

11. We prove this theorem by proving its contrapositive—namely, if neither m nor n is even (that is, if m and n are both odd), then mn is odd. Since m and n are both odd, there exist an integer r and s such that $m = 2r + 1$ and $n = 2s + 1$. Computing, we find

$$mn = (2r + 1)(2s + 1) = 4rs + 2r + 2s + 1 = 2(2rs + r + s) + 1.$$

Since the integers are closed under the operations of addition and multiplication, $2rs + r + s$ is an integer; and, therefore, mn is odd.

13. We prove this theorem by proving its contrapositive—namely, if it is not the case that either m and n are both odd integers or m and n are both even integers, then $m + n$ is an odd integer. Assume m is odd and n is even. Then there exist integers k and ℓ such that $m = 2k + 1$ and $n = 2\ell$. Hence, $m + n = 2k + 1 + 2\ell = 2(k + \ell) + 1$. Since $k + \ell$ is an integer, $m + n$ is an odd integer.

15. We prove this theorem by contradiction. Suppose mn is an odd integer and not both m and n are odd. Since m and n are not both odd, at least one is even. For definiteness, suppose m is even. Then there exists some integer k such that $m = 2k$. Multiplying, we find $mn = (2k)n = 2(kn)$. Hence, mn is even which contradicts the assumption that mn is odd.

17. We prove this theorem by contradiction. Suppose $m - n$ is an odd integer and $m + n$ is an even integer. Since $m - n$ is assumed to be odd, there exists an integer k such that (1) $m - n = 2k + 1$. And since $m + n$ is assumed to be even, there exists an integer ℓ such that (2) $m - n = 2\ell$. Adding equations (1) and (2), yields $(m - n) + (m + n) = (2k + 1) + 2\ell$. Hence, $2m = 2k + 1 + 2\ell = 2(k + \ell) + 1$ which is a contradiction, since $2m$ is an even integer but $2(k + \ell) + 1$ is an odd integer.

Exercises 2.2.2

---

1. **Proof:** We prove this theorem by considering two cases.
   Case 1. Assume n is an even integer. Thus, there exists an integer k such that $n = 2k$. Consequently, $n(n + 1) = 2k(2k + 1) = 2(k(2k + 1))$. Since $k(2k + 1)$ is an integer, $n(n + 1)$ is even.
   Case 2. Assume n is an odd integer. Hence, there exists an integer ℓ such that $n = 2\ell + 1$. Consequently, $n(n + 1) = (2\ell + 1)((2\ell + 1) + 1) = (2\ell + 1)(2\ell + 1) + 2\ell + 1 = 4\ell^2 + 4\ell + 1 + 2\ell + 1 = 2(2\ell^2 + 3\ell + 1)$. Since $2\ell^2 + 3\ell + 1$ is an integer, $n(n + 1)$ is even.

3. **Proof:** We prove this theorem by cases.
   Case 1. Assume m is an even integer. Thus, there exists an integer k such that $m = 2k$. Hence, $m^3 + m = (2k)^3 + 2k = 8k^3 + 2k = 2(4k^3 + k)$. Since $4k^3 + k$ is an integer, $m^3 + m$ is even.
   Case 2. Assume m is an odd integer. Then there exists an integer ℓ such that $m = 2\ell + 1$. Therefore, $m^3 + m = (2\ell + 1)^3 + 2\ell + 1 = 8\ell^3 + 12\ell^2 + 6\ell + 1 + 2\ell + 1 = 2(4\ell^3 + 6\ell^2 + 4\ell + 1)$. Since $4\ell^3 + 6\ell^2 + 4\ell + 1$ is an integer, $n(n + 1)$ is even.

5. **Proof:** Let m is an odd integer. By definition of odd, there exists an integer r such that $m = 2r + 1$. We prove this theorem by cases. First, we assume r is an even integer, and then we assume r is an odd integer.
   Case 1. If r is an even integer, then there exists an integer j such that $r = 2j$. Thus, for r even, $m = 2r + 1 = 2(2j) + 1 = 4j + 1$ for some integer j.
   Case 2. If r is an odd integer, then there exists an integer s such that $r = 2s + 1$. Hence, for r odd, $m = 2r + 1 = 2(2s + 1) + 1 = 4s + 2 + 1 = 4(s + 1) - 1$. Since s is an integer, $j = s + 1$ is an integer, and for r odd, $m = 4j - 1$.

7. **Proof:** First, we prove if m^3 is an odd integer, then m is an odd integer. We prove this statement by proving its contrapositive—namely, if m is even, then m^3 even. Thus, we assume m is even. Hence, there is an integer k such that $m = 2k$. Consequently, $m^3 = (2k)^3 = 2(4k^3)$ and m^3

is even. Next, we prove if m is odd, then m^3 is odd. Suppose $m = 2\ell+1$ where ℓ is an integer. Then $m^3 = (2\ell+1)^3 = 8\ell^3 + 12\ell^2 + 6\ell + 1 = 2(4\ell^3 + 6\ell^2 + 3\ell) + 1$. Since $4\ell^3 + 6\ell^2 + 3\ell$ is an integer, m^3 is odd.

9. **Proof:** We prove this theorem by cases.
   Case 1. Suppose m is an odd integer. Then there exists an integer k such that $m = 2k+1$. Computing, we find $7m - 4 = 7(2k+1) - 4 = 2(7k+1) + 1$. Since $7k+1$ is an integer, $7m-4$ is odd. We also find, $5m + 3 = 5(2k+1) + 3 = 2(5k+4)$. Since $5k+4$ is an integer, $5m+3$ is even. Hence, if m is an odd integer, then $7m-4$ is odd if and only if $5m+3$ is an even integer.
   Case 2. Suppose m is an even integer. Then there exists an integer ℓ such that $m = 2\ell$. Computing, we find $7m - 4 = 7(2\ell) = 2(7\ell)$. Hence, $7k - 4$ is an even integer. Also, $5m + 3 = 5(2\ell) + 3 = 2(5\ell + 2) + 1$. Thus, $5m+3$ is odd. Consequently, if m is an even integer, then $7m-4$ is odd if and only if $5m+3$ is an even integer.

11. **Proof:** Suppose to the contrary that (1) $111 = p + q + r$, where p, q, and r are even integers. Then there exist integers j, k, and ℓ such that $p = 2j$, $q = 2k$ and $r = 2\ell$. Substituting into (1), we find $111 = 2j + 2k + 2\ell = 2(j + k + \ell)$. Since $j + k + \ell$ is an integer, 111 is an even integer. This contradicts the fact that 111 is odd, because $111 = 2(55) + 1$.

13. **Proof:** Suppose to the contrary that there exist integers m and n such that $4m + 6n = 9$. Using the distributive property of multiplication, we see that $9 = 2(2m + 3n)$. Since $2m + 3n$ is an integer, 9 is even. However, this contradicts the fact that 9 is odd, because $9 = 2(4) + 1$.

15. $n = 3$ 17. $m = -3$, $n = 1$ 19. $x = 0$ or $x = 1$

21. **Proof:** Suppose there exist two distinct multiplicative identities 1 and 1^* for the set of integers. Thus, for all integers m, (1) $1m = m$ and (2) $1^*m = m$. Letting $m = 1^*$ in equation (1), we obtain (3) $1(1^*) = 1^*$, and letting $m = 1$ in equation (2) yields (4) $(1^*)1 = 1$. By the commutative property of multiplication $1(1^*) = (1^*)1$. Hence, from (3) and (4), $1^* = 1(1^*) = (1^*)1 = 1$. Consequently, the multiplicative identity for the set of integers is unique.

23. **Proof:** Let x be any real number. Adding y to both sides of the equation $x^2y = x - y$, we obtain (1) $x^2y + y = x$ or $(x^2+1)y = x$. Since $x^2 + 1 \neq 0$, a solution of (1) is $y = x/(x^2+1)$. Suppose there exists a distinct second solution y^*. That is, suppose (2) $x^2y^* + y^* = x$. From (1) and (2) it follows that $x^2y^* + y^* = x^2y + y$ or $(x^2+1)y^* = (x^2+1)y$. Since $x^2 + 1 \neq 0$ by the left-hand cancellation property of multiplication for real numbers, $y^* = y$.

25. **Proof:** Our proof is by cases.

    Case 1. Let n be any even integer. Then there exists an integer k such that $n = 2k$. The integers $k - 1$ and $k + 1$ are unequal and $(k - 1) + (k + 1) = 2k = n$.

    Case 2. Let n be any odd integer. Then there exists an integer ℓ such that $n = 2\ell + 1$. The integers ℓ and $\ell + 1$ are unequal and $\ell + (\ell + 1) = 2\ell + 1 = n$.

27. **Proof:** Let n and m be any integers. Since the integers are closed under the operation of subtraction, $k = n - m$ is an integer and $m + k = m + (n - m) = n$.

29. **Proof:** We prove this theorem by contradiction. Let m be any odd integer and suppose (1) $m = p + q + r$ where p, q, and r are even integers. Thus, there exist integers i, j, and k such that $p = 2i$, $q = 2j$, and $r = 2k$. Substituting into equation (1), we find $m = 2i + 2j + 2k = 2(i + j + k)$. Consequently, m is even which contradicts our assumption that m is odd.

31. **Proof:** We prove this theorem by cases.

    Case 1. Let m be any even integer. Then there exists an integer k such that $m = 2k$. Hence, $m^2 = 4k^2$ and 4 divides m^2. That is, for m even, m^2 divided by 4 has remainder 0.

    Case 2. Let m be any odd integer. Then there exists an integer ℓ such that $m = 2\ell + 1$. Hence,

    $$m^2 = (2\ell + 1)^2 = 4\ell^2 + 4\ell + 1 = 4(\ell^2 + \ell) + 1.$$

    Thus, when m^2 is divided by 4, the remainder is 1.

    Consequently, when m is any integer and m^2 is divided by 4, the remainder is 0 or 1.

Exercises 2.2.3

---

1. **Counterexample:** Let $a = 1$, $b = 3$, and $c = 4$. Then a is odd, $a + b = c$, and b is odd or c is even.

3. **Counterexample:** Let $a = 2$ and $b = -2$. Then in the set of integers a divides b, since $a(-1) = b$; b divides a, since $b(-1) = a$; and $a \neq b$.

5. **Counterexample:** Let $a = 6$, $b = 3$, and $c = 4$. Then $a = 6$ divides $bc = 12$, but 6 does not divide $b = 3$ and 6 does not divide $c = 4$.

7. **Proof:** Suppose a divides $b-c$ and a divides $c-d$. Then there exist natural numbers k and ℓ such that $ak = b-c$ and $a\ell = c-d$. Adding these two equations, we find $ak+a\ell = (b-c)+(c-d)$ or $a(k+\ell) = b-d$. Since $k+\ell$ is a natural number a divides $b-d$.

9. **Counterexample:** $2+3+4=9$.

Exercises 2.2.4

---

1. a. $14 \cdot 12 = 168 = 8 \cdot 21$ b. $6 \cdot 35 = 210 = 14 \cdot 15$ c. $6 \cdot 35 = 15 \cdot 12$

3. **Proof:** Let $\frac{a}{b}$ and $\frac{c}{d}$ be any rational numbers. By definition of addition for the rational numbers, the commutative property of multiplication for the integers, and the commutative property of addition for the integers,

$$\frac{a}{b} \oplus \frac{c}{d} = \frac{a \cdot d + b \cdot c}{b \cdot d} = \frac{d \cdot a + c \cdot b}{d \cdot b} = \frac{c \cdot b + d \cdot a}{d \cdot b} = \frac{c}{d} \oplus \frac{a}{b}$$

5. **Proof:** Let $\frac{a}{b}$, $\frac{c}{d}$, and $\frac{e}{f}$ be any rational numbers. By the definition of addition for the rational numbers, by the definition of multiplication for the rational numbers, by the left and right distributive property of multiplication over addition for the integers, by the associative and commutative properties of multiplication for the integers, and by the definition of addition and multiplication for the rational numbers, we have

$$\frac{a}{b} \odot \left(\frac{c}{d} \oplus \frac{e}{f}\right) = \frac{a}{b} \odot \left(\frac{c \cdot f + d \cdot e}{d \cdot f}\right) = \frac{a \cdot (c \cdot f + d \cdot e)}{b \cdot (d \cdot f)}$$

$$= \frac{a \cdot c \cdot f + a \cdot d \cdot e}{b \cdot d \cdot f} \odot \frac{b}{b} = \frac{(a \cdot c) \cdot (b \cdot f) + (b \cdot d) \cdot (a \cdot e)}{(b \cdot d) \cdot (b \cdot f)}$$

$$= \left(\frac{a \cdot c}{b \cdot d}\right) \oplus \left(\frac{a \cdot e}{b \cdot f}\right) = \left(\frac{a}{b} \odot \frac{c}{d}\right) \oplus \left(\frac{a}{b} \odot \frac{e}{f}\right)$$

7. **Proof:** Assume to the contrary that there exists a rational number $p = a/b$ such that a and b have no common factors and $p^2 = (a/b)^2 = a^2/b^2 = 3$. Multiplying the last equation by b^2, we have (1) $a^2 = 3b^2$. Hence, 3 divides a^2. Either 3 divides a or 3 does not divide a. Assume 3 does not divide a, then by the unique factorization theorem (The Fundamental Theorem of Arithmetic) $a = \pm 2^{n_2} 3^0 5^{n_5} \cdots$ where $n_2, n_5, \cdots$ are whole numbers. Hence $a^2 = 2^{2n_2} 3^0 5^{2n_5} \cdots$—that is, 3 does not divide

a^2. Consequently, 3 divides a. Therefore, there exists an integer k such that $a = 3k$. Squaring, we have (2) $a^2 = 9k^2$. Substituting (2) into (1), yields $9k^2 = 3b^2$. Dividing this equation by 3, we find $3k^2 = b^2$. Hence, 3 divides b^2 and by the argument above 3 divides b. Consequently, a and b have the factor 3 in common which is a contradiction.

9. **Proof:** Assume to the contrary that (1) $a = \sqrt{3} + \sqrt{5}$ is rational. Since a is rational, $a \cdot a = a^2$ is rational. Subtracting $\sqrt{3}$ from equation (1), we obtain (2) $a - \sqrt{3} = \sqrt{5}$. Squaring (2), yields $(a - \sqrt{3})^2 = (\sqrt{5})^2$ or $a^2 - 2\sqrt{3} + 3 = 5$. Solving for a^2, yields $a^2 = 2\sqrt{3} + 2$. Thus, a^2 is irrational which is a contradiction.

11. **Proof:** We prove the contrapositive: "If $x \neq 0$ and $y \neq 0$, then $xy \neq 0$." by contradiction. Hence, we assume $x \neq 0$ and $y \neq 0$ and $xy = 0$. Since every nonzero real number x has a multiplicative inverse x^{-1} with the property that $x^{-1}x = 1$, multiplication of the equation $xy = 0$ by x^{-1} results in $x^{-1}(xy) = (x^{-1}x)y = 1y = y = x^{-1}0 = 0$. That is, $y = 0$ which contradicts the hypothesis $y \neq 0$.

13. **Proof:** We prove this theorem by contradiction. Suppose x is irrational, y is rational, and $x + y$ is rational. Since the rational numbers are closed under subtraction, $(x + y) - y = x + (y - y) = x + 0 = x$ is rational. This contradicts the hypothesis that x is irrational.

15. **Proof:** This theorem is the contrapositive of property Q1.

17. **Proof:** We prove this theorem by contradiction. Hence, we assume x is irrational, $y \neq 0$ is rational, and xy is rational. Since y is rational and $y \neq 0$, there is a rational number y^{-1} with the property that $yy^{-1} = 1$. Because the rational numbers are closed under multiplication, $(xy)y^{-1} = x(yy^{-1}) = x1 = x$ is a rational number. This contradicts the hypothesis x is irrational.

19. **Proof:** Let x be any rational number, let $y = x/2 + \sqrt{2}$ which is irrational, and let $z = x/2 - \sqrt{2}$ which is irrational. Then $x = y + z$.

Review Exercises

---

1. 1. = is reflexive 2. Definition of Subtraction 3. Substitution of 2 into 1 4. Z7 5. Substitution of 4 into 3 6. Theorem 2.3 7. Substitution of 6 into 5 8. Definition of subtraction 9. Substitution of 8 into 7

3. a. **Proof:** We prove this theorem by proving its contrapositive. Let m be an even integer. Then there exists an integer k such that $m = 2k$. Squaring, we obtain $m^2 = (2k)^2 = 2(2k^2)$. Thus, m^2 is an even integer. Consequently, we have proven the theorem: "If m is an even integer, then m^2 is an even integer." Therefore, its contrapositive is a theorem also.

   b. **Proof:** We prove this theorem by proving its contrapositive—namely, "If either m is even and n is even or m is odd and n is odd, then $m - n$ is even." We prove this result by cases.

   Case 1. Assume m and n are both even integers. Then there exist integers k and ℓ such that $m = 2k$ and $n = 2\ell$. Subtraction results in $m - n = 2k - 2\ell = 2(k - \ell)$. Since $k - \ell$ is an integer, $m - n$ is an even integer.

   Case 2. Assume m and n are both odd integers. Then there exist integers k and ℓ such that $m = 2k + 1$ and $n = 2\ell + 1$. Subtraction results in $m - n = (2k + 1) - (2\ell + 1) = 2(k - \ell)$. Since $k - \ell$ is an integer, $m - n$ is an even integer.

5. a. **Proof:** We prove this theorem by cases.

   Case 1. Assume m is an even integer. Then there exists an integer k such that $m = 2k$. Computing, we find $m(m + 3) = (2k)(2k + 3) = 4k^2 + 6k = 2(2k^2 + 3k)$. Because $2k^2 + 3k$ is an integer, $m(m + 3)$ is an even integer.

   Case 2. Assume m is an odd integer. Then there exists an integer ℓ such that $m = 2\ell + 1$. Computing, we find $m(m+3) = (2\ell+1)((2\ell+1)+3) = (2\ell+1)(2\ell+1) + 3(2\ell+1) = 4\ell^2 + 4\ell + 1 + 6\ell + 3 = 4\ell^2 + 10\ell + +4 = 2(2\ell^2 + 5\ell + 2)$. Because $2\ell^2 + 5\ell + 2$ is an integer, $m(m + 3)$ is an even integer.

7. a. $n = 1$ b. $m = 2, n = 1$ c. $k = 1, m = 2, n = 0$

   d. For $x > 2$, the real number $y = -x/(x - 2)$ is negative, because the numerator is negative and the denominator is positive. The following computation proves y satisfies the equation (1) $x + xy - 2y = 0$:

$$x + xy - 2y = x + (x - 2)y = x + (x - 2)[-x/(x - 2)] = x - x = 0$$

   To prove y is the unique solution of $x + xy - 2y = 0$, suppose $y^* \neq y$ satisfies (2) $x + xy^* - 2y^* = 0$. It follows from (1) and (2) that $x + xy - 2y = x + xy^* - 2y^*$. Subtracting x and factoring, we obtain $(x - 2)y = (x - 2)y^*$. Because, $x - 2 \neq 0$, $y = y^*$ and the solution of (1) is unique.

9. a

Chapter 3

Exercises 3.1

1. a., d., e. 3. a., d., f., g., h., i., j., m., o., q., r., t.

5. **Proof:** (i) let $x \in C$. Factoring, we find $x^2 - 1 = (x-1)(x+1) = 0$. Hence, the elements of C are the integers -1 and 1. Since $|-1| = 1$, $1 \in D$ and since $|1| = 1$, $1 \in D$. Therefore, $C \subseteq D$.

   (ii) Let $n \in D$. Since n is an integer and $|n| = 1$, $n = -1$ or $n = 1$. Substituting the real number $x = -1$ into the expression $x^2 - 1$, we see $(-1)^2 - 1 = 1 - 1 = 0$, so $-1 \in C$. Likewise, substituting $x = 1$ into $x^2 - 1$, yields $1^2 - 1 = 1 - 1 = 0$, so $1 \in C$. Hence, $D \subseteq C$.

   Since $C \subseteq D$ and $D \subseteq C$, $C = D$.

7.

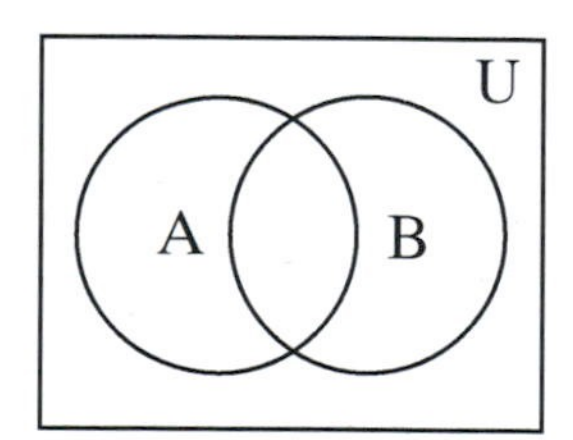

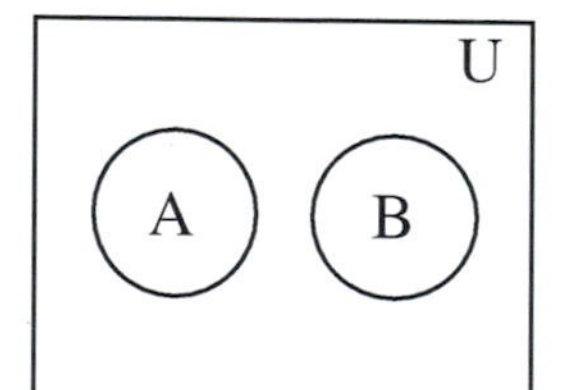

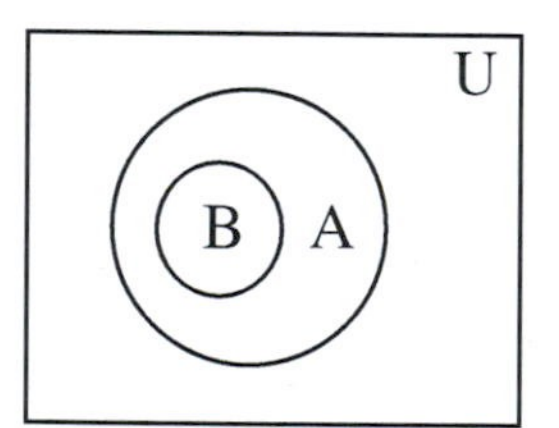

9. $2^n - 1$, $2^n - 1$, $2^n - 2$

11. a. $\emptyset$ b. $\emptyset$, $\{1\}$, $\{2\}$ c. None d. $\emptyset$

13. a. $A = \{1\}$, $B = \{1,2\}$, $C = \{1,3\}$ b. $A = \{1\}$, $B = \{1,2\}$, $C = \{3\}$

    c. $A = \{1\}$, $B = \{1\}$, $C = \{1\}$ d. $A = \{1\}$, $B = \{2\}$, $C = \{1,3\}$

    e. $A = \{1\}$, $B = \{\{1\}\}$, $C = \{1\}$

    f. $A = \{1\}$, $B = \{\{1\},1\}$, $C = \{\{1\},2\}$

Exercises 3.2

1. a. $\{1, 2, 4\}$ b. $\{2, a, e\}$ c. $\{a, e, \emptyset, 1, 4\}$
d. $\{1, 2, 4, a, e\}$ e. $\{\emptyset\}$ f. $\emptyset$
g. $\{2\}$ h. $\{1, 2, 4, a, e\}$ i. $\emptyset$
j. $\{1, 2, 4, a, e, \emptyset\} = U$ k. $\{\emptyset, 2, 4\}$ l. $\{\emptyset\}$
m. $\{a, e, \emptyset\} = A$ n. $\emptyset$

3. a. B' b. F' c. $B \cap F$ d. $B \cap F'$ e. $F \cap B'$ f. $B' \cap F' \cap T'$
g. $B \cap F \cap T$

5. **Proof:** Let A be any set and let $x \in A \cap U$. By definition, $x \in A$ and $x \in U$. Therefore, (1) $A \cap U \subseteq A$.

Let $x \in A$. Since U is the universe, $x \in U$. Hence, $x \in A \cap U$ and, therefore, (2) $A \subseteq A \cap U$. It follows from (1) and (2) that $A \cap U = A$.

7. **Proof:** Let A be any set of the universe U. By the definition of complement,

$$A' = \{x \in U \mid x \notin A\} = \{x \in U \mid \neg(x \in A)\}$$

and by double negation

$$(A')' = \{x \in U \mid \neg(\neg(x \in A))\} = \{x \in U \mid x \in A)\} = A.$$

9. **Proof:** Let $x \in A$. Then $x \in A$ or $x \in B$. That is, $x \in A \cup B$. Hence, $A \subseteq A \cup B$.

11. **Proof:** First, assume $A \subseteq B$ and suppose $x \in A \cup B$. Hence, $x \in A$ or $x \in B$. If $x \in A$, then $x \in B$, since $A \subseteq B$. In either case, $x \in B$ and (1) $A \cup B \subseteq B$. By exercise 9, (2) $B \subseteq A \cup B$. Consequently from (1) and (2), if $A \subseteq B$, then $A \cup B = B$.

Next, assume $A \cup B = B$. By exercise 9, $A \subseteq A \cup B = B$. Hence, if $A \cup B = B$, then $A \subseteq B$.

13. **Proof:** Assume $A \subseteq B$. Let $x \in B'$. Then $x \notin B$. By exercise 8. b. of exercises 3.1, we have $x \notin A$ and, therefore, $x \in A'$. Consequently, if $A \subseteq B$, then $B' \subseteq A'$.

Now assume $B' \subseteq A'$. By the first part of this proof, $(A')' \subseteq (B')'$. And by exercise 7, $(A')' = A$ and $(B')' = B$. Hence, if $B' \subseteq A'$, then $A \subseteq B$.

Consequently, $A \subseteq B \Leftrightarrow B' \subseteq A'$.

15. **Proof:** Assume $A = B$. Then $A \cup B = B \cup B = B$ and $A \cap B = B \cap B = B$. Therefore, if $A = B$, then $A \cup B = A \cap B$.

    Next, assume $A \cup B = A \cap B$. First, suppose $x \in A$. Then by definition, $x \in A \cup B$ and since $A \cup B = A \cap B$, $x \in B$. Hence, (1) $A \subseteq B$. Likewise, $x \in B$ implies $x \in A$; therefore, (2) $B \subseteq A$. From (1) and (2), it follows that $A = B$.

17. a. **Proof:** Let A be any set and assume $B \subseteq C$. Suppose $x \in A \cup B$. Then $x \in A$ or $x \in B$.

    (1) If $x \in A$, then $x \in A \cup C$.

    (2) If $x \in B$, then $x \in C$ since $B \subseteq C$. Hence, $x \in A \cup C$.

    Consequently, if A is any set and $B \subseteq C$, then $A \cup B \subseteq A \cup C$.

    b. **Proof:** Let A be any set and assume $B \subseteq C$. Suppose $x \in A \cap B$. Then $x \in A$ and $x \in B$. Because $x \in B$ and $B \subseteq C$, $x \in C$. Hence, $x \in A \cap C$. Consequently, $A \cap B \subseteq A \cap C$.

19. **Proof:** Assume $A \subseteq C$ and $B \subseteq D$ and let $x \in A \cup B$. Either (1) $x \in A$ or (2) $x \in B$.

    Case 1. If $x \in A$, then since $A \subseteq C$, $x \in C$ and $x \in C \cup D$.
    Case 2. If $x \in B$, then since $B \subseteq D$, $x \in D$ and $x \in C \cup D$.

    In either case, $A \cup B \subseteq C \cup D$.

21. **Proof:** Let $x \in B$. Since $A \cap B = \emptyset$, $x \notin A$. Thus, $x \in A'$. Since $C \subseteq A$ and $x \in A'$ by exercise 8. b. of exercises 3.1, $x \notin C$. Because $x \in B$, $x \in A \cup B$ and because $A \cup B \subseteq C \cup D$, $x \in C \cup D$. Since $x \in C \cup D$ and $x \notin C$, $x \in D$. Thus, $B \subseteq D$.

Exercises 3.3

---

1. **Proof:** By theorem 3.5, $X - Y = X \cap Y'$. Thus,

$$(A \cap B) - C = (A \cap B) \cap C' = A \cap (B \cap C') = A \cap (B - C).$$

3. **Proof:** Taking the intersection of $A - B$ and $A \cap B$, we find

$$(A - B) \cup (A \cap B) = (A \cap B') \cap (A \cap B) = A \cap (B' \cap B) = A \cap \emptyset = \emptyset.$$

   Hence, the sets $A - B$ and $A \cap B$ are disjoint.

5. **Proof:** Assume $A \cap B = \emptyset$. By exercise 2, $A = (A - B) \cup (A \cap B)$. Since $A \cap B = \emptyset$, $A = A - B$. Likewise, $B = (B - A) \cup (B \cap A)$ and since $B \cap A = A \cap B = \emptyset$, $B = B - A$.

7. **Proof:** By theorem 3.5, $X - Y = X \cap Y'$. Thus,

$$\begin{aligned}(A - C) - (B - C) &= (A \cap C') - (B \cap C') = (A \cap C') \cap (B \cap C')' \\ &= (A \cap C') \cap (B' \cup C) \\ &= [(A \cap C') \cap B'] \cup [(A \cap C') \cap C] \\ &= [A \cap (C' \cap B')] \cup [A \cap (C' \cap C)] \\ &= [A \cap (B' \cap C')] \cup [A \cap \emptyset] \\ &= [(A \cap B') \cap C')] \cup \emptyset \\ &= (A \cap B') - C = (A - B) - C\end{aligned}$$

9. a. **Proof:** $A \triangle B = (A - B) \cup (B - A) = (B - A) \cup (A - B) = B \triangle A$.

b. **Proof:** Let $x \in (A \triangle B) \triangle C = [(A \triangle B) - C] \cup [C - (A \triangle B)]$. Then (1) $x \in [(A \triangle B) - C]$ or (2) $x \in [C - (A \triangle B)]$. We consider these two cases separately.

Case 1. Suppose $x \in (A \triangle B) - C$. Then $x \in A \triangle B = (A - B) \cup (B - A)$ and $x \notin C$. There are two subcases to consider.

Case 1. a. Suppose $x \in A - B$. Then $x \in A$ and $x \notin B$. Since $x \in A$ and $x \notin B$ and $x \notin C$, $x \notin B \triangle C$ but $x \in [(A - (B \triangle C)]$; and, therefore, $x \in A \triangle (B \triangle C) = [A - (B \triangle C)] \cup [(B \triangle C) - A]$. Thus, in case 1. a., $(A \triangle B) \triangle C \subseteq A \triangle (B \triangle C)$.

Case 1. b. Suppose $x \in B - A$. Then $x \in B$ and $x \notin A$. Since $x \in B$ and $x \notin C$, $x \in B - C$ and $x \in B \triangle C = (B - C) \cup (C - B)$. Furthermore, since $x \notin A$, $x \in (B \triangle C) - A$ and $x \in A \triangle (B \triangle C) = [A - (B \triangle C)] \cup [(B \triangle C) - A]$. Hence, in case 2. b., $(A \triangle B) \triangle C \subseteq A \triangle (B \triangle C)$.

Case 2. Suppose $x \in C - (A \triangle B)$. Then $x \in C$ and $x \notin A \triangle B = (A - B) \cup (B - A)$. There are two subcases to consider. Either a. $x \in A \cap B$ or b. $x \in (A \cup B)' = A' \cap B'$.

Case 2. a. Suppose $x \in A \cap B$. Then $x \in A$ and $x \in B$. Since $x \in B$ and $x \in C$, $x \notin B \triangle C = (B - C) \cup (C - B)$. In addition, since $x \in A$, $x \in [A - (B \triangle C)]$ and $x \in [A - (B \triangle C)] \cup [(B \triangle C) - A] = A \triangle (B \triangle C)$. Thus, in case 2. a., $(A \triangle B) \triangle C \subseteq A \triangle (B \triangle C)$.

Case 2. b. Suppose $x \in A' \cap B'$. Then $x \in A'$ and $x \in B'$. Thus, $x \notin A$ and $x \notin B$. Since $x \in C$ and $x \notin B$, $x \in (C - B)$ and $x \in (B - C) \cup (C - B) = B \triangle C$. Since $x \notin A$, $x \in (B \triangle C) - A$ and $x \in [A - (B \triangle C)] \cup [(B \triangle C) - A] = A \triangle (B \triangle C)$. Hence, in case 2. b., $(A \triangle B) \triangle C \subseteq A \triangle (B \triangle C)$.

Consequently, by cases $(A \triangle B) \triangle C \subseteq A \triangle (B \triangle C)$.

To complete the proof of this theorem, one proves by cases that $A \triangle (B \triangle C) \subseteq (A \triangle B) \triangle C$.

c. **Proof:** Let A be any set. By definition,

$$A \triangle \emptyset = (A - \emptyset) \cup (\emptyset - A) = A \cup \emptyset = A.$$

d. **Proof:** Let $x \in A \cap (B \triangle C)$. Then $x \in A$ and $x \in B \triangle C = (B - C) \cup (C - B)$. Thus, $x \in A$ and (1) $x \in B - C = B \cap C'$ or (2) $x \in C - B = C \cap B'$. We consider these two cases separately.

Case 1. Suppose $x \in A$, $x \in B$, and $x \notin C$. Then $x \in A \cap B$ and $x \notin A \cap C$. Hence, $x \in [(A \cap B) - (A \cap C)]$; and, therefore, $x \in [(A \cap B) - (A \cap C)] \cup [(A \cap C) - (A \cap B)] = (A \cap B) \triangle (A \cap C)$.

Case 2. Suppose $x \in A$, $x \in C$, and $x \notin B'$. Then $x \in A \cap C$ and $x \notin A \cap B$. Hence, $x \in [(A \cap C) - (A \cap B)]$; and, therefore, $x \in [(A \cap B) - (A \cap C)] \cup [(A \cap C) - (A \cap B)] = (A \cap B) \triangle (A \cap C)$.

Consequently, $A \cap (B \triangle C) \subseteq (A \cap B) \triangle (A \cap C)$.

Next, let $x \in (A \cap B) \triangle (A \cap C)) = [(A \cap B) - (A \cap C)] \cup [(A \cap C) - (A \cap B)]$. There are two cases to consider.

Case 1. Suppose $x \in [(A \cap B) - (A \cap C)]$. Then $x \in A \cap B$ but $x \notin A \cap C$. Since $x \in A \cap B$, $x \in A$ and $x \in B$. Since $x \notin A \cap C$ and $x \in A$, $x \notin C$. Since $x \in B$ and $x \notin C$, $x \in B - C$; and, therefore, $x \in B \triangle C = (B - C) \cup (C - B)$. Furthermore, since $x \in A$, $x \in A \cap (B \triangle C)$. Hence, in this case, $(A \cap B) \triangle (A \cap C) \subseteq A \cap (B \triangle C)$.

Case 2. Suppose $x \in [(A \cap C) - (A \cap B)]$. Then $x \in A \cap C$ and $x \notin A \cap B$. Since $x \in A \cap C$, $x \in A$ and $x \in C$. Since $x \in A$ and $x \notin A \cap B$, $x \notin B$. Since $x \in C$ and $x \notin B$, $x \in C - B$; and, therefore, $x \in B \triangle C = (B - C) \cup (C - B)$. Thus, in this case, $(A \cap B) \triangle (A \cap C) \subseteq A \cap (B \triangle C)$.

Consequently, $(A \cap B) \triangle (A \cap C) \subseteq A \cap (B \triangle C)$.

In conclusion, since both $A \cap (B \triangle C) \subseteq (A \cap B) \triangle (A \cap C)$ and $(A \cap B) \triangle (A \cap C) \subseteq A \cap (B \triangle C)$, $A \cap (B \triangle C) = (A \cap B) \triangle (A \cap C)$.

e. From the definition of symmetric difference, we find

$$(1) \qquad A \triangle A = (A - A) \cup (A - A) = \emptyset \cup \emptyset = \emptyset.$$

Then it follows from parts b., (1), a., and c. that

$$A \triangle (A \triangle B) = (A \triangle A) \triangle B = \emptyset \triangle B = B \triangle \emptyset = B.$$

Hence, $X = A \triangle B$ is the solution of the equation $A \triangle X = B$.

13. **Proof:** Assume A and B are nonempty sets and assume $A \times B = B \times A$. Let $a \in A$ and let $b \in B$. Then $(a, b) \in A \times B = B \times A$. Hence, $a \in B$ and $b \in A$. Thus $A \subseteq B$ and $B \subseteq A$. Consequently, $A = B$.

Next, assume $A = B$. Then by substitution of B for A and A for B, $A \times B = B \times A$.

15. a. The given statement is false. Let $A = \{a\}$, $B = \{b\}$, and $C = \emptyset$. Then $A \times C = B \times C = \emptyset$ but $A \neq B$.

b. **Proof:** Let A and B be any sets and let $C \neq \emptyset$.

Case 1. If $A = \emptyset$, then $A \times C = \emptyset \times C = \emptyset$. By hypothesis $A \times B = B \times C$. So, $B \times C = \emptyset$. Since $C \neq \emptyset$, $B = \emptyset$ and $A = B$.

Case 2. If $B = \emptyset$, then $A \times C = B \times C = \emptyset \times C = \emptyset$. Since $C \neq \emptyset$, $A = \emptyset$ and $A = B$.

Case 3. Suppose A and B are nonempty sets. Let $a \in A$ and $c \in C$. Then $(a, c) \in A \times C = B \times C$. Hence, $a \in B$ and $A \subseteq B$. Next, let $b \in B$ and $d \in C$. Then $(b, d) \in B \times C = A \times C$. Hence, $b \in A$ and $B \subseteq A$. Because $A \subseteq B$ and $B \subseteq A$, $A = B$.

17. **Proof:** We prove this theorem by cases.

Case 1. Suppose $A = \emptyset$. Then $A \times (B \cap C) = \emptyset \times (B \cap C) = \emptyset$, $(A \times B) \cap (A \times C) = (\emptyset \times B) \cap (\emptyset \times C) = \emptyset \cap \emptyset = \emptyset$, and consequently $A \times (B \cap C) = (A \times B) \cap (A \times C)$.

Case 2. Suppose $B = \emptyset$. Then $A \times (B \cap C) = A \times (\emptyset \cap C) = A \times \emptyset = \emptyset$, $(A \times B) \cap (A \times C) = (A \times \emptyset) \cap (A \times C) = \emptyset \cap (A \times C) = \emptyset$, and consequently $A \times (B \cap C) = (A \times B) \cap (A \times C)$.

Case 3. Suppose $C = \emptyset$. Then $A \times (B \cap C) = A \times (B \cap \emptyset) = A \times \emptyset = \emptyset$, $(A \times B) \cap (A \times C) = (A \times B) \cap (A \times \emptyset) = (A \times B) \cap \emptyset = \emptyset$, and consequently $A \times (B \cap C) = (A \times B) \cap (A \times C)$.

Case 4. Assume $A \neq \emptyset$, $B \neq \emptyset$, and $C \neq \emptyset$. Then

$$\begin{aligned}
(a, b) \in A \times (B \cap C) &\Leftrightarrow (a \in A) \wedge [b \in (B \cap C)] \\
&\Leftrightarrow (a \in A) \wedge [(b \in B) \wedge (b \in C)] \\
&\Leftrightarrow [(a \in A) \wedge (b \in B)] \wedge [(a \in A) \wedge (b \in C)] \\
&\Leftrightarrow [(a, b) \in (A \times B)] \wedge [(a, b) \in (A \times C)] \\
&\Leftrightarrow (a, b) \in [(A \times B) \cap (A \times C)]
\end{aligned}$$

Therefore, $A \times (B \cap C) = (A \times B) \cap (A \times C)$.

19. **Proof:** We prove this theorem by cases.

Case 1. Suppose $A = \emptyset$, then $(A \times B) \cap (C \times D) = (\emptyset \times B) \cap (C \times D) = \emptyset \cap (C \times D) = \emptyset$ and $(A \cap C) \times (B \cap D) = (\emptyset \cap C) \times (B \cap D) = \emptyset \times (B \cap D) = \emptyset$. Hence, $(A \times B) \cap (C \times D) = (A \cap C) \times (B \cap D)$.

Case 2. Suppose $B = \emptyset$, then $(A \times B) \cap (C \times D) = (A \times \emptyset) \cap (C \times D) = \emptyset \cap (C \times D) = \emptyset$ and $(A \cap C) \times (B \cap D) = (A \cap C) \times (\emptyset \cap D) = (A \cap C) \times \emptyset = \emptyset$. Hence, $(A \times B) \cap (C \times D) = (A \cap C) \times (B \cup D)$.

Case 3. Suppose $C = \emptyset$, then $(A \times B) \cap (C \times D) = (A \times B) \cap (\emptyset \times D) = (A \times B) \cap \emptyset = \emptyset$ and $(A \cap C) \times (B \cap D) = (A \cap \emptyset) \times (B \cap D) = \emptyset \times (B \cup D) = \emptyset$. Hence, $(A \times B) \cap (C \times D) = (A \cap C) \times (B \cap D)$.

Case 4. Suppose $D = \emptyset$, then $(A \times B) \cap (C \times D) = (A \times B) \cap (C \times \emptyset) = (A \times B) \cap \emptyset = \emptyset$ and $(A \cap C) \times (B \cap D) = (A \cap C) \times (B \cap \emptyset) = (A \cap C) \times \emptyset = \emptyset$. Hence, $(A \times B) \cap (C \times D) = (A \cap C) \times (B \cap D)$.

Case 5. Assume $A \neq \emptyset$, $B \neq \emptyset$, $C \neq \emptyset$, and $D \neq \emptyset$. Then

$$\begin{aligned}(x,y) \in (A \times B) \cap (C \times D) &\Leftrightarrow [(x,y) \in A \times B] \wedge [(x,y) \in C \times D] \\ &\Leftrightarrow (x \in A) \wedge (y \in B) \wedge (x \in C) \wedge (y \in D) \\ &\Leftrightarrow [(x \in A) \wedge (x \in C)] \wedge [(y \in B) \wedge (y \in D)] \\ &\Leftrightarrow (x \in A \cap C) \wedge (y \in B \cap D) \\ &\Leftrightarrow (x,y) \in (A \cap B) \times (B \times D)\end{aligned}$$

In all cases, $(A \times B) \cap (C \times D) = (A \cap C) \times (B \cap D)$.

21. **Proof:** Suppose $(a,b) = (c,d)$. Then

$$\begin{aligned}\{\{a\},\{a,b\}\} = \{\{c\},\{c,d\}\} &\Leftrightarrow (\{a\} = \{c\}) \wedge (\{a,b\} = \{c,d\}) \\ &\Leftrightarrow (a = c) \wedge (b = d)\end{aligned}$$

That is, $(a,b) = (c,d)$ if and only if $a = c$ and $b = d$.

23. a. $\{x \mid -4 < x < 2\}$ $(-4,2)$ b. $\{x \mid -1 < x \leq 1\}$ $(-1,1]$
c. $\{x \mid -4 < x\}$ $(-4,\infty)$ d. $\{1\}$ $\{1\}$
e. $\{x \mid x \leq -1 \text{ or } 2 \leq x\}$ $(-\infty,-1] \cup [2,\infty)$
f. $\{x \mid x < 1\}$ $(-\infty,1)$ g. $\{x \mid 1 < x < 2\}$ $(1,2)$
h. $\{x \mid -4 < x < 1\}$ $(-4,1)$

Exercises 3.4

---

1. a. $\{3,4,6,8,5,9,7,11,12\}$ b. $\{8\}$

3. a. $\{2,3,4,5,6\}$ b. $\{3,4\}$ c. $\{1,7\}$ d. $\{1,7\}$
e. $\{1,2,5,6,7\}$ f. $\{1,2,5,6,7\}$ g. $\{2,3,4,5,6\}$ h. $\{3,5\}$
i. $\{2,4,6,3\}$ j. $\{2,3,4,6\}$

5. a. A_5 b. A_{10} c. A_1 d. A_6 e. $\mathbf{N} - A_1$ f. $\mathbf{N} - A_5$

7. a. (i) $[1,\infty)$ (ii) $\emptyset$ b. (i) $[1,\infty) - \mathbf{N}$ (ii) $\emptyset$
c. (i) $\mathbf{R}$ (ii) $(-1,1)$ d. (i) $(-1,2)$ (ii) $(0,1)$
e. (i) $(-1,1)$ (ii) $(0,1)$ f. (i) $[0,2)$ (ii) $[0,1]$

9. **Proof:** There are two cases to consider: $\mathcal{F} = \emptyset$ and $\mathcal{F} \neq \emptyset$.

Case 1. If $\mathcal{F} = \emptyset$, then $\bigcup_{i \in I} B_i = \emptyset$ and $\bigcup_{i \in I}(A \cap B_i) = \emptyset$. Hence,

$$A \cap (\bigcup_{i \in I} B_i) = A \cap \emptyset = \emptyset = \bigcup_{i \in I}(A \cap B_i).$$

Case 2. Suppose $\mathcal{F} \neq \emptyset$ and let $x \in U$.

$x \in A \cap (\bigcup_{i \in I} B_i)$

$\Leftrightarrow (x \in A) \wedge (x \in \bigcup_{i \in I} B_i)$	Definition of intersection
$\Leftrightarrow (x \in A) \wedge [(\exists i \in I)(x \in B_i)]$	Definition of $\bigcup_{i \in I} B_i$
$\Leftrightarrow (\exists i \in I)[(x \in A) \wedge (x \in B_i)]$	A theorem for quantifiers
$\Leftrightarrow (\exists i \in I)(x \in A \cap B_i)$	Definition of intersection
$\Leftrightarrow x \in \bigcup_{i \in I}(A \cap B_i)$	Definition of $\bigcup_{i \in I}(A \cap B_i)$

11. a. $(A_1 \cup A_2) \cap (B_1 \cup B_2 \cup B_3) =$

$[(A_1 \cup A_2) \cap B_1] \cup [(A_1 \cup A_2) \cap B_2] \cup [(A_1 \cup A_2) \cap B_3] =$

$(A_1 \cap B_1) \cup (A_2 \cap B_1) \cup (A_1 \cap B_2) \cup (A_2 \cap B_2) \cup (A_1 \cap B_3) \cup (A_2 \cap B_3) =$

$\bigcup_{j=1}^{3}(\bigcup_{i=1}^{2}(A_i \cap B_j))$

b. $\bigcup_{j=1}^{n}(\bigcup_{i=1}^{m}(A_i \cap B_j))$ c. $\bigcup_{j \in J}(\bigcup_{i \in I}(A_i \cap B_j))$

13. a. **Proof:**

$x \in B - \bigcup_{i \in I} A_i$

$\Leftrightarrow (x \in B) \wedge (x \notin \bigcup_{i \in I} A_i)$	Definition of complement
$\Leftrightarrow (x \in B) \wedge [\neg(\exists i \in I)(x \in A_i)]$	Definition of union
$\Leftrightarrow (x \in B) \wedge [(\forall i \in I)(x \notin A_i)]$	A negation theorem for quantifiers
$\Leftrightarrow (\forall i \in I)[(x \in B) \wedge (x \notin A_i)]$	A theorem for quantifiers
$\Leftrightarrow (\forall i \in I)(x \in B - A_i)$	Definition of complement
$\Leftrightarrow \bigcap_{i \in I}(B - A_i)$	Definition of intersection

b. **Proof:**

$x \in B - \bigcap_{i \in I} A_i$

$\Leftrightarrow (x \in B) \wedge (x \notin \bigcap_{i \in I} A_i)$	Definition of complement
$\Leftrightarrow (x \in B) \wedge [\neg(\forall i \in I)(x \in A_i)]$	Definition of intersection
$\Leftrightarrow (x \in B) \wedge [(\exists i \in I)(x \notin A_i)]$	A negation theorem for quantifiers
$\Leftrightarrow (\exists i \in I)[(x \in B) \wedge (x \notin A_i)]$	A theorem for quantifiers
$\Leftrightarrow (\exists i \in I)(x \in B - A_i)$	Definition of complement
$\Leftrightarrow \bigcup_{i \in I}(B - A_i)$	Definition of union

Review Exercises

1. b., d. 3. a., d., e., g., i., j.

5. a. $\{3, 5, 2, 6\}$ b. $\emptyset$ c. U d. $\emptyset$ e. $\{3, 6\}$ f. C g. $\{2\}$
h. $\{1, 3\}$ i. $\{4, 7\}$ j. $\{(3, 1), (3, 3), (3, 6), (5, 1), (5, 3), (5, 6)\}$
k. $\{(1, 3), (1, 5), (3, 3), (3, 5), (6, 3), (6, 5)\}$ l. $\{(3, 2), (5, 2)\}$

9. a. A_1 b. $\emptyset$

Chapter 4

Exercises 4.1

1. a. 3 $\{(1, a), (2, a), (3, a)\}$
b. 8 $\{\emptyset, \{(1, a)\}, \{(2, a)\}, \{(3, a)\}, \{(1, a), (2, a)\}, \{(1, a), (3, a)\}, \{(2, a), (3, a)\}, A \times B\}$
c. 8

3. $\{(2, 2), (2, 6), (2, 12), (3, 6), (3, 12), (4, 12)\}$

5. a. $\{c, b, 3, 2, a\}$ b. $\{1, 3, e, b, f, 6\}$
c. $\{(1, c), (3, b), (e, 3), (b, 2), (f, a), (6, b)\}$ d. S

9. a. $\{(5,4),(3,2)\}$ b. $\{(1,4)\}$ c. $\{(2,2),(3,3)\}$ d. $\{(1,1),(2,2)\}$
e. $\emptyset$ f. $\{(2,4)\}$ g. $\{(4,5),(2,3)\}$ h. $\{(4,3),(3,2)\}$

11. a. $\{(x,y) \mid y$ is the paternal grandfather of $x.\}$

b. $\{(x,y) \mid y$ is the paternal grandmother of $x.\}$

c. $\{(x,y) \mid y$ is the maternal grandfather of $x.\}$

d. $\{(x,y) \mid y$ is the mother of a brother of $x.\}$

e. $\{(x,y) \mid y$ is the uncle of x on the mother's side of the family.$\}$

f. $\{(x,y) \mid y$ is the father of a sister of $x.\}$

g. $\{(x,y) \mid y$ is the aunt of x on the mother's side of the family.$\}$

h. $\{(x,y) \mid y$ is the mother of a sister of $x.\}$

13. a. Let $R = \{(1,2)\}$ and $S = \{(3,4)\}$. Then $S \circ R = \emptyset$, $\text{Dom}(S \circ R) = \emptyset$, $\text{Dom}(R) = \{1\}$, and $\text{Dom}(R) \not\subseteq \text{Dom}(S \circ R)$.

b. Let $R = \{(1,2)\}$ and $S = \{(3,4)\}$. Then $S \circ R = \emptyset$, $\text{Rng}(S \circ R) = \emptyset$, $\text{Rng}(R) = \{2\}$, and $\text{Rng}(R) \not\subseteq \text{Rng}(S \circ R)$.

Exercises 4.2

---

1. T1. **Proof:** $x < 0 \Leftrightarrow x - 0$ is positive. $\Leftrightarrow x$ is positive.

T3. **Proof:** Let $x, y \in \mathbf{R}$. Then $y - x \in \mathbf{R}$ and by axiom O1, exactly one of the following statements is true:

(1) $y - x = 0$, in which case $x = y$.

(2) $y - x$ is positive. Thus, by T1, $0 < y - x$ and $x < y$.

(3) $-(y - x)$ is positive. Hence, by T1, $0 < x - y$ and $y < x$.

T5. **Proof:** By definition, $x < 0 \Leftrightarrow 0 - x = -x$ is positive. $\Leftrightarrow$ x is negative.

T7. **Proof:** By T5, since $x < 0$ and $y < 0$, x and y are negative; and, therefore, $-x$ and $-y$ are positive. By axiom O3, $(-x)(-y) = xy$ is positive. Consequently, by T1, $xy > 0$.

T9. **Proof:** By T3 with $x = 0$ and $y = 1$, exactly one of the following statements is true (1) $0 = 1$, (2) $0 < 1$, or (3) $1 < 0$. Clearly, (1) is false. Suppose (3) is true. By T7 with $x = y = 1$, $1 \cdot 1 = 1 > 0$, which contradicts theorem T3, because $1 < 0$ and $1 > 0$ are both true.

T10. First, we prove two lemmas.

Lemma 1. Let x and y be real numbers. If $x > 0$ and $y < 0$, then $xy < 0$.

Proof: Since $y < 0$, by T5, y is negative and by definition $-y$ is positive. Since $x > 0$, by T1, x is positive. By axiom O3, $x(-y) = -(xy)$ is positive. Thus, by definition, xy is negative and by T5, $xy < 0$.

Lemma 2. Let z be a real number. If $z > 0$, then $\frac{1}{z} > 0$.

Proof: By T3, exactly one of the following is true: (1) $\frac{1}{z} = 0$, (2) $\frac{1}{z} < 0$, or (3) $0 < \frac{1}{z}$. Because $z \cdot \left(\frac{1}{z}\right) = 1 \neq 0$, statement (1) is false. Suppose $\frac{1}{z} < 0$. Then, by lemma 1, $\left(\frac{1}{z}\right) \cdot z = 1 < 0$ which is false by T9. Hence, $0 < \frac{1}{z}$.

Proof of T10: Assume $0 < x < y$. Because $0 < x$ and $0 < y$, x and y are positive and by axiom O3, xy is positive. Hence, by T1, $xy > 0$ and by lemma 2, $\frac{1}{xy} > 0$. Since $y > 0$, by lemma 2, (4) $0 < \frac{1}{y}$. Since $x < y$, we have $y - x > 0$. Because $\frac{1}{xy} > 0$ and $y - x > 0$, $\frac{1}{xy}(y - x) > 0$. That is, $\frac{1}{x} - \frac{1}{y} > 0$ and (5) $\frac{1}{x} > \frac{1}{y}$. It follows from (4) and (5) that $0 < \frac{1}{y} < \frac{1}{x}$.

T11. **Proof:** Since $x < y$, $y - x$ is positive. By T1, z is positive, because $0 < z$. By axiom O3, $(y - x)z = yz - xz$ is positive, and by T1, $yz - xy > 0$. Hence, $yz > xz$ and $xz < yz$.

T13. **Proof:** Assume $x < y$. Then $y - x > 0$. Since $-(-y) = y$, by substitution $-(-y) - x = (-x) - (-y) > 0$. Hence, $-y < -x$.

T15. **Proof:** Since $z < w$, $w - z > 0$. And since $x > 0$, $x(w - z) = xw - xz > 0$. Therefore, (1) $xz < xw$. Since $x < y$, $y - x > 0$, and since $w > 0$, $(y - x)w = yw - xw > 0$. Hence, (2) $xw < yw$. It follows from (1) and (2) by T2 that $xz < yw$.

Because $zw < yw$, $yw - xz > 0$. Since $z > 0$ and $w > 0$, $zw > 0$ and by lemma 2 in the proof of T10, $\frac{1}{zw} > 0$. Multiplication and cancellation yields, $\frac{1}{zw}(yw - xz) = \frac{y}{z} - \frac{x}{w} > 0$. Therefore, $\frac{x}{w} < \frac{y}{z}$.

T17. **Proof:** Assume $x \leq y$ and $y \leq x$. Since $x \leq y$, either $x < y$ or $x = y$. And since $y \leq x$, either $y < x$ or $y = x$. There are four cases to consider.

Case 1. Suppose $x < y$ and $y < x$. By T2, $x < x$. This is a contradiction.
Case 2. Suppose $x < y$ and $y = x$. By substitution $x < x$. This is a contradiction.
Case 3. Suppose $x = y$ and $y < x$. By substitution $x < x$. This is a contradiction.
Case 4. Suppose $x = y$ and $y = x$. Then $x = y$.
Consequently, if $x \leq y$ and $y \leq x$, then $x = y$.

T19. **Proof:** Assume $x \leq y$ and $z < 0$. Since $x \leq y$, either $x < y$ or $x = y$.
Case 1. Suppose $x < y$. Then $y - x > 0$ and since $z < 0$, $(y - x)z = yz - xz < 0$. Therefore, $yz < xz$.
Case 2. Suppose $x = y$. Then $xz = yz$.
From cases 1 and 2, it follows that if $x \leq y$ and $z < 0$, then $yz \leq xz$.

3.

	Least Element	Greatest Element	Greatest Lower Bound	Least Upper Bound
a.	$\sqrt{2}$	e	$\sqrt{2}$	e
b.		5	-3	5
c.	-7		-7	
d.		5	-4	5
e.	-4		-4	3
f.			$-\sqrt{50}$	$\sqrt{50}$
g.		3	2	3
h.	1	3	1	3
i.	-2	$3/2$	-2	$3/2$
j.	-2	$3/2$	-2	$3/2$
k.	3		3	6
l.	1	6	1	6

5. **Proof:** We prove this theorem by contradiction. Assume S is a nonempty set of real numbers with greatest lower bound m and suppose there exists an $\epsilon_1 > 0$ such that for every $x \in S$, $x > m + \epsilon_1$. Hence, $m + \epsilon_1$ is a lower bound for S and since $\epsilon_1 > 0$, $m + \epsilon_1 > m$. Therefore, m is not the greatest lower bound for S. Contradiction.

Exercises 4.3

---

1. a. transitive b. symmetric c. symmetric d. reflexive, symmetric and transitive e. reflexive, symmetric and transitive

3. a.

i	$\text{Dom}(T_i)$	$\text{Rng}(T_i)$
1	$\{1, 2, 3, 4\}$	$\{2, 3, 4, 5\}$
2	A	A
3	A	A
4	A	A
5	$\{2, 3, 5\}$	$\{2, 3, 4, 5\}$

c. $T_1 = \{(1,2), (1,3), (1,4), (1,5), (2,3), (2,4), (2,5), (3,4), (3,5), (4,5)\}$
$T_2 = \{(1,1), (1,2), (1,3), (1,4), (1,5), (2,2), (2,3), (2,4), (2,5), (3,3),$
$(3,4), (3,5), (4,4), (4,5), (5,5)\}$
$T_3 = \{(1,1), (2,2), (3,3), (4,4)\}$
$T_4 = \{(1,1)(1,2), (1,3), (1,4), (1,5), (2,2), (2,4), (3,3), (4,4), (5,5)\}$
$T_5 = \{(2,2), (2,4), (3,3), (5,5)\}$

d. T_1 is transitive.

T_2 is reflexive and transitive.

T_3 is reflexive, symmetric, and transitive.

T_4 is reflexive, symmetric, and transitive.

T_5 is transitive.

e. T_3 and T_4 are equivalence relations on A.

5. **Proof:** Assume $a \in A$. Since $a^2 > 0$, $(a, a) \in P$ and P is reflexive on A. Assume $(a, b) \in P$. Then $ab > 0$. Since multiplication is commutative in the set of real numbers, $ba > 0$ and $(b, a) \in P$. Hence, P is symmetric on A. Assume $(a, b), (b, c) \in P$. Then $ab > 0$ and $bc > 0$. By axiom O3, $ab^2c > 0$. By T8, $b^2 > 0$ and by lemma 2 in the proof of T10, $1/b^2 > 0$. By T11, $(ab^2c)/b^2 = ac > 0$. Hence, $(a, c) \in P$ and P is transitive on A. Since P is reflexive, symmetric, and transitive on A, P is an equivalence relation on A.

7. **Proof:** Suppose $x \in \mathbf{Z}$. Since the sum of two odd integers is an even integer and since the sum of two even integers is an even integer, $(x, x) \in R$. Assume $(x, y) \in R$. Then by definition of R, $x + y = y + x$ is even. Hence, $(y, x) \in R$. Assume $(x, y), (y, z) \in R$. Then $x + y$ and $y + z$ are even integers. Adding, we find $x + 2y + z$ is an even integer. Since $2y$ is an even integer, and since the difference of two even integers is even, $(x + 2y + z) - 2y = x + z$ is even and $(x, z) \in R$. Consequently, R is an equivalence relation on $\mathbf{Z}$.

9. **Proof:** Let $x \in \mathbf{Z}$. Then $x - x = 0 = 3 \cdot 0$. Hence, $(x, x) \in T$ and T is reflexive. Suppose $(x, y) \in T$. By definition of T, there exists a $k \in \mathbf{Z}$ such that $x - y = 3k$. Multiplying this equation by -1 and rearranging, we find $y - x = 3(-k)$. Since $-k \in \mathbf{Z}$, $(y, x) \in T$ and T is symmetric.

Suppose $(x, y), (y, z) \in T$. Then there exist integers k and ℓ such that $x - y = 3k$ and $y - z = 3\ell$. Addition yields, $(x - y) + (y - z) = 3k + 3\ell$ or $x - z = 3(k + \ell)$. Since k and ℓ are integers, $k + \ell$ is an integer, $(x, z) \in T$, and T is transitive. Therefore, T is an equivalence relation on $\mathbf{Z}$.

11. **Proof:** Let $x \in \mathbf{N}$. Since $x^2 = x^2$, $(x, x) \in V$. Suppose $(x, y) \in V$. Then $x^2 = y^2$. Since equality is a symmetric relation, $y^2 = x^2$ and $(y, x) \in V$. Suppose $(x, y), (y, z) \in V$. Then $x^2 = y^2$ and $y^2 = z^2$. Since equality is a transitive relation $x^2 = z^2$ and $(x, z) \in V$. Therefore, V is an equivalence relation on $\mathbf{N}$.

13. **Proof:** Let $(a, b) \in A$. Since multiplication is commutative on the set of integers, $ab = ba$ and $((a, b), (a, b)) \in X$. Therefore, X is reflexive on A. Suppose $((a, b), (c, d)) \in X$. Then $ad = bc$. Since equality is a symmetric relation $bc = ad$ and since multiplication is commutative $cb = da$. Therefore, $((c, d), (a, b)) \in X$ and X is symmetric. Assume $((a, b), (c, d)) \in X$ and $((c, d), (e, f)) \in X$. Then $ad = bc$ and $cf = de$. Multiplying the first equation by f and multiplying the second equation by b, we find $adf = bcf$ and $bcf = bde$. Hence, $adf = bde$. Dividing by $d \neq 0$, we obtain $af = be$. Hence, $((a, b), (e, f)) \in X$ and X is transitive. Consequently, X is an equivalence relation on A.

15. **Proof:** Let (a, b) be an element of the equivalence relation E^{-1}. Then $(b, a) \in E$. Since E is an equivalence relation, by symmetry, $(a, b) \in E$. Hence, $E^{-1} \subseteq E$. Next, let $(a, b) \in E$. Then $(b, a) \in E^{-1}$. Since E^{-1} is an equivalence relation, by symmetry, $(a, b) \in E$. Hence, $E \subseteq E^{-1}$. Since $E^{-1} \subseteq E$ and $E \subseteq E^{-1}$, $E^{-1} = E$

17. a. **Proof:** Assume R and S are reflexive relations on the set A and let $x \in A$. Since R is a reflexive relation on A, $(x, x) \in R$ and since S is a reflexive relation on A, $(x, x) \in S$. Therefore, $(x, x) \in R \cap S$ and $R \cap S$ is a reflexive relation on A.

    b. **Proof:** Assume R and S are symmetric relations on the set A and let $(x, y) \in R \cap S$. Hence, $(x, y) \in R$ and $(x, y) \in S$. Since R is a symmetric relation, $(x, y) \in R$ implies $(y, x) \in R$ and since S is a symmetric relation, $(x, y) \in S$ implies $(y, x) \in S$. Therefore, $(y, x) \in R \cap S$ and $R \cap S$ is symmetric.

    c. **Proof:** Assume R and S are transitive relations on the set A and let $(x, y), (y, z) \in R \cap S$. Hence, $(x, y), (y, z) \in R$ and $(x, y), (y, z) \in S$. Since R is a transitive relation, $(x, y), (y, z) \in R$ implies $(x, z) \in R$ and since S is a transitive relation, $(x, y), (y, z) \in S$ implies $(x, z) \in S$. Therefore, $(x, z) \in R \cap S$ and $R \cap S$ is transitive.

    d. **Proof:** Assume R and S are equivalence relations on the set A. Then R and S are reflexive, symmetric, and transitive relations on A.

By parts a, b, and c above $R \cap S$ is a reflexive, symmetric, and transitive relation on A. Consequently, $R \cap S$ is an equivalence relation on A.

Exercises 4.4

1. a. $[a] = [c] = \{a, c\}$, $[b] = [e] = \{b, e\}$, $[d] = \{d\}$, $[e] = \{e\}$

   b. $\{\{a, c\}, \{b, e\}, \{d\}, \{e\}\}$

3. a. $E = \{n \in \mathbf{Z} \mid n \text{ is even}\}$ b. $O = \{n \in \mathbf{Z} \mid n \text{ is odd}\}$ c. $\{E, O\}$

5. a. $[(2,3)]_T = \{(x, y) \in \mathbf{R} \times \mathbf{R} \mid y = x + 1\}$. The line $y = x + 1$ in the Cartesian plane.

   b. For $r \in \mathbf{R}$ define $L_r = \{(x, y) \in \mathbf{R} \times \mathbf{R} \mid y = x + r\}$. $P = \{L_r \mid r \in \mathbf{R}\}$ is the partition of $\mathbf{R} \times \mathbf{R}$ corresponding to the relation T.

7. a. (1) $[0] = \mathbf{Z}$ (2) $\{[0]\}$

   b. (1) $[0] = \{\ldots, -4, -2, 0, 2, 4, \ldots\}$, $[1] = \{\ldots, -5, -3, -1, 1, 3, 5, \ldots\}$ (2) $\{[0], [1]\}$

   c. (1) $[0] = \{\ldots, -6, -3, 0, 3, 6, \ldots\}$, $[1] = \{\ldots, -5, -2, 1, 4, 7, \ldots\}$, $[2] = \{\ldots, -4, -1, 2, 5, 8, \ldots\}$ (2) $\{[0], [1], [2]\}$

   d. (1) $[0] = \{\ldots, -10, -5, 0, 5, 10, \ldots\}$, $[1] = \{\ldots, -9, -4, 1, 6, 11, \ldots\}$, $[2] = \{\ldots, -8, -3, 2, 7, 12, \ldots\}$ $[3] = \{\ldots, -7, -2, 3, 8, 13, \ldots\}$ $[4] = \{\ldots, -6, -1, 4, 9, 14, \ldots\}$ (2) $\{[0], [1], [2], [3], [4]\}$

9. a. $E = \{n \in \mathbf{Z} \mid n \text{ is even}\}$ b. $O = \{n \in \mathbf{Z} \mid n \text{ is odd}\}$ c. $\{E, O\}$

11. **Proof:** For $n \in \mathbf{Z}$ let $I_n = [n, n + 1)$. To prove $\mathcal{P} = \{I_n \mid n \in \mathbf{Z}\}$ is a partition of $\mathbf{R}$, we must prove

    (1) For $n \in \mathbf{Z}$, $I_n \neq \emptyset$.

    (2) If $I_n, I_m \in \mathcal{P}$, then either $I_n = I_m$ or $I_n \cap I_m = \emptyset$.

    (3) $\bigcup_{n \in \mathbf{Z}} I_n = \mathbf{R}$.

    (1) By definition, $n \in I_n$ and, therefore, $I_n \neq \emptyset$.

    (2) Assume $I_n, I_m \in \mathcal{P}$. Then either $n = m$ or $n \neq m$. If $n = m$, then $I_n = I_m$. So, suppose $n \neq m$ and assume that $I_n \cap I_m \neq \emptyset$. Without loss of generality, we may assume that $n < m$ and that $x \in I_n \cap I_m$. Since $x \in I_n$, $n \leq x < n + 1$ and since $x \in I_m$, $m \leq x < m + 1$. Since $n < m$, $n < m \leq x < n + 1$ and, consequently, there exists an integer m between the consecutive integers n and $n + 1$, which is a contradiction.

    (3) Let $x \in \mathbf{R}$. By theorem 4.9, there exists a unique integer $n \in \mathbf{Z}$ such that $n \leq x < n + 1$. Hence, $x \in I_n$ and $\bigcup_{n \in \mathbf{Z}} I_n = \mathbf{R}$.

13. **Proof:** Let $S_n = \{(x, y) \in \mathbf{R} \times \mathbf{R} \mid n - 1 \leq \sqrt{x^2 + y^2} < n\}$. To prove $\mathcal{P} = \{S_n \mid n \in \mathbf{N}\}$ is a partition of $\mathbf{R} \times \mathbf{R}$, we must prove

    (1) For $n \in \mathbf{N}$, $S_n \neq \emptyset$.

    (2) If $S_n, S_m \in \mathcal{P}$, then either $S_n = S_m$ or $S_n \cap S_m = \emptyset$.

    (3) $\bigcup_{n \in \mathbf{N}} S_n = \mathbf{R} \times \mathbf{R}$.

    (1) Let $n \in \mathbf{N}$. Then $(n - 1, 0) \in S_n$ because $n - 1 = \sqrt{(n-1)^2 + 0^2}$.

    (2) Assume $S_n, S_m \in \mathcal{P}$. Then either $n = m$ or $n \neq m$. If $n = m$, then $S_n = S_m$. So, suppose $n \neq m$ and assume that $S_n \cap S_m \neq \emptyset$. Without loss of generality, we may assume that $n < m$ and that $(x, y) \in S_n \cap S_m$. Then $n - 1 \leq \sqrt{x^2 + y^2} < n$ and $m - 1 \leq \sqrt{x^2 + y^2} < m$. Since $n < m$, $n - 1 < m - 1 < \sqrt{x^2 + y^2} < n$. That is, there exists an integer, $m - 1$, between the consecutive integers $n - 1$ and n, which is a contradiction.

15. **Proof:** First, suppose $[a]_R = [b]_R$ and let $x \in [a]_R$. Then $(a, x) \in R$. Since $[a]_R = [b]_R$, $x \in [b]_R$ and $(b, x) \in R$. Because R is symmetric, $(x, b) \in R$; and because R is transitive, $(a, b) \in R$.

    Next, suppose $(a, b) \in R$. By definition $b \in [a]_R$ and by theorem 4.14, $[b]_R = [a]_R$.

17. **Proof:** We prove $[a]_R \cap [b]_R = \emptyset \Rightarrow (a, b) \neq R$ by proving its contrapositive: $(a, b) \in R \Rightarrow [a]_R \cap [b]_R \neq \emptyset$. By theorem 4.15, $(a, b) \in R \Rightarrow [a]_R = [b]_R$. Since $a \in [a]_R$ and $[a]_R = [b]_R$, $a \in [b]_R$ and $[a]_R \cap [b]_R \neq \emptyset$.

    And we prove $(a, b) \notin R \Rightarrow [a]_R \cap [b]_R = \emptyset$ by proving its contrapositive: $[a]_R \cap [b]_R \neq \emptyset \Rightarrow (a, b) \in R$. By theorem 4.16, $[a]_R \cap [b]_R \neq \emptyset \Rightarrow [a]_R = [b]_R$. It follows from theorem 4.15 that $[a]_R = [b]_R \Rightarrow (a, b) \in R$.

Review Exercises

---

1. $\text{Dom}(R) = \{a, b, 1, x, 3\}$ $\quad$ $\text{Rng}(R) = \{2, 1, c, d, y\}$

    $R^{-1} = \{(2, a), (1, b), (c, 1), (x, d), (y, 3)\}$

3. $T^{-1} = \{(x, y) \mid 3y + 2x = 4\}$

5. a. $W \circ X = \{(x, y) \mid y = \cos(\ln x)\}$ $\quad$ $X \circ W = \{(x, y) \mid y = \ln(\cos x)\}$

    b. $\text{Dom}(W \circ X) = (0, \infty)$ $\quad$ $\text{Rng}(W \circ X) = [-1, 1]$

    $\text{Dom}(X \circ W) = \bigcup_{n \in \mathbf{Z}} ((4n-1)\pi/2, (4n+1)\pi/2)$ $\quad$ $\text{Rng}(X \circ W) = (-\infty, 0]$

7. a. not reflexive, not symmetric, transitive, and not an equivalence relation

    b. reflexive, symmetric, transitive, and an equivalence relation

c. reflexive, not symmetric, not transitive, and not an equivalence relation

d. not reflexive, symmetric, not transitive, and not an equivalence relation

9. $R = \{(1,1),(3,3),(1,3),(3,1),(2,2),(4,4),(6,6),(2,4),(4,2),(2,6),$ $(6,2),(4,6),(6,4),(5,5)\}$

11. No

13. **Proof:** Assume R is a symmetric and transitive relation on A and assume $\text{Dom}(R) = A$. Since $\text{Dom}(R) = A$, for every $x \in A$ there exists some $y \in A$ such that the ordered pair $(x,y) \in R$. Because R is symmetric, $(y,x) \in R$. Since R is transitive and because $(x,y) \in R$ and $(y,x) \in R$, we have $(x,x) \in R$. Therefore, R is reflexive on A.

15. **Proof:** To prove R is an equivalence relation on $\mathbf{Z}$, we prove R is reflexive, symmetric and transitive on $\mathbf{Z}$.

    Suppose $n \in \mathbf{Z}$. Then because $n = n + 5 \cdot 0$ and $0 \in \mathbf{Z}$, the ordered pair $(n,n) \in R$ and R is reflexive on $\mathbf{Z}$.

    Suppose $m, n \in \mathbf{Z}$ and $(m,n) \in R$. Then there exists a $k \in \mathbf{Z}$ such that $n = m + 5k$. Hence, $m = n + 5(-k)$. Since $-k \in \mathbf{Z}$, the ordered pair $(n,m) \in R$ and R is symmetric on $\mathbf{Z}$.

    Suppose $\ell, m, n \in \mathbf{Z}$, $(\ell, m) \in R$, and $(m,n) \in R$. Since $(\ell,m) \in R$, there exists a $k \in \mathbf{Z}$ such that (1) $m = \ell + 5k$. Also, since $(m,n) \in R$, there exists a $j \in \mathbf{Z}$ such that (2) $n = m + 5j$. Substituting for m in equation (2) from equation (1), we find $n = (\ell + 5k) + 5j = \ell + 5(k+j)$. Since $k + j$ is an integer, $(\ell, n) \in R$ and R is transitive on $\mathbf{Z}$.

Chapter 5

Exercises 5.1

---

1. a. {January, February, March, April, May, June, July, August, September, October, November, December}

   b. $\{28, 30, 31\}$ c. (i) 31 (ii) 28 (iii) 30

   d. (i) February (ii) April, June, September, November (iii) January, March, May, July, August, October, December (iv) None

3. b., c. 5. c., d., e., f., h. 9. a. No b. Yes

11.

	Domain	Range
a	$\mathbf{N}$	$\{9\}$
b	$\mathbf{N}$	$\{n \mid n = m^2 - 5 \text{ and } m \in \mathbf{N}\}$
c	$\mathbf{Z}$	The set of odd integers
d	$\mathbf{Z} - \{3\}$	$\mathbf{Z} - \{6\}$
e	$\mathbf{Q}$	$\mathbf{Q}$
f	$\mathbf{Q}$	$\{s \mid s = 4 - r^2 \text{ and } r \in \mathbf{Q}\}$
g	$\mathbf{R}$	$[2, \infty)$
h	$\mathbf{R} - \{n\pi/2 \mid n \in \mathbf{Z}\}$	$(-\infty, -3] \cup [3, \infty)$

13. a. $2^3 = 8$ b. $3^2 = 9$ c. $n(Y)^{n(X)}$

15. a. (i) well-defined (ii) $f \circ g = \{(1,4), (2,1), (3,4)\}$
 b. (i) not well-defined
 c. (i) well-defined (ii) $(f \circ g)(n) = 2n - 17$
 d. (i) not well-defined
 e. (i) well-defined (ii) $(f \circ g)(n) = n^2 + 6$
 f. (i) well-defined (ii) $(f \circ g)(n) = 7 - \sqrt{n + n^2}$
 g. (i) not well-defined

17. **Proof:** Suppose $(a, b) \in f \cap g$. Since f and g are functions, $b = f(a) = g(a)$. Hence, on $\text{Dom}(f \cap g)$, $f = g$. Next, suppose $(a, b) \in f \cap g$ and $(a, c) \in f \cap g$. Then $(a, b) \in f$ and $(a, c) \in f$. Since f is a function, $b = c$. Therefore, by the definition of a function, $f \cap g$ is a function.

Exercises 5.2

---

1. a. $B = \{1, 2, 3\}$ $f = \{(a, 1), (b, 1), (c, 2)\}$
 b. $B = \{1, 2\}$ $f = \{(a, 1), (b, 1), (c, 2)\}$
 c. $B = \{1, 2, 3, 4\}$ $f = \{(a, 1), (b, 2), (c, 3)\}$
 d. $B = \{1, 2, 3\}$ $f = \{(a, 1), (b, 2), (c, 3)\}$

3. a. $A = \{1\}$, $B = \{a, b\}$, $C = \{x, y\}$, $f = \{(1, a)\}$, $g = \{(a, x), (b, y)\}$
 b. $A = \{1\}$, $B = \{a, b\}$, $C = \{x, y\}$, $f = \{(1, a)\}$, $g = \{(a, x)\}$
 c. $A = \{1\}$, $B = \{a, b\}$, $C = \{x\}$, $f = \{(1, a)\}$, $g = \{(a, x), (b, x)\}$
 d. $A = \{1, 2\}$, $B = \{a\}$, $C = \{x\}$, $f = \{(1, a), (2, a)\}$, $g = \{(a, x)\}$
 e. $A = \{1, 2\}$, $B = \{a, b\}$, $C = \{x\}$, $f = \{(1, a), (2, b)\}$, $g = \{(a, x), (b, x)\}$
 f. $A = \{1\}$, $B = \{a, b\}$, $C = \{x\}$, $f = \{(1, a)\}$, $g = \{(a, x), (b, x)\}$

5. a. $[2, \infty)$ b. $[0, \pi]$ c. $(-\pi/2, \pi/2)$ d. $[-2/3, \infty)$

7. **Proof:** Assume $f : A \xrightarrow{1\text{-}1} B$ and $g : B \xrightarrow{1\text{-}1} C$. And suppose there exists $a, b \in A$ such that $(g \circ f)(a) = (g \circ f)(b)$. Thus, $g(f(a)) = g(f(b))$. Since g is one-to-one, $f(a) = f(b)$. And since f is one-to-one, $a = b$. Therefore, $g \circ f$ is one-to-one.

9. a. The solution of the equation $m + 2 = 1$ is $m = -1 \notin \mathbf{N}$. Therefore, there is no natural number m which satisfies the equation $f(m) = m + 2 = 1$. Hence, f is not onto $\mathbf{N}$.

   b. The solutions of the equation $m^3 + m = 1$ are $.6823278 \notin \mathbf{N}$, $-.3411639 + 1.161541i \notin \mathbf{N}$, and $-.3411639 - 1.161541i \notin \mathbf{N}$. Thus, there is no natural number m which satisfies the equation $f(m) = m^3 + m = 1$. Hence, f is not onto $\mathbf{N}$.

   c. For $x \in \mathbf{R}$, $e^x > 0$, so there is no real number x such that $e^x = 0$. Hence, f is not onto $\mathbf{R}$.

   e. Since $|x| \geq 0$, there is no real number x such that $|x| = -1$. Hence, f is not onto $\mathbf{R}$.

   g. Because $\sqrt{x^2 - 3} \geq 0$, there is no real number x such that $\sqrt{x^2 - 3} = -1$. Hence, f is not onto $\mathbf{R}$.

   i. Since $-1 \leq |\sin x| \leq 1$, there is no real number x such that $\sin x = 2$. Hence, f is not onto $\mathbf{R}$.

   k. For $x \in [-\pi/2, \pi/2]$, $-1 \leq |\sin x| \leq 1$. Therefore, there is no real number x such that $\sin x = 2$. Hence, f is not onto $\mathbf{R}$.

11. e. Since $-2 \in \mathbf{R}$, $2 \in \mathbf{R}$ and $f(-2) = |-2| = 2 = |2| = f(2)$, the function f is not one-to-one.

    f. Since $-2 \in \mathbf{R}$, $2 \in \mathbf{R}$, $f(-2) = |-2| - 4 = -2$, and $f(2) = |2| - 4 = -2 = f(-2)$, the function f is not one-to-one.

    i. Since $0 \in \mathbf{R}$, $\pi \in \mathbf{R}$, $f(0) = \sin 0 = 0$, and $f(\pi) = \sin \pi = 0 = f(0)$, the function f is not one-to-one.

    j. See i.

    n. Since $(1, 2) \in \mathbf{R} \times \mathbf{R}$, $(2, 1) \in \mathbf{R} \times \mathbf{R}$, $f((1, 2)) = 2$, and $f((2, 1)) = 2 = f((1, 2))$, the function f is not one-to-one.

13. a., b., c., g., k. 15. e., i.

Exercises 5.3

1. **Proof:** Let $a, b \in \mathbf{R}$ and assume $f(a) = f(b)$. Then $3a - 2 = 3b - 2$, $3a = 3b$, and $a = b$. Hence, f is one-to-one. Let b be in the codomain $\mathbf{R}$ and let $a = (b+2)/3$ which is in $\mathbf{R}$. Calculating, we find $f(a) = 3(b+2)/3 - 2 = b + 2 - 2 = b$. Therefore, for every b in the codomain $\mathbf{R}$ there exists an a in the domain $\mathbf{R}$ such that $f(a) = b$ and f is onto. Since f is one-to-one and onto, f is a one-to-one correspondence.

   The function $f^{-1}(y) = (y+2)/3$ is the inverse of f.

3. **Proof:** Assume $b \in [1, \infty)$. Then $b - 1 \geq 0$ and $\sqrt{b-1} \geq 0$. So $a = -\sqrt{b-1} \in (-\infty, 0]$ and $f(a) = (-\sqrt{b-1})^2 + 1 = b - 1 + 1 = b$ and f is onto $[1, \infty)$. Next, assume there exists $a, b \in (-\infty, 0]$ such that $f(a) = f(b)$. Then $a^2 + 1 = b^2 + 1$ and $a^2 = b^2$. So either (1) $a = b$ or (2) $a = -b$. For $a = -b$ and $a, b \in (-\infty, 0]$, the only solution of $a^2 = b^2$ is $a = b = 0$. Therefore, $a = b$ and f is one-to-one. Thus, f is a one-to-one correspondence.

   The function $f^{-1}(y) = -\sqrt{y-1}$ is the inverse of f.

5. a. $f^{-1}(y) = (y-3)/(-2)$ $\qquad g^{-1}(y) = (y+5)/4$

   b. $(g \circ f)(x) = -8x + 7$ $\qquad (g \circ f)^{-1}(y) = (y-7)/(-8) = (f^{-1} \circ g^{-1})(y)$

   c. Yes

7. a. $(0, \infty)$

   b. Let $y \in (0, \infty)$. Then $y > 0$, $y + 1 > 1 > 0$, and $x = \dfrac{y}{1+y} > 0$. Computing, we find

$$f(x) = f\left(\frac{y}{1+y}\right) = \frac{\dfrac{y}{1+y}}{1 - \dfrac{y}{1+y}} = \frac{y}{1+y-y} = y.$$

   Therefore, f is onto $(0, \infty)$.

   Let $x_1, x_2 \in (0, 1)$ and assume $f(x_1) = f(x_2)$. Then $\dfrac{x_1}{1-x_1} = \dfrac{x_2}{1-x_2}$ which implies $x_1 - x_1x_2 = x_2 - x_1x_2$ which implies $x_1 = x_2$. Consequently, f is one-to-one.

   c. The function $f^{-1}(y) = \dfrac{y}{1+y}$ is the inverse of f.

   d. Let $x \in (0, 1)$. Then

$$(f^{-1} \circ f)(x) = f^{-1}(f(x)) = f^{-1}\left(\frac{x}{1-x}\right) = \frac{\dfrac{x}{1-x}}{1 + \dfrac{x}{1-x}} = \frac{x}{1-x+x} = x.$$

Hence, $f^{-1} \circ f = I_{(0,1)}$.

Let $y \in (0, \infty)$. Then

$$(f \circ f^{-1})(y) = f(f^{-1}(y)) = f\left(\frac{y}{1+y}\right) = \frac{\frac{y}{1+y}}{1 - \frac{y}{1+y}} = \frac{y}{1+y-y} = y.$$

Hence, $f \circ f^{-1} = I_{(0,\infty)}$.

9. Define f by $f(x) = x$ and g by $g(x) = -x$. Then the function $(f+g)(x) = f(x) + g(x) = 0$ is not one-to-one and not onto; and, therefore, not a one-to-one correspondence. Furthermore, $(fg)(x) = f(x)g(x) = -x^2$ is not one-to-one and not onto.

11. Assume $g \circ f = I_A$. By theorem 5.12 since f is a one-to-one correspondence from A to B, $f^{-1} \circ f = I_A$. Thus $f^{-1} \circ f = g \circ f$ which implies $(f^{-1} \circ f) \circ f^{-1} = (f \circ f) \circ f^{-1}$. By associativity of composition $f^{-1} \circ (f \circ f^{-1}) = f \circ (f \circ f^{-1})$. Hence $f^{-1} \circ I_B = g \circ I_B$ and $f^{-1} = g$. When we assume $f \circ g = I_B$, the proof is similar.

13. a. $f|_A = \{(-2, 12), (0, 2), (1, -3), (3, -13)\}$

b. $\text{Rng}(f|_{[-2,3]}) = [-13, 12]$

15. a. $[0, 1]$ b. $[-1, 0]$ c. $[0, 1]$ d. $[1, \infty)$

Exercises 5.4

---

1. a. $\{2\}$ b. $\{3\}$ c. $\{x, y\}$ d. $\emptyset$ e. $\{v, x, y, z\}$ f. $\{3, 4, 5\}$

3. a. $\{15, 75, 45, 225, 135, 675\}$ b. $\{(1, 1), (2, 1), (1, 2)\}$

5. a. **Proof:** By definition, $f(\emptyset) = \{y \in B \mid y = f(x) \text{ for some } x \in \emptyset\}$. Suppose $y \in f(\emptyset)$. Then there exists some $x \in \emptyset$ such that $y = f(x)$. Since there are no x in the empty set, we have a contradiction. Hence, there are no $y \in f(\emptyset)$ and, therefore, $f(\emptyset) = \emptyset$.

b. **Proof:** By definition, $f^{-1}(\emptyset) = \{x \in A \mid f(x) \in \emptyset\}$. Since there are no $f(x) \in \emptyset$, there are no $x \in f^{-1}(\emptyset)$—that is, $f^{-1}(\emptyset) = \emptyset$.

c. **Proof:** The set $f(\{x\}) = \{y \in B \mid y = f(x) \text{ for some } x \in \{x\}\}$. Since x is the only element in the set $\{x\}$ and since f is a function, the only element in the set $f(\{x\})$ is $f(x)$. Thus, $f(\{x\}) = \{f(x)\}$.

d. **Proof:** Let $x \in C$. Since $f(C) = \{y \in B \mid y = f(x) \text{ for some } x \in C\}$ and since $x \in C$, it follows that $f(x) \in f(C)$. That is, $x \in C$ which implies $f(x) \in f(C)$.

e. **Proof:** The set $f^{-1}(D) = \{x \in A \mid f(x) \in D\}$. Consequently, $x \in f^{-1}(D)$ implies that $f(x) \in D$.

f. **Proof:** The set $f^{-1}(D) = \{x \in A \mid f(x) \in D\}$. Suppose $f(x) \in D$. Then, by definition of $f^{-1}(D)$, we have $x \in f^{-1}(D)$.

7. **Proof:** Assume that $C \subseteq D$ and suppose $y \in f(C)$. It follows from the definition of $f(C)$ that there exists an $x \in C$ such that $y = f(x)$. By theorem 5.16.d., $x \in C$ implies $f(x) \in f(C)$. Since $C \subseteq D$, we have $x \in D$ and by theorem 5.15.d., $y = f(x) \in f(D)$. Therefore, $f(C) \subseteq f(D)$.

9. **Proof:** Suppose $x \in C$. Then by theorem 5.16.d., $f(x) \in f(C)$. In addition, by theorem 5.16.f., $x \in f^{-1}(f(C))$. Therefore, $C \subseteq f^{-1}(f(C))$.

11. **Proof:** Suppose $y \in f(f^{-1}(E))$. Then there exists an $x \in f^{-1}(E) \subseteq A$ such that $y = f(x)$. Therefore, $y \in f(A) = \text{Rng}(A)$. Since $x \in f^{-1}(E)$, by theorem 5.16.e., $f(x) = y \in E$. Hence, $y \in f(A) \cap E$ and the set $f(f^{-1}(E)) \subseteq f(A) \cap E$.

    Now suppose that $y \in f(A) \cap E$. Hence, $y \in f(A)$ which implies there exists an $x \in A$ such that $y = f(x)$. It follows from the definition of $f^{-1}(E)$ that $x \in f^{-1}(E)$. Then by theorem 5.16.e., we have $y = f(x) \in f(f^{-1}(E))$. Hence, $f(A) \cap E \subseteq f(f^{-1}(E))$.

    Consequently, $f(f^{-1}(E)) = f(A) \cap E$, because $f(f^{-1}(E)) \subseteq f(A) \cap E$ and $f(A) \cap E \subseteq f(f^{-1}(E))$.

13. **Proof:** Because $f^{-1}(B) = A$ by theorem 5.15 and substitution,

$$f^{-1}(B - E) = f^{-1}(B) - f^{-1}(E) = A - f^{-1}(E).$$

15. **Proof:** Assume that $f(f^{-1}(E)) = E$ for all $E \subseteq B$. Let $y \in B$. For $E = \{y\}$, we have $f(f^{-1}(\{y\})) = \{y\}$. By theorem 5.16.c., $f^{-1}(\{y\}) = \{f^{-1}(y)\}$ and $f(\{f^{-1}(y)\}) = \{f(f^{-1}(y))\}$. Hence, $f(f^{-1}(\{y\})) = \{y\} = \{f(f^{-1}(y))\}$. That is, for all $y \in B$, we have $f(f^{-1}(y)) = y$. Hence, $f \circ f^{-1} = I_B$ and by theorem 5.14 the function f is onto.

17. **Proof:** From problem 9, we have $C \subseteq f^{-1}(f(C))$. Thus, to prove this theorem, we must prove that if f is one-to-one, then $f^{-1}(f(C)) \subseteq C$. Let $x \in f^{-1}(f(C))$. Then by theorem 5.16.e., $f(x) \in f(C)$. Consequently, for some $x' \in C$, we have $f(x) = f(x')$. Since f is one-to-one, $x' = x$ and $x \in C$. Hence, $f^{-1}(f(C)) \subseteq C$ and, therefore, $f^{-1}(f(C)) = C$.

19. **Proof:** By theorem 5.17, $f(C \cap D) \subseteq f(C) \cap f(D)$. Therefore, we must prove that if f is one-to-one, then $f(C) \cap f(D) \subseteq f(C \cap D)$. Suppose $y \in f(C) \cap f(D)$. Then $y \in f(C)$ and $y \in f(D)$. Hence, there exists an $x_1 \in C$ such that $f(x_1) = y$ and there exists an $x_2 \in D$ such that $f(x_2) = y$. Thus, $f(x_1) = f(x_2)$ and since f is one-to-one, $x_1 = x_2$.

Therefore, $x_1 \in C \cap D$ and $y = f(x_1) \in f(C \cap D)$. Consequently, $f(C) \cap f(D) \subseteq f(C \cap D)$.

21. **Proof:** By theorem 5.16.d., if $x \in C$, then $f(x) \in f(C)$. So, assume that f is one-to-one and suppose $f(x) \in f(C)$. By theorem 5.16.f. there exists a $z \in f^{-1}(f(C))$ such that $f(z) = f(x)$. Since f is one-to-one, $z = x$. Also since f one-to-one, $f^{-1}(f(C)) = C$ by problem 17. Consequently, $x \in f^{-1}(f(C)) = C$.

23. **Proof:** Since f is one-to-one, $f(A - C) = f(A) - f(C)$ by problem 20. And since f is onto, $f(A) = B$. Therefore, $f(A - C) = B - f(C)$.

25. Let $A = \{a, b\}$, $B = \{1\}$, $C = \{a, b\}$, $D = \{a\}$, and $f = \{(a, 1), (b, 1)\}$. Then $C - D = \{b\}$, $f(C - D) = \{1\}$, and $f(C) = \{1\} = f(D)$, so $f(C) - f(D) = \emptyset$. Consequently, $f(C - D) = \{1\} \neq \emptyset = f(C) - f(D)$.

27. Let $f : \mathbf{R} \to \mathbf{R}$ be defined by $f(x) = x^2$. Let $C = \{3\}$. Then $f(C) = \{9\}$ and $f^{-1}(f(C)) = f^{-1}(\{9\}) = \{-3, 3\} \neq C$.

29. Let $A = \{x\}$, $B = \{1, 2\}$, $E = \{2\}$ and $f = \{(x, 1)\}$. Then $E \neq \emptyset$ but $f^{-1}(E) = \emptyset$.

Review Exercises

---

1. a. not a function, $\text{Dom}(f) \neq A$ b. function

   c. not a function, $(a, b), (a, c) \in h$ and $b \neq c$

   d. function e. function

3. $\text{Dom}(f) = (-\infty, -3) \cup (-3, 3) \cup (3, \infty)$ $\text{Rng}(f) = (-\infty, -1] \cup (1, \infty)$

5. $\pm 2\sqrt{2}$

7. $R = I_A$, because for all $x \in A$, we have $(x, x) \in R$ since R is reflexive.

9. a. (1) $f(x) = \sqrt{x}$, $g(x) = 2x - 5$ (2) $f(x) = \sqrt{x - 5}$, $g(x) = 2x$

   b. (1) $f(x) = \cos|x|$, $g(x) = 3x + 2$ (2) $f(x) = \cos x$, $g(x) = |3x + 2|$

11. a. $f^{-1}(y) = (y + 4)/3$ d. $f^{-1}(y) = -\sqrt{(y - 1)/2}$

    f. $f^{-1}(y) = \text{Tan}^{-1}(y) = \text{Arctan}\, y$ j. $f^{-1}(y) = \dfrac{1}{2}\ln y$

13. a. $A_1 = (-\infty, 3)$ $A_2 = (3, \infty)$ b. $A_1 = (-\pi, 0)$ $A_2 = (0, \pi)$

15. a. $\{e, a\}$ b. $\{3, 2\}$ c. $\{2, 1, 4\}$ d. $\{e\}$

17. **Proof:** Since $h \circ h$ is a one-to-one correspondence, $h \circ h$ is onto. From theorem 5.4, if f and g are functions on S and if $g \circ f$ is onto, then g is onto. Therefore, h is onto. Since $h \circ h$ is a one-to-one correspondence, $h \circ h$ is one-to-one. From theorem 5.6, if f and g are functions on S and if $g \circ f$ is one-to-one, then f is one-to-one. Therefore, h is one-to-one; and, consequently, h is a one-to-one correspondence.

Chapter 6

Exercises 6.1

---

1. Let $P(n)$ be the statement

$$(1) \qquad 1 + 3 + 5 + \cdots + (2n - 1) = n^2.$$

(i) For $n = 1$, the left-hand side of the statement $P(1)$ is 1 and the right-hand side is $1^2 = 1$. So, $P(1)$ is true.

(ii) Let $n \in \mathbf{N}$ and assume that the statement $P(n)$ is true. That is, for $n \in \mathbf{N}$ assume that (1) is true. The statement $P(n+1)$ is the statement

$$1 + 3 + 5 \cdots + (2n - 1) + (2n + 1) = (n + 1)^2.$$

Substituting (1) into the left-hand side of $P(n + 1)$, we obtain

$$[1 + 3 + 5 \cdots + (2n - 1)] + (2n + 1) = n^2 + 2n + 1 = (n + 1)^2.$$

By the Principle of Mathematical Induction, for all $n \in \mathbf{N}$,

$$1 + 3 + 5 + \cdots + (2n - 1) = n^2.$$

5. Let $P(n)$ be the statement

$$(1) \qquad 1 + 2^1 + 2^2 + \cdots + 2^{n-1} = 2^n - 1$$

(i) For $n = 1$, the left-hand side of the statement $P(1)$ is 1 and the right-hand side is $2^1 - 1 = 1$. So, $P(1)$ is true.

(ii) Let $n \in \mathbf{N}$ and assume that the statement $P(n)$ is true. That is, for $n \in \mathbf{N}$ assume that (1) is true. The statement $P(n+1)$ is the statement

$$1 + 2^1 + 2^2 + \cdots + 2^{n-1} + 2^n = 2^{n+1} - 1.$$

Substituting (1) into the left-hand side of $P(n+1)$, we obtain

$$[1 + 2^1 + 2^2 + \cdots + 2^{n-1}] + 2^n = 2^n - 1 + 2^n = 2 \cdot 2^n - 1 = 2^{n+1} - 1.$$

By induction, for all $n \in \mathbf{N}$,

$$1 + 2^1 + 2^2 + \cdots + 2^{n-1} = 2^n - 1.$$

11. Let $P(n)$ be the statement

$$(1) \qquad 2 \cdot 6 \cdot 10 \cdots (4n-2) = \frac{(2n)!}{n!}$$

(i) For $n = 1$, the left-hand side of the statement $P(1)$ is 2 and the right-hand side is $\frac{(2 \cdot 1)!}{1!} = 2$. So, $P(1)$ is true.

(ii) Let $n \in \mathbf{N}$ and assume that the statement $P(n)$ is true. That is, for $n \in \mathbf{N}$ assume that (1) is true. The statement $P(n+1)$ is the statement

$$2 \cdot 6 \cdot 10 \cdots (4n-2) \cdot (4n+2) = \frac{(2n+2)!}{(n+1)!}$$

Substituting (1) into the left-hand side of $P(n+1)$, we obtain

$$[2 \cdot 6 \cdot 10 \cdots (4n-2)] \cdot (4n+2) = \frac{(2n)!}{n!} \cdot (4n+2) =$$

$$\frac{(2n)!}{n!} \cdot 2(2n+1)\frac{n+1}{n+1} = \frac{(2n+2)!}{(n+1)!}$$

13. Let $P(n)$ be the statement $3^n > n^2$.

(i) For $n = 1$, the left-hand side of the statement $P(1)$ is $3^1 = 3$, the right-hand side is $1^2 = 1$, and $3 > 1$. So, $P(1)$ is true.

For $n = 2$, the left-hand side of the statement $P(2)$ is $3^2 = 9$, the right-hand side is $2^2 = 4$, and $9 > 4$. So, $P(2)$ is true.

(ii) Now, assume that for $n \in \mathbf{N}$ where $n \geq 2$ the statement $P(n)$ is true. That is, for $n \in \mathbf{N}$ and $n \geq 2$ assume that (1) $3^n > n^2$ is true. The statement $P(n+1)$ is $3^{n+1} > (n+1)^2$. Since $n \geq 2$, we have $n - 1 \geq 1 > 0$ and $2n(n-1) \geq 2n \geq 4 > 1$. Hence, $2n^2 - 2n > 1$ or (2) $2n^2 > 2n + 1$. Multiplying (1) by 3 and using both (1) and (2), we obtain

$$3^{n+1} = 3 \cdot 3^n > 3n^2 = n^2 + 2n^2 > n^2 + 2n + 1 = (n+1)^2.$$

That is, for all natural numbers $n \geq 2$ the statement $P(n+1)$ is true.

Hence, by the Generalized Principle of Mathematical Induction, $3^n > n^2$ for all $n \in \mathbf{N}$.

17. Let $P(n)$ be the statement (1) $n! > 2^n$.

   (i) For $n = 4$, we have $4! = 24 > 16 = 2^4$. Therefore, $P(4)$ is true.

   (ii) Assume for $n \in \mathbf{N}$ and $n \geq 4$ that the statement (1) is true. The statement $P(n+1)$ is $(n+1)! > 2^{n+1}$. Since $(n+1)! = (n+1)n!$, it follows from (1) that $(n+1)! > (n+1)2^n$. Since $n \geq 4$, we have $n+1 > 2$ and $(n+1)! > 2 \cdot 2^n = 2^{n+1}$. Hence, the statement $P(n+1)$ is true.

   Hence, by the Generalized Principle of Mathematical Induction, $n! > 2^n$ for $n \geq 4$.

19. Let $P(n)$ be the statement $(n+1)! > 2^{n+3}$.

   (i) For $n = 5$, we find $(5+1)! = 720 > 256 = 2^8$. Hence, the statement $P(5)$ is true.

   (ii) The statement $P(n+1)$ is $(n+2)! > 2^{n+4}$. Assume for $n \in \mathbf{N}$ and $n \geq 5$ that (1) $(n+1)! > 2^{n+3}$ is true. From the definition of factorial and the inequality (1), we obtain

$$(n+2)! = (n+2)(n+1)! > (n+2)2^{n+3} = n2^{n+3} + 2^{n+4} > 2^{n+4}.$$

   Consequently, by the Generalized Principle of Mathematical Induction, $(n+1)! > 2^{n+3}$ for $n \geq 5$.

Exercises 6.2

---

1. a. (i) $112 = 2^4 \cdot 7$, $320 = 2^6 \cdot 5$ (ii) 16 (iii) $x = 3$, $y = -1$

   b. (i) $2387 = 7 \cdot 11 \cdot 31$, $7469 = 7 \cdot 11 \cdot 97$ (ii) 77 (iii) $x = -25$, $y = 8$

   c. (i) $-4174 = -2 \cdot 2087$, $10672 = 2 \cdot 23 \cdot 29$ (ii) 2

   (iii) $x - -2319$, $y = -907$

   d. (ii) $2^3 \cdot 3^2 \cdot 5 \cdot 7 = 2520$ (iii) $x = -9$, $y = 29$

3. a. **Proof:** Since $D = \gcd(a, b)$, by theorem 6.1 there exist integers x and y such that (1) $D = ax + by$. Substituting for a and b in (1), we obtain $D = drx + dsy = d(rx + sy)$. Hence, d divides D.

   b. **Proof:** Since $d = \gcd(a, b)$, by theorem 6.1 there exist integers x and y such that (1) $d = ax + by$. Substitution for a and b yields $da'x + db'y = d$. Cancelling, we find $a'x + b'y = 1$ and, therefore, $\gcd(a', b') = 1$.

   c. **Proof:** Let $D = \gcd(na, nb)$ and let $d = \gcd(a, b)$. By the definition of the greatest common divisor, d divides a and d divides b. That is, there exist integers a' and b' such that $a = da'$ and $b = db'$. It follows from part b. that $\gcd(a', b') = 1$. Substituting, we have $D = \gcd(nda', ndb')$.

By part a. the natural number n divides D and d divides D. Since a' and b' are relatively prime, they have no common factor other than 1. Hence, $D = nd$. That is, $\gcd(na, nb) = n \gcd(a, b)$.

5. Formula: $x_n = n^2$ for $n \geq 1$.

   Proof: (i) For $n = 1$, we have $x_1 = 1^2 = 1$ and for $n = 2$, we have $x_2 = 2^2 = 4$. So, the formula is correct for $n = 1$ and $n = 2$.

   (ii) Assume for some natural number $n \geq 2$ that $x_k = 2^k$ for $k = 1, 2, \ldots, n$. We must show that $x_{n+1} = 2^{n+1}$. Using the recursive definition for x_{n+1} and the inductive hypothesis, we obtain

$$\begin{aligned} x_{n+1} &= 2x_n - x_{n-1} + 2 = 2n^2 - (n-1)^2 + 2 \\ &= 2n^2 - (n^2 - 2n + 1) + 2 = n^2 + 2n + 1 = (n+1)^2 \end{aligned}$$

   Hence, by induction, $x_n = n^2$ for all $n \in \mathbf{N}$.

7. **Proof:** (i) For $n = 1$, we are given $x_1 = \sqrt{5}$. Obviously, $4 < 5 < 9$. Taking square roots of this inequality, we obtain $2 < \sqrt{5} < 3$.

   (ii) Let n be any natural number and assume $2 < x_n < 3$. Adding 5 to this inequality, yields $7 < 5 + x_n < 8$. Taking square roots, we obtain $2 < \sqrt{7} < \sqrt{5 + x_n} < \sqrt{8} < 3$. Since x_{n+1} is defined recursively by $x_{n+1} = \sqrt{5 + x_n}$, we have $2 < x_{n+1} < 3$. Thus, by induction, $2 < x_n < 3$ for all $n \in \mathbf{N}$.

Chapter 7

Exercises 7.1

---

1. a., b., f., g., h.

3. **Proof:** Our proof is by contradiction. Hence, we assume that the set A is infinite, $A \subseteq B$, and the set B is finite. By theorem 7.8, since $A \subseteq B$ and B is finite, the set A is finite, which contradicts the hypothesis A is infinite.

5. **Proof:** Because the function $f : \mathbf{N} \to E$ defined by $f(n) = 2n$ is a one-to-one correspondence, $\mathbf{N} \sim E$.

7. Let $A = \{a, b\}$ and $B = \{b\}$. Then $|A| = 2$, $|B| = 1$, and $A \cup B = \{a, b\}$. So $|A \cup B| = 2$ but $|A| + |B| = 2 + 1 = 3$. That is, $|A \cup B| = 2 \neq 3 = |A| + |B|$.

11. **Proof:** If A or B is the empty set, then $A \times B = \emptyset$, $B \times A = \emptyset$, and $A \times B \sim B \times A$. Assume neither A nor B is the empty set. For $(a, b) \in A \times B$ define $f : A \times B \to B \times A$ by $f((a, b)) = (b, a)$. Suppose $(a, b), (c, d) \in A \times B$ and that $f((a, b)) = f((c, d))$. By definition of f, $(b, a) = (d, c)$ which implies $b = d$ and $a = c$. Therefore, $(a, b) = (c, d)$ and f is one-to-one. Suppose $(b, a) \in B \times A$. Then $b \in B$, $a \in A$, and $(a, b) \in A \times B$. By definition, $f((a, b)) = (b, a)$. Hence, f is onto and a one-to-one correspondence from $A \times B$ onto $B \times A$.

Exercises 7.2

---

1. **Proof:** We prove that f is a one-to-one function by cases. Suppose n_1 and n_2 are integers and for definiteness that $n_1 \leq n_2$.

   Case 1. If $n_1 \leq n_2 \leq 0$ and $f(n_1) = f(n_2)$, then $1 - 2n_1 = 1 - 2n_2$ which implies $n_1 = n_2$.

   Case 2. If $n_1 \leq 0 < n_2$ and $f(n_1) = f(n_2)$, then $1 - 2n_1 = 2n_2$ which implies $n_1 + n_2 = 1/2$. This is a contradiction, because the sum of two integers is an integer.

   Case 3. If $0 < n_1 \leq n_2$ and $f(n_1) = f(n_2)$, then $2n_1 = 2n_2$ which implies $n_1 = n_2$.

   Consequently, the function f is one-to-one.

   Suppose $m \in \mathbf{N}$. If m is even, then $m = 2n$ for some $n \in \mathbf{N} \subset \mathbf{Z}$. If m is odd, then $m - 1$ is even and $m - 1 = 2k$ for some $k \in \mathbf{N} \cup \{0\}$. Therefore, $m = 1 - 2(-k)$ where $-k$ is a negative integer or zero. Hence, f is onto; and, therefore, f is a one-to-one correspondence from $\mathbf{Z}$ onto $\mathbf{N}$.

3. **Proof:** Let $P(n)$ be the statement: "If A is a denumerable set and if B is a finite set with $|B| = n$, then $A \cup B$ is a denumerable set."

   (i) For $n = 1$, the statement $P(n)$ is "If A is a denumerable set and if $B = \{x\}$, then $A \cup B$ is a denumerable set." By theorem 7.14, this statement is true.

   (ii) Suppose for some natural number n the statement $P(n)$ is true. Let A be a denumerable set and let B be a finite set with $|B| = n+1$. Since B has $n+1$ elements, B is nonempty and there exists some $x \in B$. The set $A \cup B = [A \cup (B - \{x\})] \cup \{x\}$. By theorem 7.10, the set $B - \{x\}$ is finite and $|B - \{x\}| = |B| - 1 = n + 1 - 1 = n$. Hence, the set $A \cup (B - \{x\})$ is the union of the denumerable set A and the finite set $B - \{x\}$ which has n elements. By the induction hypothesis, $A \cup (B - \{x\})$ is denumerable and by theorem 7.14, the set $A \cup (B - \{x\})] \cup \{x\}$ is a denumerable set. That is, the statement $P(n + 1)$ is true and by induction the statement $P(n)$ is true for all natural numbers.

5. a. **Proof:** The set F^+ is denumerable, because the function $f : \mathbf{N} \to F^+$ defined by $f(n) = 4n$ is a one-to-one correspondence. You should verify this fact.

   b. **Proof:** Let $F^- = \{\ldots, -12, -8, -4\}$. The function $g : \mathbf{N} \to F^-$ defined by $g(n) = -4n$ is a one-to-one correspondence. By theorem 7.16, the set $F^+ \cup F^-$ is denumerable, since F^+ and F^- are denumerable and $F^+ \cap F^- = \emptyset$. Hence, by theorem 7.14, the set $F = (F^+ \cup F^-) \cup \{0\}$ is denumerable.

   c. **Proof:** The function $h : \mathbf{N} \to G$ defined by $h(n) = n + 100$ is a one-to-one function from $\mathbf{N}$ onto G. Verify this.

   d. **Proof:** The function $f : \mathbf{N} \to G$ defined by $f(n) = -50 - n$ is a one-to-one function from $\mathbf{N}$ onto H. Verify this.

   e. **Proof:** The function $g : \mathbf{N} \to \mathbf{N} - \{1, 3, 5, 7, 9\}$ defined by

$$f(n) = \begin{cases} 2n, & \text{if } n \leq 5 \\ n + 5, & \text{if } n \geq 6 \end{cases}$$

   is a one-to-one correspondence. Verify this fact.

7. **Proof:** Suppose to the contrary that the set $A - B$ is not denumerable. Thus, $A - B$ is a finite set and by problem 8 of exercises 7.1, since B is a finite set, the set $(A - B) \cup B = A$ is a finite set. This contradicts the hypothesis that A is denumerable.

Exercises 7.3

---

3. **Proof:** We prove this theorem by contradiction. Let A be an uncountable set and let $A \subseteq B$. Assume that the set B is a countable set. Since B is countable, either (1) B is finite or (2) B is denumerable.

   Case 1. By theorem 7.8, since $A \subseteq B$ and B is a finite set, the set A is a finite set which contradicts the hypothesis that A is uncountable.

   Case 2. Since A is an uncountable set, the set A is an infinite set. By theorem 7.19, since A is an infinite subset of the denumerable set B, the set A is denumerable, which contradicts the hypothesis that A is uncountable.

5. a. **Proof:** By definition, $|A| = |B|$ implies $A \sim B$ and $|B| = |C|$ implies $B \sim C$. By transitivity, $A \sim C$ which implies $|A| = |C|$.

   b. **Proof:** Since the identity function $I_A : A \to A$ is one-to-one, we have $|A| \leq |A|$.

   c. **Proof:** Because $|A| \leq |B|$, there exists a function $f : A \xrightarrow{1\text{-}1} B$ and because $|B| \leq |C|$, there exists a function $g : B \xrightarrow{1\text{-}1} C$. The composition $g \circ f : A \to C$ is one-to-one; therefore, $|A| \leq |C|$.

   d. **Proof:** Let $A \subseteq B$. The function $f : A \to B$ defined by $f(a) = a$ is one-to-one. Hence, $|A| \leq |B|$.

   e. **Proof:** Since $A \cap B \subseteq B$, we have $|A \cap B| \leq |B|$ by exercise 5 d.

   f. **Proof:** Since $B \neq \emptyset$, there exists some $b \in B$. Let the function $f : A \to A \times B$ be defined by $f(a) = (a, b)$. This function is one-to-one. Therefore, $|A| \leq |A \times B|$.

   g. **Proof:** By exercise 5 d, since $A \subseteq B$, we have $|A| \leq |B|$. Also since $B \subseteq C$, we have $|B| \leq |C| = |A|$ by hypothesis. Consequently, by the Schröder-Bernstein Theorem $|A| = |B|$.

7. a. **Proof:** By hypotheses $|A| \leq |B|$ and $|B| \leq |C|$. So by exercise 5 c, we have $|A| \leq |C|$. Also by hypothesis $|C| \leq |A|$. Hence, by the Schröder-Bernstein Theorem, $|A| = |C|$ and $A \sim C$. By hypotheses $|B| \leq |C|$ and $|C| \leq |A|$. So by exercise 5 c, we have $|B| \leq |A|$. Also, by hypothesis, $|A| \leq |B|$ and by the Schröder-Bernstein Theorem, $|A| = |B|$ and $A \sim B$. Hence, $A \sim B \sim C$.

   b. The function $f : \mathbf{R} \to \mathbf{R} \times \mathbf{R}$ defined by $f(x) = (x, 0)$ is one-to-one. Hence, $|\mathbf{R}| \leq |\mathbf{R} \times \mathbf{R}|$. The set $\mathbf{R}$ and the interval $(0, 1)$ both have cardinality $\mathbf{c}$. To prove that $|\mathbf{R} \times \mathbf{R}| \leq |\mathbf{R}|$, we prove that there exists a one-to-one function from $(0, 1) \times (0, 1)$ into $(0, 1)$. For each $x \in (0, 1)$, let x be written in normalized decimal form. The function $g : (0, 1) \times (0, 1) \to (0, 1)$ defined by

$$g((x, y)) = g((0.x_1x_2x_3\ldots,\ 0.y_1y_2y_3\ldots)) = 0.x_1y_1x_2y_2x_3y_3\ldots$$

   is one-to-one. Hence, $|\mathbf{R} \times \mathbf{R}| \leq |\mathbf{R}|$ and by the Schröder-Bernstein Theorem $|\mathbf{R} \times \mathbf{R}| \leq |\mathbf{R}|$ and $\mathbf{R} \times \mathbf{R} \sim \mathbf{R}$.

Chapter 8

Exercises 8.1

1. a. $a_n = \dfrac{2n+2}{2n+1}$ b. $a_n = \dfrac{2n+1}{2n}$

c. $a_n = \dfrac{(-1)^n}{n!}$ d. $a_n = \dfrac{n - \left(\frac{1+(-1)^n}{2}\right)}{2n-1}$

e. $a_n = \dfrac{(-1)^{n+1}n}{3n-1}$ f. $a_n = \dfrac{(-1)^n \sin(2n+1)}{(n+1)^2}$

3. a. $\lim_{n\to\infty} a_n = -7$

Proof: Let $\epsilon > 0$ be given. By the Archimedean property of real numbers, there exists a natural number $N > 1/\epsilon$. For all $n > N$,

$$\left|(-7 + \frac{1}{n}) - (-7)\right| = \left|\frac{1}{n}\right| = \frac{1}{n} < \frac{1}{N} < \epsilon.$$

Hence, $\lim_{n\to\infty}(-7 + \dfrac{1}{n}) = -7.$

e. $\lim_{n\to\infty} a_n = 1/2$

Proof: Let $\epsilon > 0$ be given. By the Archimedean property of real numbers, there exists a natural number $N > \sqrt{1/(12\epsilon)}$. For all $n > N$,

$$\left|\frac{3n^2+1}{6n^2+1} - \frac{1}{2}\right| = \left|\frac{6n^2+2-6n^2-1}{12n^2+2}\right| = \frac{1}{12n^2+2} < \frac{1}{12n^2} < \frac{1}{12N^2} < \epsilon.$$

Hence, $\lim_{n\to\infty} \dfrac{3n^2+1}{6n^2+1} = \dfrac{1}{2}.$

g. $\lim_{n\to\infty} a_n = 1/4$

Proof: Let $\epsilon > 0$ be given. By the Archimedean property of real numbers, there exists a natural number $N > 1/(256\epsilon^2)$. For all $n > N$,

$$\left|\frac{\sqrt{n}+2}{4\sqrt{n}+7} - \frac{1}{4}\right| = \left|\frac{4\sqrt{n}+8-4\sqrt{n}-7}{16\sqrt{n}+28}\right|$$
$$= \frac{1}{16\sqrt{n}+28} < \frac{1}{16\sqrt{n}} < \frac{1}{16\sqrt{N}} < \epsilon.$$

Hence, $\lim_{n\to\infty} \dfrac{\sqrt{n}+2}{4\sqrt{n}+7} = \dfrac{1}{4}.$

5. $a_n = -1/n \to 0$ which is nonnegative.

7. a. Let $a_n = (-1)^{n+1}$ and let $b_n = (-1)^n$. The sequences $< a_n >$ and $< b_n >$ both diverge; however, the sequence $< a_n + b_n >=< 0 >$ converges.

   b. Let $< a_n >= 1, 0, 1, 0, \ldots$ and $b_n >= 0, 1, 0, 1, \ldots$. Then the sequence $< a_n b_n >= 0, 0, 0, \ldots$ converges.

9. **Proof**: Our proof is by contradiction. Suppose that $a_n < b_n$ for all $n \in \mathbf{N}$, $a_n \to A$, $b_n \to B$, and $A > B$. Let $\epsilon = (A - B)/2 > 0$. Since $a_n \to A$, there exists a natural number N_1 such that $|a_n - A| < (A - B)/2$ for all $n > N_1$. That is, for $n > N_1$,

$$\frac{A+B}{2} = A - \frac{A-B}{2} < a_n < A + \frac{A-B}{2}. \tag{1}$$

   Since b_n converges to B, there exists a natural number N_2 such that $|b_n - B| < (A - B)/2$ for all $n > N_2$. That is, for $n > N_2$,

$$B - \frac{A-B}{2} < b_n < B + \frac{A-B}{2} = \frac{A+B}{2}. \tag{2}$$

   Since $a_n < b_n$ for all $n \in \mathbf{N}$, we have from equations (1) and (2) for $n > \max(N_1, N_2)$ that

$$\frac{A+B}{2} < a_n < b_n < \frac{A+B}{2}$$

   which is a contradiction.

11. Let

$$a_n = 1 - \frac{1}{6n^2}, \qquad b_n = \frac{\sin \dfrac{1}{n}}{2n\left(1 - \cos \dfrac{1}{n}\right)}, \qquad \text{and} \qquad c_n = 1.$$

   Since $a_n < b_n < c_n$ for all $n \in \mathbf{N}$, we have $a_n \leq b_n \leq c_n$ for all $n \in \mathbf{N}$. By theorem 8.3, since $a_n \to 1$ and $c_n \to 1$, we conclude $b_n \to 1$.

13. **Proof**: Suppose that $a_n \to 0$ and let $\epsilon > 0$ be given. Since $a_n \to 0$, there exists a natural number N such that if $n > N$, then $|a_n - 0| < \epsilon$. Because $||a_n| - 0| = ||a_n| - |0|| \leq |a_n - 0|$, we conclude that if $n > N$, then $||a_n| - 0| < \epsilon$. Therefore, $\lim_{n\to\infty} |a_n| \to 0$.

   Now, suppose that $|a_n| \to 0$ and let $\epsilon > 0$ be given. Since $|a_n| \to 0$, there exists a natural number N such that if $n > N$, then $||a_n| - 0| < \epsilon$. Because $|a_n - 0| = |a_n| = |a_n| - 0 = ||a_n| - 0|$, we conclude if $n > N$, then $|a_n - 0| < \epsilon$. Hence, $\lim_{n\to\infty} a_n = 0$.

15. a. The given sequence is bounded above by 5, is bounded below by 0, and is bounded by 5.

    b. The given sequence is bounded above by 3, is bounded below by 0, and is bounded by 3.

    c. The given sequence is bounded above by 9/16, is bounded below by $-3/4$, and is bounded by 3/4.

    d. The given sequence is not bounded above, is not bounded below, and is not bounded.

    e. The given sequence is bounded above by 1, is bounded below by 0, and is bounded by 1.

    f. The given sequence is bounded above by 1, is bounded below by 0, and is bounded by 1.

    g. The given sequence is not bounded above, is bounded below by 0, and is not bounded.

    h. The given sequence is bounded above by $\sqrt{2}$, is bounded below by 0, and is bounded by $\sqrt{2}$.

17. **Proof**: Suppose that $< a_n >$ and $< b_n >$ are bounded sequences. By definition, there exist positive real numbers A and B such that $|a_n| \le A$ and $|b_n| \le B$ for all $n \in \mathbf{N}$.

    (i) By the triangle inequality, for all $n \in \mathbf{N}$, we have

$$|a_n + b_n| \le |a_n| + |b_n| \le A + B.$$

    That is, the sequence $< a_n + b_n >$ is bounded.

    (ii) Also, by the triangle inequality, for all $n \in \mathbf{N}$,

$$|a_n - b_n| = |a_n + (-b_n)| \le |a_n| + |-b_n| = |a_n| + |b_n| \le A + B.$$

    Therefore, the sequence $< a_n - b_n >$ is bounded.

    (iii) For all $n \in \mathbf{N}$, we have $|a_n b_n| = |a_n||b_n| \le AB$. Therefore, the sequence $< a_n b_n >$ is bounded.

Exercises 8.2

---

1. **Proof**: Let $\epsilon > 0$ be given. Since $a_n \to A$ there exists a natural number N_1 such that $n > N_1$ implies $|a_n - A| < \epsilon/2$. Since $b_n \to B$ there exists a natural number N_2 such that $n > N_2$ implies $|b_n - B| < \epsilon/2$. Let $N = \max\{N_1, N_2\}$. For $n > N$, we have

$$\begin{aligned}|(a_n - b_n) - (A - B)| &= |(a_n - A) + (-(b_n - B))| \\ &\le |a_n - A| + |-(b_n - B)| < \frac{\epsilon}{2} + \frac{\epsilon}{2} = \epsilon.\end{aligned}$$

3. **Proof**: Assume that the sequence $< a_n >$ converges to A and that k is a real constant. Since the sequence $< k >$ converges to k, the sequence $< ka_n >$ converges to kA by theorem 8.7.

5. **Proof**: Our proof is by contradiction. Assume that the sequence $< a_n >$ converges, the sequence $< b_n >$ diverges, and the sequence $< a_n + b_n >$ converges. Since $< a_n >$ converges and $< a_n + b_n >$ converges, by theorem 8.6 the sequence $< (a_n + b_n) - a_n > = < b_n >$ converges which contradicts the hypothesis that the sequence $< b_n >$ diverges.

7. **Proof**: Suppose that $a_n \to A \neq 0$ and the sequence $< a_n b_n >$ converges. Suppose that $A > 0$ and let $\epsilon = A/2 > 0$. Since $a_n \to A$, there exists a natural number N such that for $n > N$, we have

$$0 < \frac{A}{2} = A - \frac{A}{2} < a_n < A + \frac{A}{2}.$$

That is, $0 < a_n$ for all $n \geq N$. Hence, the sequence $< a_n >_{n=N}^{\infty}$ is a sequence for which $a_n \neq 0$. Since the sequence $< a_n b_n >$ converges by hypothesis, the sequence $< a_n b_n >_{n=N}^{\infty}$ converges. By theorem 8.8, the sequence $< a_n b_n / a_n >_{n=N}^{\infty} = < b_n >_{n=N}^{\infty}$ converges. The sequence $< b_n >$ converges, because it only contains the finite number of terms $b_1, b_2, \ldots, b_{N-1}$ preceeding the terms of the convergent sequence $< b_n >_{n=N}^{\infty}$. The proof for the case in which $A < 0$ is similar. Let $\epsilon = |A|/2 > 0$ and proceed as above.

9. **Proof**: Suppose that $a_n \geq 0$ for all $n \in \mathbf{N}$ and that $a_n \to A > 0$. Let $\epsilon > 0$ be given. Since $a_n \to A$, there exists a natural number N such that $|a_n - A| < \sqrt{A}\epsilon$ for all $n > N$. Hence, for all $n > N$, we have

$$\begin{aligned} |\sqrt{a_n} - \sqrt{A}| &< \left| (\sqrt{a_n} - \sqrt{A}) \left(\frac{\sqrt{a_n} + \sqrt{A}}{\sqrt{a_n} + \sqrt{A}} \right) \right| \\ &= \frac{|a_n - A|}{\sqrt{a_n} + \sqrt{A}} < \frac{|a_n - A|}{\sqrt{A}} < \epsilon. \end{aligned}$$

That is, $\sqrt{a_n} \to \sqrt{A}$.

11. $2B/5$

Exercises 8.3

---

1. $< (-1)^n >$ 2. $< n >$ 3. $< (-1)^n n >$ 4. $< \dfrac{(-1)^n}{n} >$

5. $< n >$ 6. $< (-1)^n >$

7. **Proof**: We prove by induction that $a_n < 5/3$ for all $n \in \mathbf{N}$. Let $P(n)$ be the statement "$a_n < 5/3$." The statement $P(1)$ is $1 < 5/3$ which is true. Now, assume that $P(n)$ is true for some natural number n. That is, we assume that $a_n < 5/3$ for some $n \in \mathbf{N}$. From the recursive definition, we have $a_{n+1} = (a_n + 5)/4 < (5/3 + 5)/4 = 5/3$. Hence, the statement $P(n + 1)$ is true and $a_n < 5/3$ for all $n \in \mathbf{N}$. That is, the sequence $< a_n >$ is bounded above by $5/3$.

   Next, we prove by induction that the sequence $< a_n >$ is monotone increasing. Let $Q(n)$ be the statement "$a_n \leq a_{n+1}$." The statement $Q(1)$ is $1 \leq 3/2$ which is true. Now, we assume that $Q(n)$ is true for some natural number n. That is, we assume that $a_n \leq a_{n+1}$ for some $n \in \mathbf{N}$. Adding 5 to the last inequality and then dividing by 4, we obtain $(a_n + 5)/4 \leq (a_{n+1} + 5)/4$. From the recursive definition, it follows that $a_{n+1} \leq a_{n+2}$. Therefore, the statement $Q(n+1)$ is true, and by mathematical induction the sequence $< a_n >$ is monotone increasing. Hence, the sequence is bounded below by $a_1 = 1$.

   By theorem 8.9 (the monotone convergence theorem), the sequence $< a_n >$ converges to some real number A. Taking the limit of the recursive definition of a_{n+1} as n approaches infinity, we find $A = (A + 5)/4$. Solving for A, we obtain $A = 5/3$.

11. **Proof**: First, we prove that $a_n > 3$ for all $n \in \mathbf{N}$. Let $P(n)$ be the statement "$a_n > 3$." The statement $P(1)$ is true, because $a_1 = 4 > 3$. Suppose for some natural number n that $a_n > 3$. Adding 6 to this inequality, we obtain $6 + a_n > 9$ which implies $a_{n+1} = \sqrt{6 + a_n} > 3$. Thus, the statement $P(n + 1)$ is true. Hence, by induction, $a_n > 3$ for all $n \in \mathbf{N}$ and the sequence $< a_n >$ is bounded below by 3.

    Next, we prove that $a_n > a_{n+1}$ for all $n \in \mathbf{N}$. Since $a_n > 3$, we have $a_n - 3 > 0$ and $a_n + 2 > 0$. Therefore, $0 < (a_n - 3)(a_n + 2) = a_n^2 - a_n - 6$. Solving for a_n^2, we find $a_n^2 > 6 + a_n$ which implies $a_n > \sqrt{6 + a_n} = a_{n+1}$ by the recursive definition. Therefore, the sequence $< a_n >$ is monotone decreasing and bounded above by $a_1 = 4$.

    By theorem 8.9, the sequence $< a_n >$ converges to a real number A. From the recursive definition, we see that A satisfies the equation $A = \sqrt{6 + A}$. Squaring this equation, we find $A^2 = 6 + A$. Hence, $A^2 - A - 6 = (A - 3)(A + 2) = 0$ and $A = 3$ or $A = -2$. Since $3 < a_n < 4$ for all $n \in \mathbf{N}$, we conclude $a_n \to 3$.

15. $< a_{2n+1} > = < 2\cos\dfrac{(2n+1)\pi}{2} > = < 0 > \to 0$

    $< a_{4n} > = < 2\cos(2n\pi) > = < 2 > \to 2$

    $< a_{4n+2} > = < 2\cos(2n+1)\pi > = < -2 > \to -2$

The sequence $< a_n >$ diverges.

17. a. $0, 1, 1, 2, 2, 3, 3, 4, 4, \ldots$

    b. $\frac{1}{2}, \frac{1}{3}, \frac{1}{4}, \ldots$

Exercises 8.4

---

1. $< n >$ 2. $< \frac{(-1)^n}{n} >$ 3. $< (-1)^n >$

Chapter 9

Exercises 9.1

---

1. The following $*$ are not binary operations.

   a. Since $1, 2 \in \mathbf{N}$ and $1 * 2 = \frac{1}{2} \notin \mathbf{N}$, $*$ is not a binary operation.

   e. Since $1 \in \mathbf{N}$ and $|1 - 1| = 0 \notin \mathbf{N}$, $*$ is not a binary operation.

   g. Since $0, 1 \in \mathbf{Q}$ and $1*0 = 1/0$ is undefined, $*$ is not a binary operation.

   i. Since $1, 2 \in \mathbf{Q}^+$ and $\sqrt{2} \notin \mathbf{Q}^+$, $*$ is not a binary operation.

3. Table 9.4 is associative and commutative. There is no identity element; and, therefore, there are no inverses.

   Table 9.5 is associative and commutative. The identity element is 0. The inverse of 0 is 0 and the inverse of 1 is 1.

   Table 9.6 is associative and commutative. There is no identity element; and, hence, there are no inverses.

   Table 9.7 is associative but not commutative. There is no identity element; and, hence, there are no inverses.

Exercises 9.2

---

1. d., e., g., i., j.

3. d. $(U, *)$ is an abelian group with identity 1. The inverse of 1 is 1, the inverse of 2 is 4, and the inverse of 4 is 2.

   e. $(V, +)$ is an abelian group with identity 0. The inverse of 0 is 0, the inverse of 2 is 4, and the inverse of 4 is 2.

   g. $(E, +)$ is an abelian group with identity 0. The inverse of each $x \in E$ is $-x$.

   i. $(X, *)$ is an abelian group with identity 0. The inverse of each $a \in X$ is $-a/(1-a)$.

   j. $(\mathbf{Z}, \circ)$ is an abelian group with identity -1. The inverse of each $a \in \mathbf{Z}$ is $-2-a$.

5.

*	**0**	**1**	**2**
0	1	0	2
1	2	1	0
2	0	2	1

The operation $*$ is not associative, because $(0 * 0) * 2 = 1 * 2 = 0$ and $0 * (0 * 2) = 0 * 2 = 2$. Also, there is no identity element.

7.

*	**e**	**a**	**b**	**c**
e	e	a	b	c
a	a	e	c	b
b	b	c	e	a
c	c	b	a	e

This group is abelian.

9. **Proof**: Let $P(n)$ be the statement "$(a_1a_2\cdots a_n)^{-1} = a_n^{-1}\cdots a_2^{-1}a_1^{-1}$."
   (i) For $n = 1$, the statement $P(1)$ is $a_1^{-1} = a_1^{-1}$ which is true.

   (ii) Assume for some natural number n that the statement $P(n+1)$ is true. That is, assume that $(a_1a_2\cdots a_n)^{-1} = a_n^{-1}\cdots a_2^{-1}a_1^{-1}$ is true. For $a, b \in G$ we have $(ab)^{-1} = b^{-1}a^{-1}$ by theorem 9.4. Hence, using the induction hypothesis, we find

$$\begin{aligned}(a_1a_2\cdots a_na_{n+1})^{-1} &= ((a_1a_2\cdots a_n)a_{n+1})^{-1}\\ &= a_{n+1}^{-1}(a_1a_2\cdots a_n)^{-1} = a_{n+1}^{-1}a_n^{-1}\cdots a_2^{-1}a_1^{-1}.\end{aligned}$$

   Therefore, the statement $P(n+2)$ is true and by induction for all $n \in \mathbf{N}$, $(a_1a_2\cdots a_n)^{-1} = a_n^{-1}\cdots a_2^{-1}a_1^{-1}$.

11. **Proof**: Let e be the identity of the group G and let a^{-1} be the inverse of a. Consider $y = ba^{-1} \in G$. By associativity, we have $ya = (ba^{-1})a = b(aa^{-1}) = be = b$. Hence, y is one solution of $ya = b$. Suppose that y_1 and $y_2 \neq y_1$ are both solutions of the equation $ya = b$. Then $y_1a = y_2a$ and by the right cancellation law $y_1 = y_2$ which contradicts $y_2 \neq y_1$. Therefore, $y = ba^{-1}$ is the unique solution of $ya = b$.

13. **Proof**: First, we assume that G is an abelian group. Let $a, b \in G$. By theorem 9.4, we have (1) $(ab)^{-1} = b^{-1}a^{-1}$. Since G is abelian, (2) $b^{-1}a^{-1} = a^{-1}b^{-1}$. From (1) and (2), we see that $(ab)^{-1} = a^{-1}b^{-1}$.

    Next, suppose that $(ab)^{-1} = a^{-1}b^{-1}$ for all $a, b \in G$. Let e be the identity element of G. Since $a, b \in G$, the elements a^{-1}, b^{-1}, $a^{-1}b^{-1}$, $(ab)^{-1}$ are elements of G. By definition of inverse (1) $(ab)(ab)^{-1} = e$. By hypothesis (2) $(ab)^{-1} = a^{-1}b^{-1}$. Substituting (2) into (1), we obtain (3) $(ab)a^{-1}b^{-1} = e$. Multiplying (3) on the right by b, yields $aba^{-1}b^{-1}b = eb$ or (4) $aba^{-1} = b$. Then, multiplying (4) on the right by a, we obtain $ab = ba$. Therefore, G is abelian.

15. a. $\begin{pmatrix} 0 & 0 \\ 0 & 0 \end{pmatrix}$ b. $\begin{pmatrix} -a & -b \\ -c & -d \end{pmatrix}$

    c. **Proof**:

$$\begin{pmatrix} a_1 & b_1 \\ c_1 & d_1 \end{pmatrix} + \begin{pmatrix} a_2 & b_2 \\ c_2 & d_2 \end{pmatrix} = \begin{pmatrix} a_1 + a_2 & b_1 + b_2 \\ c_1 + c_2 & d_1 + d_2 \end{pmatrix}$$
$$= \begin{pmatrix} a_2 + a_1 & b_2 + b_1 \\ c_2 + c_1 & d_2 + d_1 \end{pmatrix} = \begin{pmatrix} a_2 & b_2 \\ c_2 & d_2 \end{pmatrix} + \begin{pmatrix} a_1 & b_1 \\ c_1 & d_1 \end{pmatrix}$$

    d. Let $\begin{pmatrix} a_1 & b_1 \\ c_1 & d_1 \end{pmatrix}$ and $\begin{pmatrix} a_2 & b_2 \\ c_2 & d_2 \end{pmatrix}$ be in $GL(2, \mathbf{R})$. Then

$$\begin{pmatrix} a_1 & b_1 \\ c_1 & d_1 \end{pmatrix} \cdot \begin{pmatrix} a_2 & b_2 \\ c_2 & d_2 \end{pmatrix} = \begin{pmatrix} a_1a_2 + b_1c_2 & a_1b_2 + b_1d_2 \\ c_1a_2 + d_1c_2 & c_1b_2 + d_1d_2 \end{pmatrix}$$

    The determinant of the matrix on the right hand side is

$$(a_1a_2 + b_1c_2)(c_1b_2 + d_1d_2) - (c_1a_2 + d_1c_2)(a_1b_2 + b_1d_2)$$
$$= a_1a_2b_2c_1 + a_1a_2d_1d_2 + b_1b_2c_1c_2 + b_1c_2d_1d_2$$
$$-a_1a_2b_2c_1 - a_2b_1c_1d_2 - a_1b_2c_2d_1 - b_1c_2d_1d_2$$
$$= (a_1d_1 - b_1c_1)(a_2d_2 - b_2c_2) \neq 0,$$

    since $a_1d_1 - b_1c_1 \neq 0$ and $a_2d_2 - b_2c_2 \neq 0$.

    e. $I = \begin{pmatrix} 1 & 0 \\ 0 & 1 \end{pmatrix}$

    f. Verify that (1) $AA^{-1} = A^{-1}A = I$ and (2) $\det A^{-1} \neq 0$.

    g. Let $A = \begin{pmatrix} 1 & 1 \\ 0 & 1 \end{pmatrix}$ and $B = \begin{pmatrix} 0 & 1 \\ 1 & 1 \end{pmatrix}$. Then

$$AB = \begin{pmatrix} 1 & 1 \\ 0 & 1 \end{pmatrix}\begin{pmatrix} 0 & 1 \\ 1 & 1 \end{pmatrix} = \begin{pmatrix} 1 & 2 \\ 1 & 1 \end{pmatrix}$$

    while

$$BA = \begin{pmatrix} 0 & 1 \\ 1 & 1 \end{pmatrix}\begin{pmatrix} 1 & 1 \\ 0 & 1 \end{pmatrix} = \begin{pmatrix} 0 & 1 \\ 1 & 2 \end{pmatrix}$$

Exercises 9.3

1. No 2. No 3. Yes

5. **Proof**: Let G be a group with identity e and let H and K be subgroups of G. By theorem 9.6, $e \in H$ and $e \in K$. Therefore, $e \in H \cap K$ and $H \cap K$ is a nonempty set. Suppose that $a, b \in H \cap K$. Then $a, b \in H$ and $a, b \in K$. Since H and K are subgroups of G, the element $b^{-1} \in H$ and $b^{-1} \in K$ by theorem 9.6. Since H and K are groups, $ab^{-1} \in H$ and $ab^{-1} \in K$. Thus, $ab^{-1} \in H \cap K$ and by theorem 9.7, the set $H \cap K$ is a subgroup of G.

9. $\{e\}$ 10. $\{0, 1, 2, 3\}$

11. **Proof**: Let G be a group with identity e and let a be any element of G. Since $ea = ae$, we have $e \in C(a)$; and, therefore, $C(a)$ is nonempty. Suppose that $g \in C(a)$. Then (1) $ga = ag$. Multiplying equation (1) on the left by g^{-1}, we obtain

$$g^{-1}(ga) = (g^{-1}g)a = ea = a = g^{-1}ag.$$

Multiplying $a = g^{-1}ag$ on the right by g^{-1}, we find

$$ag^{-1} = (g^{-1}ag)g^{-1} = (g^{-1}a)(gg^{-1}) = (g^{-1}a)e = g^{-1}a.$$

Therefore $g^{-1} \in C(a)$. Now, suppose that $g, h \in C(a)$. Then $ga = ag$ and $ha = ah$. Computing, we find

$$(gh)a = g(ha) = g(ah) = (ga)h = (ag)h = a(gh).$$

Hence, $gh \in C(a)$ and by theorem 9.8, the set $C(a)$ is a subgroup of G.

13. **Proof**: Let G be a group with identity e and let a be a fixed element of G. Let H be a subgroup of G and let $K = \{a^{-1}ha \mid h \in H\}$. Since H is a subgroup of G, the identity e is an element of H and $a^{-1}ea = e \in K$. That is, K is a nonempty set. Suppose that $b, c \in K$. Then there is an $h_1 \in H$ such that $b = a^{-1}h_1a$ and there is an $h_2 \in H$ such that $c = a^{-1}h_2a$. Computing, we find

$$\begin{aligned} bc^{-1} = (a^{-1}h_1a)(a^{-1}h_2a)^{-1} &= (a^{-1}h_1a)(a^{-1}h_2^{-1}a) \\ &= a^{-1}h_1(aa^{-1})h_2^{-1}a = a^{-1}(h_1h_2^{-1})a \end{aligned}$$

Since $h_1, h_2 \in H$ and since H is a group, $h_1h_2^{-1} \in H$ and, therefore, $bc^{-1} \in H$. Hence, by theorem 9.7, the set K is a subgroup of G.

15. Cyclic: $C = \{0, 1, 2, 3\}$ Noncyclic: $N = \{0, 2, 4, 5\}$

References

Boyer, C. B. 1968. *A History of Mathematics.* Princeton: Princeton University Press.

Bunt, L. N. H., Jones, P. S., and Bedient, J. D. 1976. *The Historical Roots of Elementary Mathematics.* Englewood Cliffs: Prentice-Hall, Inc.

Chartrand, G., Polimeni, A. D., and Zhang, P. 2003. *Mathematical Proofs: A Transition to Advanced Mathematics.* Boston: Addison Wesley.

Copi, I. M. 1982. *Introduction to Logic.* New York: Macmillan Publishing Co., Inc.

Eves, H. 1983. *Great Moments in Mathematics (After 1650).* The Mathematical Association of America.

Eves, H. 1983. *Great Moments in Mathematics (Before 1650).* The Mathematical Association of America.

Fujii, J. N. 1961. *An Introduction to the Elements of Mathematics.* New York: John Wiley & Sons, Inc.

Gallian, J. A. 2002. *Contemporary Abstract Algebra.* Boston: Houghton Mifflin Company.

Gordon, R. A. 2002. *Real Analysis: A First Course.* Boston: Addison Wesley.

Lin, S. T. and Lin, Y. 1974. *Set Theory: An Intuitive Approach.* Boston: Houghton Mifflin Company.

McCoy, N. H. 1960. *Introduction to Modern Algebra.* Boston: Allyn and Bacon, Inc.

Morash R. P. 1987. *Bridge to Abstract Mathematics: Mathematical Proof and Structures.* New York: Random House, Inc.

Olmsted, John M. H. 1959. *Real Variables, An Introduction to the Theory of Functions.* New York: Appleton-Century-Crofts, Inc.

Smith, D., Eggen, M., and St. Andre, R. 2006. *A Transition to Advanced Mathematics.* Belmont: Thomson Brooks/Cole.

Smith, D. E. 1958. *History of Mathematics, Volume I.* New York: Dover Publications, Inc.

Smith, D. E. 1958. *History of Mathematics, Volume II.* New York: Dover Publications, Inc.

Stahl, S. 1999. *Real Analysis: A Historical Approach.* New York: John Wiley & Sons, Inc.

Stoll, M. 2001. *Introduction to Real Analysis.* Boston: Addison Wesley.

Stoll, R. R. 1961. *Sets, Logic, and Axiomatic Theories.* San Francisco: W. H. Freemen and Company.

Suzuki, J. 2002. *A History of Mathematics.* Upper Saddle River: Prentice-Hall, Inc.

Index

D